An Introduction to Neural Information Processing

Peiji Liang • Si Wu • Fanji Gu

An Introduction to Neural Information Processing

Springer

Peiji Liang
School of Biomedical Engineering
Shanghai Jiao Tong University
Shanghai, China

Si Wu
Beijing Normal University
Beijing, China

Fanji Gu
School of Life Sciences
Fudan University
Shanghai, China

ISBN 978-94-017-7391-1 ISBN 978-94-017-7393-5 (eBook)
DOI 10.1007/978-94-017-7393-5

Library of Congress Control Number: 2015955554

Springer Dordrecht Heidelberg New York London

Printed on acid-free paper

Springer Science+Business Media B.V. Dordrecht is part of Springer Science+Business Media (www.springer.com)

Preface

The human brain might be the most complicated system in the universe. Neuroscience, the aim of which is to elucidate the mechanisms underlying brain functions, is one of the scientific frontiers in the twenty-first century. To understand how brains work, neuroscience researches involve various approaches.

It has long been understood that anatomical architecture of the brain, morphological structure of cells, and even chemical structure of membrane proteins are essential for brain functions, and knowledge about how brain functions are realized has been accumulated via traditional approaches of neuroscience including neuroanatomy, neurophysiology, neurochemistry, cellular and molecular neuroscience, etc. However, brain functions mostly involve neural information processing, which relies on dynamic processes at various levels; computational approaches are therefore indispensable for understanding the process and mechanism underlying neural information processing.

Although a brain should have somewhat stable anatomical connectivity among its different areas, the functional connectivity among relevant brain areas is dynamically changing corresponding to the brain function it is undertaking; at the meantime, the activation of a certain brain area is characterized by the firing activity of the neurons in the area, while the temporal pattern of neuronal firing sequence and the concerted activity of neuronal population are crucial for neural signal transmission and neural coding; furthermore, biophysical properties of ion channels located on the plasma membrane and their dynamic modulation under various physiological and functional conditions determine the activity of a neuron and also influence the synaptic connectivity between neurons which is responsible for signal transmission. To understand these aspects, experimental observations made by approaches such as neuroimaging and neuroelectrophysiology are doubtlessly necessary, but quantitative analyses or even theoretical analyses via modeling and simulation using mathematical tools contribute significantly for getting more insight into such processes in detail. Such demands in neuroscience research give rise to the branch of computational neuroscience, which is developed based on the interactions between neuroscience and other disciplines such as mathematics, physics, as well as cybernetics and information science and makes it a growing field during the past decades.

However, to solve biological problems by approaches from mathematical and/or relevant disciplines is not simply for biologists to pass their data to mathematicians for computation. To make such interdisciplinary collaboration efficient and productive, it is necessary for researchers with different backgrounds to understand the language of his/her collaborators. It should be good for people conducting computation to understand relevant background of the biological problems they are dealing with, so that they can have an insight into their biological significance. At the meantime, it would also be helpful for his/her partner-biologist to understand the philosophy underlying the mathematical tools adopted, so as to make the communication effective.

On the other hand, given that the human brain is the most complicated and delicate system for information processing, people have long been exploring to develop brain-like machine or system to mimic brain functions. Actually, artificial intelligence has persistently been inspired by the functions of biological neural systems. Therefore, to acquire some basic knowledge about biological neural systems should also be beneficial for those who are working in the field of neural engineering, neuromorphic engineering, neural robotics, and so on and so forth.

Having backgrounds in relevant fields such as mathematics, physics, and biomedical engineering and having been working in this multidisciplinary area of computational neuroscience for decades, the authors of this book are well experienced in combining knowledge from these different disciplines into the field of computational neuroscience. They well understand the difficulties that may be encountered by those with background other than biology but are practicing in or preparing themselves for neuroscience research. The authors wish that this book can provide some basic knowledge about neurobiology at a proper level, neither too detailed nor too simple.

The potential readers for this book are mainly supposed to be those with mathematics, physics, or informatics background, who are willing to join neuroscience research and apply their knowledge in these fields to solve neurobiological problems. For those with neurobiology background who are interested in employing mathematical tools in neuroscience studies, this book may also provide some guidance by introducing some pioneer works as well as some recent advances in the field. Some of the authors' own works are also introduced.

The main parts of this book are organized as follows:

Chapter 1 is an introduction about the interdisciplinary essence of computational neuroscience. An overview of this particular branch of neuroscience is presented by giving a historical review on the development of this field together with a number of pioneers' contribution to this field.

In Chap. 2, the necessary neuroscience background is introduced in more depth than is generally the case in monographs with similar topics. In addition, the materials presented there were selected from a point of view of neural information processing, so that the readers would have enough knowledge to read monographs and original papers without too much difficulties.

In Chap. 3, models of single neurons—typical examples of neural information processing studies—are elucidated in detail. This is not only for the reason that neurons are considered as both anatomical and functional elements of nervous systems, understanding neurons is essential for further reading, but also the classical studies on neuron modeling give readers good examples how to develop neural models based on questions raised by neuroscientists and how to solve such problems using mathematics and informatics approaches verified by biological experiments. This chapter gives readers many classical examples to show how to build models firmly grounded on experimental results and in turn guide further experiments. Both reductionist and dynamic approaches are used. The topics discussed in this chapter are simple enough to be understood without too much background knowledge and detailed enough to make readers realize the essence of interdisciplinary studies. Some arguments unsolved until now are also mentioned for reader's own consideration and judgment.

In Chap. 4, neural coding theories are introduced. Given that the main task of the nervous system is to process and transmit the information of the environment stimulations, as well as conduct the action of motor system, one of the essential issues of neuroscience research is to understand the "code" that the nervous system applies for information processing and transmission. During the past decades, neuroscientist investigated the relationships between the environmental stimulation/motor actions and the properties of neuronal activities, such as firing rate, the temporal structure of the firing sequence, as well as the coordination of neuronal activities in neuronal assembly, which are termed as rate coding, precise time coding (temporal coding), and population coding, respectively. These aspects will be introduced together with Bayesian inference.

In Chap. 5, a number of advanced fields in sensory information processing are introduced, including some aspects in visual information processing and olfactory information processing. Among the sensory systems, the visual system is the most complex and at the meantime the best understood one. Visual information processing occurs in the retina which is the peripheral part of the system, as well as in the central visual regions. Neurons in different parts of the visual pathway have different response properties, which can be measured using well-designed laboratory stimulations together with proper computational tools. Besides, adaptation to the environment is one of the important features of visual neurons; such property allows the visual neurons to well encode the visual stimulations which changes among a wide range via their limited range of activity. Coding during visual adaptation forms another topic of this chapter. The olfactory system is another sensory system which is important for animals to survive in the natural environment. Olfactory information processing is a hot topic that has been intensively studied in recent years.

In Chap. 6, a number of neural network models aiming for modeling brain functions are introduced. The Hopfield model is such a classical one which played a very important role in the early time of applying network models to mimic brain functions. It assumes very simple and abstract forms of neurons and synapses, but nevertheless captures some fundamental features of neural information retrieving. Continuous attractor neural network is a one-step advance of the Hopfield model. It incorporates more biological elements and describes the encoding of continuous variables in neural system successfully. Reservoir networks, including liquid state machine and echo state machine, emphasize on the fact that a neural system typically contains a huge number of neurons. It exploits the enormous variations of network states and simplifies the effort of information decoding significantly. Special network models are also proposed to implement decision-making in the brain. Recent studies have explored how the short-term plasticity of synapses enriches the dynamics and computational capacities of a neural network.

Among these, Fan-Ji Gu drafted Chaps. 1 and 3; Chap. 2 was drafted by Fanji Gu and Peiji Liang; Chaps. 4 and 5 were coauthored by Pei-Ji Liang and Si Wu, and Chap. 6 was by Si Wu.

Although numerous textbooks on neurobiology and neuroscience are available, textbooks dedicated to computational neuroscience are comparatively scarce. By providing knowledge on both biological and computational sides, as well as the combination of these knowledge in the field of computational neuroscience, we wish this book provides a good guidance for those researchers and students who are preparing themselves for this interdisciplinary field. Of course, it is unrealistic to pack all aspects of computational neuroscience in one book; further readings are encouraged. Reading materials including more detailed textbook on neurobiology and relevant references about computational neuroscience are recommended in the last sections of related chapters.

Last but not least, we would like to express our heartiest gratitude to Prof. Walter J. Freeman for his kind permission to quote his words in our personal communication and the figures published in *Scientific American*, which he owes the copyright. We are also indebted to Profs. Tiande Shou, Hongbo Yu, Hans Liljenstöm, and Hans Braum for discussing some ideas published in this book. We acknowledge Drs. Danke Zhang, Yuanyuan Mi, Mrs. Wenhao Zhang, Luozheng Li, Liutao Yu, Xiaolong Zhou, Gang Wang, Tao Wang, and Miss Yan Xia for their hardworking in preparing the documents for the book. We also wish to thank Dr. Peter Butler and Dr. Peng Zhang for encouraging us to publish this book in Springer and helping us with many things.

Peiji Liang
Si Wu
Fanji Gu

Contents

Chapter 1
Introduction

Abstract In its essence, the brain is an information processing system, which is the most complicated system people have ever known. Mathematical modeling and computer simulation may play a vital role in integrating data from different levels of the brain and thus to elucidate the underlying mechanisms of the brain function. Many interdisciplinary fields have been developed for different aspects of this topic. However, the core of these different branches is the same—neural information processing, which is the main topic of this book.

A brief history of the study of neural information processing is given. Close combining theoretical analysis, modeling and simulation, and biological experiments is emphasized. This book is mainly for those readers who want to know the necessary knowledge about the brain with adequate depth, learn how to use mathematical modeling and computer simulation to solve neurobiological problems, and learn some latest progress in this field.

Keywords Neural information processing • Mathematical modeling • Computer simulation • Interdisciplinary study

People are used to using the most advanced machine at their time as a metaphor of the brain, such as a water-powered mill, a steam engine, a telephone switchboard, a digital computer, and the Internet. However, the brain is the most complicated system in the universe we have ever known; no machine mentioned above is a precise metaphor of the brain, which contains 10^{11} neurons, a number that can be compared with the one of all the stars in our known universe. In addition, there are 10^{15} connections called synapses between neurons. Thus, to study the brain and mind is the biggest challenge scientists have ever met; the task is a little like imaging an elephant by a group of blind men where each man can only touch a part of the animal. The man who touches its nose may say that it is like a tube and the one touching the foot may say that it is a column, while the one touching its trunk announces that it is just like a wall. Only summarizing all the impressions together may give a quite realistic picture. So does brain research. What makes the thing more complicated is the fact that the brain is a multilevel hierarchical system, where each level has its own emergent properties which its composing elements do not have. No single technology can solve all the problems; a variety of different technologies are needed to study the brain and mind from different aspects and at

P. Liang et al., *An Introduction to Neural Information Processing*,
DOI 10.1007/978-94-017-7393-5_1

different levels, and only synthesizing all the results together may give an approximately realistic view of the brain and its function, for which mathematical modeling and computer simulation may play a vital role.

One thing now we definitely know is that brains are information processing systems. Of course, energy and substances are important for normal functions of the brain; however, these are so-called enabling factors of the brain function. The essential role which the brain plays for the survival of the animal is to receive, transfer, process, store, and retrieve information, based on which the animal can decide its action, e.g., fight or flight.

A quantitative approach should be used for such studies. This idea can be traced back to the 1940s. American neurophysiologist Warren McCulloch and mathematical biologist and logician Walter Harry Pitts, Jr. (1943), are among the first to describe neuronal function with mathematics. They are the founders of neural network theories. Their 1943 paper is the first important one to inspire others to study neural information processing quantitatively, although some mathematical models of neurons had been developed even long ago before them (Lapicque 1907).

In 1948, American mathematician Norbert Wiener, who had been interested in biology since his childhood, published his classical book *Cybernetics: Or Control and Communication in the Animal and Machine* (Wiener 1948), in which he used the word "communication" to denote information coding, processing, storage, and retrieving; he emphasized the key role of the concept "information" in understanding an animal's behavior. He is also a pioneer in encouraging mathematicians, physicists, electrical engineers, and experts from other fields to study nervous systems. In the introduction of his book, he emphasized:

> It is these boundary regions of science which offer the richest opportunities to the qualified investigator. They are at the same time the most refractory to the accepted techniques of mass attack and the division of labor. If the difficulty of a physiological problem is mathematical in essence, ten physiologists ignorant of mathematics will get precisely as far as one physiologist ignorant of mathematics, and no further. If a physiologist, who knows no mathematics, works together with a mathematician who knows no physiology, the one will be unable to state his problem in terms that the other can manipulate, and the second will be unable to put the answers in any form the first can understand. Dr. Rosenblueth has always insisted that a proper exploration of these blank spaces on the map of science could only be made by a team of scientists, each a specialist in his own field, but each pocessing a thoroughly sound and trained acquaintance with the fields of his neighbors; all in the habit of working together, of knowing one another's intellectual suggestion before it has taken on a full formal expression. The mathematician need not have the skill to conduct a physiological experiment, but he must have the skill to understand one, to criticize one, and to suggest one. The physiologist need not be able to prove a certain mathematical theorem, but he must be able to grasp its physiological significance and to tell the mathematician for what he should look.

Although these words were written more than 65 years ago, they are still as instructive today as at that time. Wiener pointed out that the concept of energy, which people had paid much attention to before, was not the key to understand control and communication. Energy is only an enabling factor for information coding and processing. The core concept is information! He and Hungarian–American mathematician John von Neumann, who is also one of the founders of digital computers and wrote a book titled *The Computer and the Brain* (von Neumann 1958) to compare these two systems just before his death in 1957, organized a series of symposiums on control and communication since 1943. At his later years, Wiener had paid more attention to a branch of cybernetics, biological cybernetics, especially to neural information processing (Wiener and Schade 1964).

Although McCulloch and Pitts proposed a mathematical model of neurons and elucidated that their neural network could do any job a Turing machine could do, their network lacks the ability to learn. However, brains can learn. Canadian psychologist Donald Hebb is the first to denote that synaptic plasticity may underlie learning. He wrote in his famous book *The Organization of Behavior: A Neuropsychological Theory* (Hebb 1949) the following hypothesis: "When an axon of cell A is near enough to excite a cell B and repeatedly or persistently takes part in firing it, some growth process or metabolic change takes place in one or both cells such that A's efficiency, as one of the cells firing B, is increased." This hypothesis is usually summarized as "Cells that fire together, wire together." And it is used to being named as the Hebbian learning rule. His idea was later used by neural network modelers to build networks which can learn. Although the idea was only a hypothesis without experimental evidence when Hebb proposed it, now it has been strongly supported by physiological experiments (Paulsen and Sejnowski 2000).

American psychologist Frank Rosenblatt's perceptron is the first neural network, which can learn with try-and-error approach (Rosenblatt 1962). After many times of training, the network can learn to classify some inputs, such as handwriting digits, into categories defined by regular printed types, with slight modification of connection weights based on its output if it is the one expected by the trainer or not; Rosenblatt thought brains would work in a similar way.

The success of the perceptron absorbed many artificial intelligence (AI) students to shift their researches to neural networks, and part of the financial support to AI flowed to neural network studies. This annoyed AI researchers. AI was the mainstream in technologies trying to duplicate human intelligence at that time. They insisted that machine intelligence could be realized only by the way of finding some formal representation of knowledge, reasoning, and programming. It seemed to them that any idea that a machine could create its own knowledge representation and logic with changing its own connection weights would only lead to a dead end. In 1969, American cognitive scientist and mathematician Marvin Minsky, one of AI pioneers, and mathematician Seymour Papert published their famous book *Perceptrons: An Introduction to Computational Geometry*, in which they proved rigorously that a one-layer perceptron cannot even classify some very simple inputs. This was a heavy blow on the neural network movement, which was in an awkward predicament in the 1970s. There were only a few scientists who still insisted on neural network studies. However, Minsky's proof is only true for a neural network with one layer. For neural networks with multilayers, they only said: "We consider it to be an important research problem to elucidate (or reject) our intuitive judgment that the extension is sterile" (Minsky and Papert 1969). No detailed explanation was given, but they were wrong this time. Neural networks with multilayers can solve problems which Minsky and Papert thought they could not. Although Rosenblatt himself had thought about such possibilities, he could not find a way to train the intermediate layers.

In 1982, American biophysicist John Hopfield developed a recurrent neural network rather than a feedforward network such as a perceptron; he proved that, under certain conditions, the state of his network would converge to certain local minimum, depending on its connection weights, which could be considered as stored memory in wildly distributed synaptic strengths, and the stored memory (local minimum) could be retrieved by partial information (the initial state), a phenomenon similar to associative memory of the brain (Hopfield 1982). Hopfield's paper is one of the landmarks of the rejuvenation of the neural network movement. For details, please see Sect. 5.1 of this book. Others also found that a network with multilayers could do jobs that Minsky thought it could not. However, the progress of artificial neural network studies just as traditional AI studies met its bottleneck around the turn of this century. These techniques have been found very difficult or even failed to carry out many important functions of brains such as cognition or acting in a new environment; the key problem is that they just try to mimic human behavior without understanding the underlying mechanism of the brain. For artificial neural networks, although its elements are similar to biological neurons in some way, they usually are oversimplified, and the networks are not organized in similar ways that biological neural circuits do. Now neural network modelers are paying more attention to the underlying mechanism of their biological prototypes and their works become one important part of the field of neural information processing. We will give some important examples in Chap. 5.

The rise and fall of traditional AI and artificial neural network studies has given people an important lesson: To understand human intelligence, create an intelligent machine, and treat mental disorders, the behaviorist approach is far from enough; digging deeply into the underlying mechanism of brains is necessary. It is essential to understand how brains process information, using all available techniques, including mathematical modeling and computer simulation. Developing new technologies is also important for this purpose, as what the US BRAIN Initiative suggested.

As a matter of fact, exploring the biological mechanism of nervous systems using mathematical model and simulation can be traced back to British physiologists Alan Hodgkin and Andrew Huxley's seminar papers (Hodgkin and Huxley 1952) on the generation and propagation of action potentials in the 1950s (for details, see Sect. 2.1 of this book). By the way, although Hodgkin took physiology as his main major in the university when he was a college student, he also took mathematics and physics as elective courses, as suggested by his zoology professor, and Huxley was a physics student in his college days, taking physiology as an elective course.

In the 1950s and 1960s, there were several other important works in this field.

American biophysicist Haldan Keffer Hartline found that there was mutual inhibition between neighboring eccentric cells (a kind of bipolar neurons; their axons form optical nerves to its brain) in the retina of *Limulus* compound eyes, which could enhance the spatial contrast of the edge and details in a scene (Hartline 1974). He and his colleagues established a nonlinear algebra equation—Hartline–Ratliff equation—to model such a lateral inhibitory network and used it to elucidate the mechanism underlying the Mach band, an optical illusion discovered by Austrian scientist Ernst Mach in the nineteenth century. Now lateral inhibition has been proven to be one of the universal principles in sensory information processing. Hartline is also the one who firstly proposed the concept of a visual receptive field of a given neuron—an area in the retina, on which a proper optical stimulus could change the pattern of this neuron's activities.

In 1959, American neuroscientist Jerome Lettvin and his colleagues including McCulloch and Pitts published a paper titled "What the frog's eye tells the frog's brain" (Lettvin et al. 1959), in which four different types of ganglion cells in frog's retina were reported that could detect different features in visual stimuli. A feature detection theory was proposed for visual information processing.

In the same year, Canadian–American and Swedish neuroscientists David Hubel and Torsten Wiesel found that, in the primary cortex, there were neurons sensitive to special orientations of bars in their receptive fields. Such function may be the basis of form perception (Hubel 1981).

After these findings, a lot of papers on receptive field modeling were published. Recent progresses in this direction are given in Chap. 4 in detail.

American neuroscientist Wilfrid Rall, who was a physics student, treated every branch of dendrites as a kind of cable (Rall 1959, 1960, 1962, 1964); he analyzed electrical conduction in branched dendrites by the cable theory, which was developed by engineers to study the similar problems of undersea cables across oceans. Although anatomists had already found complicatedly arborescent structures of dendrites, neuron modelers used to neglect such complexity and treat them just as a node. Rall is the first to describe the dendrite tree with a mathematical approach to explain how synaptic inputs are integrated to control spiking. In his model, he even predicted that there might be dendrodendritic synapses (Rall et al. 1966), which were verified experimentally a few months later by his colleagues. In the following years, he not only considered passive properties of dendrites but also active properties and suggested that the latter may contribute to local amplification of the input signals.

During the Second World War, Bernhard Hassenstein, a biology student, and Werner Reichardt, a high school graduate, were sent to the field. They agreed that they would organize an institute combining biology and physics if they could survive. Fortunately, both survived. After the war, Hassenstein got his Ph.D. degree in biology with a thesis on beetle optomotor responses, and Reichardt got a Ph.D. degree in physics and they met together again; the latter found that he could model the former's results, and a model was developed (Hassenstein and Reichardt 1956; Reichardt 1961), which can not only mimic Hassenstein's biological experiment results but also predict that the response would not increase monotonically with the stimulus speed forever; instead, it would reach a peak at certain speed and then decline, and the optimal speed depended on stimulus spatial frequency. These predictions were verified by further biological experiments. Their researches promoted the trend to study nervous system activities quantitatively with mathematical modeling. Intuition is not always true; mathematical analysis may uncover the truth, as what was said that "a pen might be wiser than the person who holds it."

They established a biological cybernetics group in the Max Planck Institute of Biology in Tübingen, which has been developed into an independent institute—the Max-Planck-Institut für biologische Kybernetik in Tübingen. Reichardt's model started studies on movement detectors and he launched a journal *Kybernetik* (now *Biological Cybernetics*), which is the first international journal devoting itself to topics about biological cybernetics. Although having graduated in physics, Reichardt advised every young person that joined in his institute that theories must always combine with experiments closely, and he did not believe that any brain theory without biological foundation would have any chance to be successful.

About the same period, British neuroscientist David Marr, who obtained his BA in mathematics and Ph.D. in physiology, developed a model of the cerebellum based on Australian neurophysiologist John Eccles' anatomical and physiological data in his Ph.D. thesis. He shifted his focus on visual information processing after that. His classical book *Vision: A Computational Investigation into the Human Representation and Processing of Visual Information* (Marr 1982) summarizing his studies was completed by his colleagues after his passing away, in which he proposed that there are three analysis levels in visual information processing: theory, algorithm, and hardware levels. He emphasized the independence between these levels; however, later studies showed that such independence was not as strong as what he thought. In fact, brain function is heavily constrained by its structure. And ironically he himself paid much attention to visual physiology and psychology in his book. His works inspired studies on computer vision and became a classical example for visual information processing research using mathematical tools.

Although feature detection and receptive field theories tell us how visual systems analyze different features in a scene, they have not answered how visual systems synthesize these features into a unified object or combine different local features into different objects in a scene—the so-called binding problem. In the 1980s, German computer scientist Christoph von der Malsburg proposed a synchronization theory to answer it—neurons giving response to the same object fire synchronously (von der Malsburg 1981, 1985)—and his theory were supported by German neuroscientists Wolf Singer and his colleagues (Singer 1993); their results were confirmed by another German group headed by Reinhard Eckhorn (1988).

American neuroscientist Walter J. Freeman, who also graduated in physics and mathematics, is one of the pioneers who proposed a mesoscopic dynamic approach to model the olfactory system; he is also one of the pioneers to study chaotic dynamics in nervous systems (Freeman 2000). German physicist Hermann Haken used a synergetic approach developed by him to study brain activity and its dynamics (Haken 1996, 2002).

The above has only given an uncompleted list of pioneering works on neural information processing, which promote researchers to study nervous systems with mathematical modeling and computer simulation. In 1985, Eric L. Schwartz, a cognitive neuroscientist and computer scientist, coined the term "computational neuroscience" and organized the first international conference on this topic (Schwartz 1990). In 1994, the first international journal titled *Journal of Computational Neuroscience* devoting to this topic was published. In the editorial of its first issue, it is emphasized that the goal of computational neuroscience is to study "how the nervous system processes information to produce meaningful behavior in animals and humans," i.e., to understand the rules how the nervous system encodes, processes, or "computes" the information, and how neural machinery implements the corresponding computational algorithms (Rinzel et al. 1994).

Computational neuroscience has been one of the frontiers in twenty-first century's sciences. In a supplement to *Nature Neuroscience* focused on computational approaches to brain function organized by the NIH, it is said that "Clearly, computational neuroscience is, and will continue to be a growing field" (NIH 2000).

In the recent years, more related disciplines such as neuroinformatics, neuroengineering, neurodynamics, computational neuroethology (computational neuroecology), neurorobotics, neuromorphic engineering, etc., have emerged and grown quickly. Although their emphasized aspects are different in some way, they share a core concept—neural information processing—and they are all interdisciplinary fields between neuroscience and informatics/IT. For such researches, theoretical analysis, modeling and simulation, and biological experiments must go hand in hand; now a lot of mathematicians, physicists, computer scientists, and IT engineers are rushing into these golden mine areas. Just as what Norbert Wiener said more than half a century ago: "For many years, Dr. Rosenblueth and I had shared the conviction that the most fruitful areas for the growth of the sciences were those which had been neglected as a no-man's land between various established fields" (Wiener 1948). British molecular biologist and neuroscientist Francis Crick, who also graduated in physics and called as the greatest biologist in the twentieth century by Nobel laureate Eric Kandel, also pointed out that: "In nature hybrid species are usually sterile, but in science the reverse is often true. Hybrid subjects are often astonishingly fertile, whereas if a scientific discipline remains too pure it usually wilts" (Crick 1988).

From the above brief review of the history of this field, it is easy to notice that all the pioneers are people like what Wiener demanded. Although they focused their studies on neuroscience, they had sound foundation of mathematics, physics, or informatics. However, as what is stated in the quotation we cited from Wiener, it is not easy to master all these different disciplines. Every discipline has its own approach and its own thinking habit. Pure mathematicians are good at logical reasoning based on some axioms; they don't care what concrete object they are treating, so that a joke says that a pure mathematician is the person who does not know what he or she is talking about. Traditional biologists emphasized observation and experiment while physicists are used to idealizing an object into a model, deducing results from the model, and comparing them with experimental ones. Every approach has its own merit and deficit. People working with neural information processing should combine all these approaches into one and be accustomed to discussing with colleagues from other fields. For doing this, one must have background knowledge with some depth in his neighboring fields and understand others' thinking habit.

This book is for readers who are interested in brain research with backgrounds other than neuroscience, just like the authors themselves of this book. At the very beginning, when they try to work in this field, they often find that they lack enough knowledge about nervous systems. Neuroscience textbooks are often too detailed for them to quickly grasp the essence of the neuroscience background for studying, just as what American neuroscientist John Dowling (2001) said: "Textbooks in neuroscience (and most other branches of biology) have become more and more encyclopedic and, in my view, less and less useful for the beginning student or for others interested in the principles of a field." On the other hand, monographs on computational neuroscience usually do not give enough pages to introduce the neuroscience background, which makes the beginners have difficulty to really understand. Even worse, some popular science books on these topics just give a cartoon-like description of the brain, and this may mislead readers to think that their descriptions are the realistic view of the nervous systems, thus underestimating the complexity of the living brain. Some readers may develop their "brain models" which almost share nothing common to a living brain, just like the model proposed by a scientist visiting Francis Crick, who thought that his model was pretty and worked well. However, the comment given by Crick is "My dear chap, that's a criterion you would use for selling a vacuum cleaner – I don't see what it has to do with the brain" (Ramachandran 2004). Thus, the authors have to make some trade-off in introducing the neuroscience background neither too detailed nor too simplified.

Of course, this book may also be useful for neuroscience students, who have some mathematical background and are interested in using mathematics, physics, and informatics approaches for their own studies. This book will give those readers a review of basic neuroscience background related to information processing and give them classical examples on how to use mathematical and physical ideas and approaches to solve biological problems and some recent important advances in this field so that they may learn some lessons to solve their own problems. For those who don't have enough mathematical background, reading some concise mathematics textbook, especially those for neuroscience students, is necessary (Gabbiani and Cox 2010).

As an introductory book, of course, it is impossible to cover all achievements in neural information processing, which are so broad, that the first edition of a handbook on this topic published in 1995 which has 1,118 pages and a weight over 3 kg (Arbib 1995); however, the main criticism about the handbook in a book review is that many important achievements hadn't been included in that book (Aleksander 1995)! Further reading is necessary and lists of them are recommended in the last sections of the related chapters.

From the authors' view, studies on neural information processing are still in its infant period; the mysteries of the brain and mind could not be solved in the near future. Just as what Walter J. Freeman said (Personal communication 2006):

> Fifty years ago, enthused by successes in creating digital computers and the DNA model of heredity, scientists were confident that solutions to the problems of understanding biological intelligence and creating machine intelligence were within their grasp. Progress at first seemed rapid. Giant 'brains' that filled air-conditioned rooms were shrunk into briefcases. The speed of computation doubled every 2 years.
>
> What these advances revealed is not the solutions but the difficulties of the problems. We are like the geographers who 'discovered' America, not as a collection of islands but as continents seen only at shores and demanding exploration. We are astounded less by the magnitude of our discoveries about how brains cogitate than by the enormity of the tasks we have undertaken, to explain and replicate the higher functions of brains.

There is still a long way to explore these mysteries; it seems unlikely that the goal could be reached by one or two giant projects in the next decade or even a little longer (Gu 2013). However, as what Francis Crick pointed out more than 20 years ago (Crick 1988): "The present state of the brain sciences reminds me of the state of molecular biology and embryology in, say, 1920s and 1930s. Many interesting things have been discovered, each year steady progress is made on many fronts, but the major questions are still largely unanswered and are unlikely to be without new techniques and new ideas ...... The brain sciences have still a long way to go, but the fascination of the subject and the importance of the answers will inevitably carry it forward. It is essential to understand our brain in some detail, if we are access correctly our place in this vast and complicated universe we see all around us."

The future to understand the brain is bright; the way ahead is still long. For young readers, we will present them Shakespeare's verse with all respect:

> There is a tide in the affairs of men,
> Which taken at the flood leads on to fortune;
> Omitted, all the voyage of their life
> Is bound in shallows and in miseries.
> (William Shakespeare: Julius Caesar, Act IV, Scene 3, Line 217)

Enjoy reading.

References

Aleksander I. Fleshing out intelligent machines. Science. 1995;376:564.

Arbib MA, editor. The handbook of brain theory and neural networks. Cambridge, MA: MIT Press; 1995.

Crick F. What mad pursuit – a personal view of scientific discovery. New York: Basic Books; 1988.

Dowling JE. Neurons and networks – an introduction to neuroscience. 2nd ed. Cambridge, MA/-London: The Belknap Press of Harvard University Press; 2001.

Eckhorn R, Bauer R, Jordan W, et al. Coherent oscillations: a mechanism of feature linking in the visual cortex? Multiple electrode and correlation analysis in the cat. Biol Cybern. 1988;60:121–30.

Freeman WJ. Neurodynamics – an exploration in mesoscopic brain dynamics. London: Springer; 2000.

Gabbiani F, Cox SJ. Mathematics for neuroscientists. Amsterdam: Academic; 2010.
Gu F. The human brain project EU is unlikely to create an artificial whole-brain in a decade. Brain-Mind Mag. 2013;2:4–6.
Haken H. Principles of brain functioning – a synergetic approach to brain activity, behavior and cognition. Berlin: Springer; 1996.
Haken H. Brain dynamics – synchronization and activity patterns in pulse-coupled neural nets with delays and noise. Berlin: Springer; 2002.
Hartline HK. Visual receptors and retinal interaction. In: Ratliff F, editor. Studies on excitation and inhibition in the retina. New York: The Rockefeller University Press; 1974.
Hassenstein B, Reichardt W. Systemtheoretische analyse bewegungsperzeption des ruessel – kaefers chlorophanus. Z Naturforsch. 1956;11B:513–24.
Hebb DO. The organization of behavior. New York: Wiley & Sons; 1949.
Hodgkin AL, Huxley AF. A quantitative description of membrane current and its application to conduction and excitation in nerve. J Physiol. 1952;117:500–44.
Hopfield JJ. Neural networks and physical systems with emergent collective computational abilities. Proc Natl Acad Sci U S A. 1982;79:2554–8.
Hubel D. Evolution of ideas on the primary visual cortex, 1955–1978: a biased historical account. Nobel lecture, 8 Dec 1981. In: Hubel DH, Wiesel TN, editors. Brain and visual perception: the story of a 25-year collaboration. Oxford: Oxford University Press; 2005. p. 659–86.
Lapicque L. Recherches quantitatives sur l'excitation électrique des nerfs traitée comme une polarisation. J Physiol Pathol Gen. 1907;9:620–35.
Lettvin YJ, Maturana HR, McCulloch WS, Pitts WH. What the frog's eye tells the frog's brain. Proc IRE. 1959;47:1940–51.
Marr D. Vision. A computational investigation into the human representation and processing of visual information. New York: W.H. Freeman and Company; 1982.
McCulloch W, Pitts W. A logical calculus of the ideas immanent in nervous activity. Bull Math Biophys. 1943;5:115–33.
Minsky M, Papert S. Perceptrons: an introduction to computational geometry. Cambridge, MA: The MIT Press; 1969.
NIH. Computational neuroscience at the NIH. Nat Neurosci. 2000;3:1161–64.
Paulsen O, Sejnowski TJ. Natural patterns of activity and long-term synaptic plasticity. Curr Opin Neurobiol. 2000;10(2):172–9.
Rall W. Branching dendritic trees and motoneuron membrane resistivity. Exp Neurol. 1959;1:491–527.
Rall W. Membrane potential transients and membrane time constant of motoneurons. Exp Neurol. 1960;2:503–32.
Rall W. Theory of physiological properties of dendrites. Ann NY Acad Sci. 1962;96:1071–92.
Rall W. Theoretical significance of dendritic trees for neuronal input-output relations. In: Reiss RF, editor. Neural theory and modeling. Stanford: Stanford University Press; 1964. p. 73–97.
Rall W, Shepherd GM, Reese TS, Brightman MW. Dendro-dendritic synaptic pathway for inhibition in the olfactory bulb. Exp Neurol. 1966;14:44–56.
Ramachandran VS. The astonishing Francis Crick. Perception. 2004;33:1151–4.
Reichardt W. Autocorrelation, a principle for the evaluation of sensory information by the central nervous system. In: Rosenblith WA, editor. Sensory communication. New York: MIT Press/ John Wiley and Sons; 1961. p. 303–17.
Rinzel J, et al. Editorial. J Comput Neurosci. 1994;1:9–10.
Rosenblatt F. Principles of neurodynamics: perceptrons and the theory of brain mechanisms. Washington, DC: Spartan; 1962.
Schwartz EL, editor. Computational neuroscience. Cambridge, MA: The MIT Press; 1990.
Singer W. Synchronization of cortical activity and its putative role in information processing and learning. Annu Rev Physiol. 1993;55:349–74.
von der Malsburg C. The correlation theory of brain function. Internal report 81–2. Göttingen: Dept. of Neurobiology, Max-Planck Institute for Biophysical Chemistry; 1981.

von der Malsburg C. Nervous structure with dynamical links. Berichte der Bunsengesellschaft fuer Phys Chem. 1985;89:703–10.
von Neumann J. The computer and the brain. New Haven: Yale University Press, Inc; 1958.
Wiener N. Cybernetics – or control and communication in the animal and machine. New York: The Technology Press, John Wiley & Sons, Inc.; 1948.
Wiener N, Schade JP (Ed.) (1964) *Progress in biocybernetics Vol. 1* Elsevier Pub. Co.

Chapter 2
Neurobiological Basis Underlying Neural Information Processing

Abstract Although modeling and calculation from a perspective of neural information processing are sometimes based on some necessary assumptions to make the nervous system a simplified one, which enables the computation feasible, the nervous system is itself a very complicated one. Besides, different parts of the nervous system have their own characteristics. Therefore, it should be very helpful for those people who are naïve of neuroscience but wish to work on topics related to neural information processing to start with some fundamental ideas about neuroscience. This chapter provides the necessary biological background of the brain with adequate details. Gross anatomy of nervous systems, structural and functional properties of nerve cells, functional organization of various sensory systems and somatomotor system, as well as structure and function of cerebellum and hippocampus are introduced from a point of view of information processing.

Keywords Nervous system • Nerve cells • Sensory and somatomotor systems • Cerebellum • Hippocampus

Pioneers of traditional AI thought that machine intelligence, which might be compared with human intelligence or even surpass, could be realized by sophisticated programming and symbolic processing. In 1967, Marvin Minsky, one of the AI founders, concluded optimistically that the task could essentially be solved within one generation. However, his prediction failed. In 1982, he had to acknowledge that the problem of artificial intelligence was one of the most difficult one which science had ever met. The traditional AI approach met its bottleneck. For creating intelligent machines or treating mental disorders, in addition to understanding human beings ourselves, we must know mechanisms underlying neural information processing in the living brain. Thus, at the very beginning of this book, the necessary and essential background about neuroanatomy and neurophysiology for understanding neural information processing should be given.

As what we have mentioned in the introduction, brain functions are constrained by their "hardware." Marr's three levels for analyzing brain function, computational theory, representation algorithm, and hardware implementation (Marr 1982), are not absolutely independent to each other as what he declared in his book. In biology, functions always depend on structures. "If you want to understand function, seek to understand structure." (Koch 2004) American neurophilosopher

P. Liang et al., *An Introduction to Neural Information Processing*,
DOI 10.1007/978-94-017-7393-5_2

Patricia S. Churchland and computational neuroscientist Terrence J. Sejnowski indicated (Churchland and Sejnowski 1992): "In contrast to the doctrine of independence......Knowledge of brain architecture, far from being irrelevant to the project, can be the essential basis and invaluable catalyst for devising likely and powerful algorithms—algorithms that have a reasonable shot at explaining how in fact the neurons do the job."

Therefore, before discussing neural information processing, it is necessary to give fundamental facts about neural anatomy and physiology in this chapter. We will first describe gross structures of nervous systems, introduce its functional and structural units—neurons, and its most important composing elements for information processing—synapses, and ionic channels. Its electrical properties will also be explained, as the electrical signal is one of the essential carriers of neural information the readers must understand. As what we have mentioned in the introduction, nervous systems are hierarchical systems; there are many levels[1] in them: from ionic channels or even below (say, biological macromolecules), synapses, neurons, neural circuits, maps, specific systems such as the visual system, the auditory system, etc. until the whole brain or the whole nervous system. For brevity, only some of such specific systems which are most important to neural information processing and their subsystems are introduced in the next sections. In addition to the organization of the cerebral cortex which is the organ for the higher brain functions, the anatomical structure and physiological function of sensory systems including visual systems, auditory systems, olfactory systems, and somatosensory systems are introduced in details. The functional organization of the somatomotor system is also described. Besides, cerebellums and the hippocampus are introduced, respectively—not only for their own functions involved in motor control and in learning and memory but also that these two systems provide nice models for studying synaptic plasticity. Of course, in a short chapter as this one, it is impossible to cover all the fields in neurobiology; further readings are recommended in the last section for those readers who wish to learn more.

2.1 Gross Anatomy of Nervous Systems

Nervous systems have inhomogeneous 3D structures. To determine the location and shape of some structure within it, one way is to view its projection on some section plan with special orientation. The following three special section plans are generally accepted to be the reference ones: The sagittal plane is the plane which divides the head (or body) into right and left parts (Fig. 2.1d); the coronal plane is the one which divides the head into front and back parts (Fig. 2.1c), usually people view it from behind; and the horizontal plane is the one which divides the head into upper and lower parts (Fig. 2.1b), usually people look down to it. Generally speaking, a central sagittal section, which is the sagittal section dividing the body into equal

[1] Here the term "level" means organization level, which is different from Marr's analysis level.

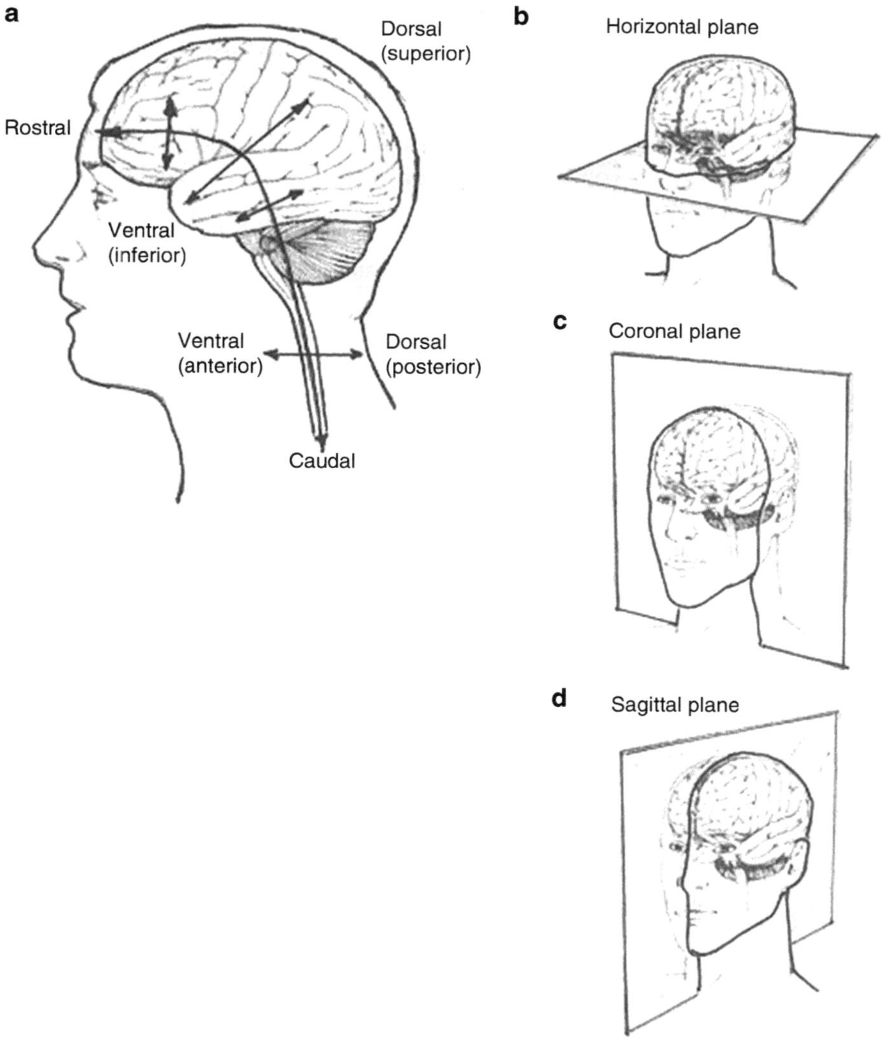

Fig. 2.1 Terms of reference. (**a**) Direction terms along the body axes and cerebrum axis, (**b**) horizontal plane, (**c**) coronal plane, (**d**) sagittal plane (Adapted with permission from Steward O (2000) *Functional Neuroscience,* Springer, Fig. 1.1, from Martin JH (1989) Neuroanatomy Text and Atlas, Elsevier, and Kandel ER Schwartz JH (1985) Principles of Neural Science, Elsevier)

halves, is taken first to locate where the structure is along the body axis; then a coronal section is taken to show the projection of location and shape of that structure on the plane.

For convenience of talking about the location of some structure related to others, it may be necessary to define some terms to denote the location relationship between them. It is shown in Fig. 2.1a that the structures of the nervous system are almost placed in a straight line along the body axis below the cerebrum; however, the line turns off for 120° approximately in the cerebrum, as shown in this figure. It is clear for the meaning of "superior," "inferior," "anterior," and "posterior" for all structures, as shown in the figure. However, the meaning of "rostal," "caudal," "dorsal," or "ventral" is a little bit different for different cases. In the case of subcerebral structures, "dorsal" means the direction to the back, "ventral" to the belly, "rostal" to the top of the head, and "caudal" to the opposite direction, while in the case of cerebrum, "dorsal" means the direction to the top of the head, "ventral" to the opposite direction, "rostal" to the face, and "caudal" to its opposite direction. In addition, "medial" is used to denote the direction toward the middle line of the body, while "lateral" away from it.

Laymen are apt to confuse brains with nervous systems. As a matter of fact, the brain is only a part of the nervous system. The nervous system is composed of the central nervous system and the peripheral nervous system, while the central nervous system is composed of the brain and the spinal cord (Fig. 2.2).

The brain is the organ which is responsible for all the higher functions, such as perception, initiation of movements, learning, memory, emotion, and consciousness, and other basic functions for survival such as breathing, regulating heart rate, blood pressure, and so on and so forth. The spinal cord is responsible for reflex activities and rhythmic actions and is the pathway both for transmitting peripheral sensory information from sensory receptors to the brain and action instructions from the brain to muscles and glands.

A more detailed division of nervous systems could be listed as follows:

- Nervous system
  - Central nervous system
    - Brain
      - Forebrain
        - Cerebral hemisphere
          - Neocortex
          - Basal ganglia
          - Limbic system
      - Diencephalon
        - Thalamus
        - Hypothalamus
      - Midbrain (mesencephalon)
      - Hindbrain
        - Metencephalon
          - Cerebellum
          - Pons
        - Medulla oblongata

Spinal cord

Peripheral nervous system

Somatic nervous system
Autonomic nervous system

Sympathetic division
Parasympathetic division

The above jargons may make beginners feel dizzy. However, it would be much clearer if one looks for the structures corresponding to these terms in Figs. 2.2 and 2.3, while reads the above list.

Central Nervous Systems Mainly, the information processing in central nervous systems is discussed in this book, while only a very brief introduction to peripheral nervous system is given. The locations of a variety of structures in the central nervous systems in sagittal section are shown in Fig. 2.3a. The terms forebrain, midbrain, and hindbrain are named from their locations in the early stage of the development of the brain.

Cerebrum The cerebrum is composed of two hemispheres: the left hemisphere and the right hemisphere, separated by a longitudinal fissure (Fig. 2.4). Sensation and movement of the trunk and limbs are controlled by the contralateral hemisphere, while sensation and movement of the head and face are controlled by the ipsilateral hemisphere.

There is a giant bundle of nerve fibers—corpus callosum—interconnecting the two hemispheres at the bottom of the longitudinal fissure (the section of this bundle is shown in Fig. 2.3a). Besides corpus callosum, another bundle of nerve fibers—fornix (its fimbria is shown in Fig. 2.4c)—interconnects bilateral hippocampus and hypothalamus between the two hemispheres. It is these fibers that make information be exchanged between the two hemispheres and make it possible to coordinate their functions.

The surface of cerebrum is separated by many sulci (only very deep sulcus was named as fissure) into many gyri. The main sulci and gyri on the cerebrum surface and their names are shown in Fig. 2.4b. According to its location and function, the surface of cerebral hemisphere can be separated by some sulci or fissures into four main lobes—frontal lobe, parietal lobe, temporal lobe, and occipital lobe (Fig. 2.4b); sometimes, a part of the medial cerebral surface encircling the upper brain stem is called limbic lobe as the fifth lobe (Fig. 2.5). Lateral sulcus (Sylvian fissure) separates temporal lobe from the other areas (Fig. 2.6). There is also an area called insula at the bottom of this fissure, which is related to tasting. Central sulcus (Rolandic fissure) separates front lobe from parietal lobe. Parietooccipital sulcus is the border between parietal lobe and occipital lobe.

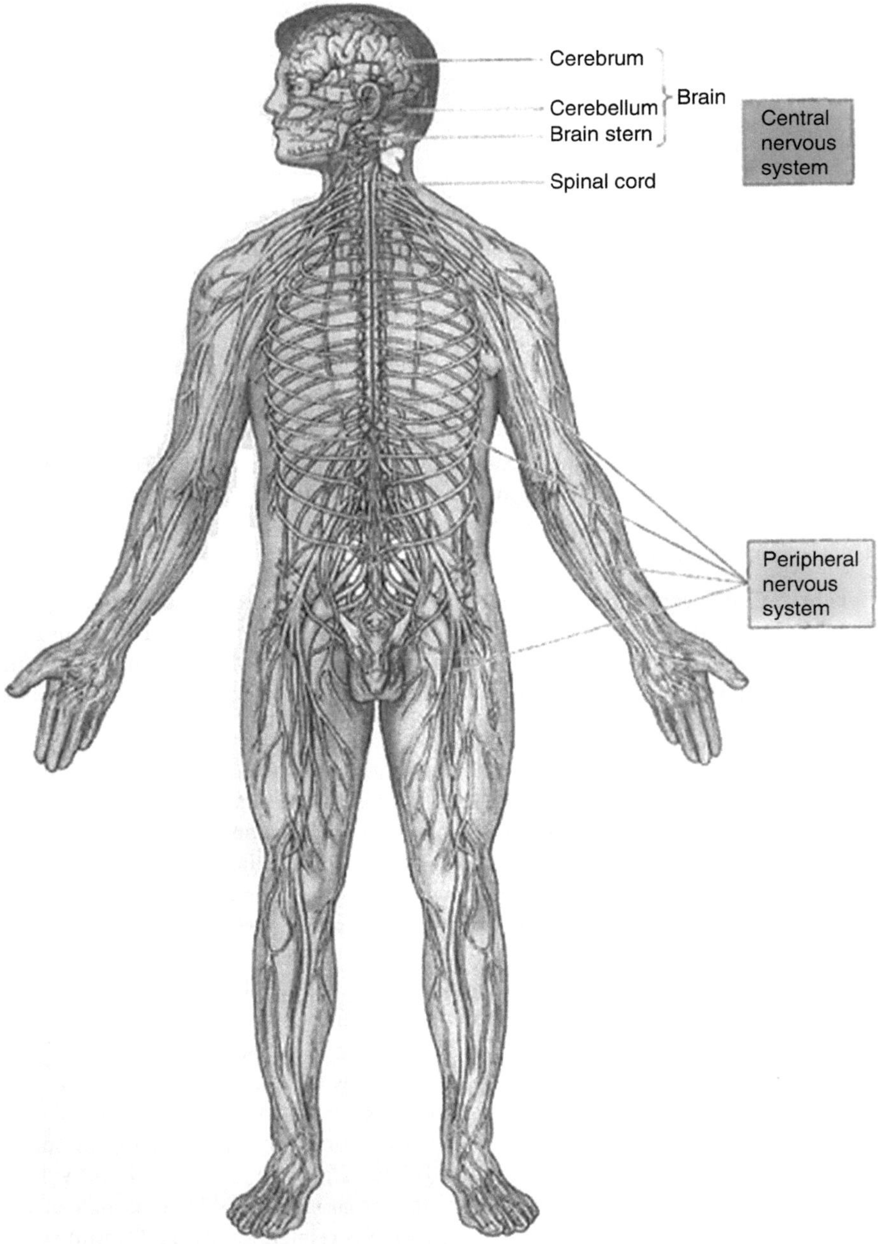

Fig. 2.2 The nervous system. The nervous system is composed of central nervous system (CNS) and peripheral nervous system (PNS), while the former consists of the brain and the spinal cord, and the latter consists of the nerves and nerve cells outside CNS (Reproduced with permission from Bear MF, Connors BW, Paradiso MA (2007) Neuroscience: Exploring the Brain (3rd Edition). Lippincott Williams & Wilkins. Fig. 1.7)

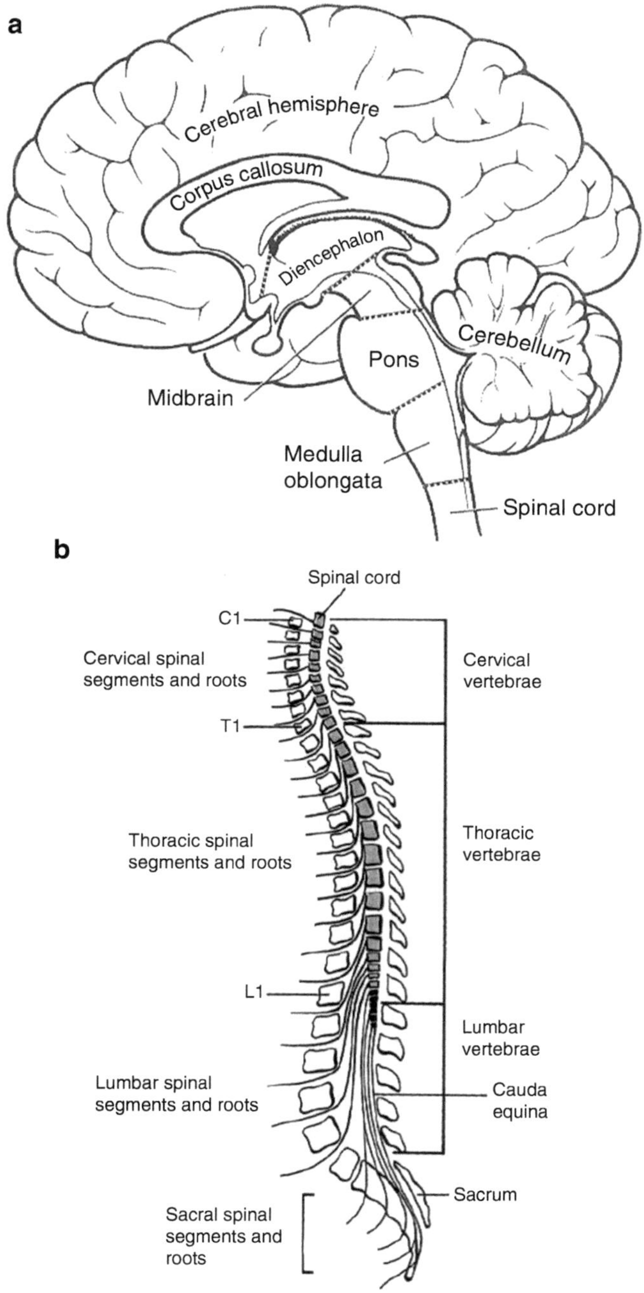

Fig. 2.3 The sagittal section of the central nervous system. (**a**) A sagittal section of the brain, which consists of cerebral hemispheres, diencephalon, midbrain, pons, medulla oblongata, and cerebellum (Reproduced with permission from Strominger NL, Demarest RJ, Laemle LB (2012) *Noback's Human Nervous System Seventh Edition: Structure Function*. Humana Press. Fig. 1.1). (**b**) A sagittal section of the spinal cord, which consists of cervical, thoracic, lumbar, and sacral segments (Reproduced with permission from Conn PM (Ed) (2008) *Neuroscience in Medicine (3rd Edition)* Humana Press (Murray M. Organization of the spinal cord. Fig. 1))

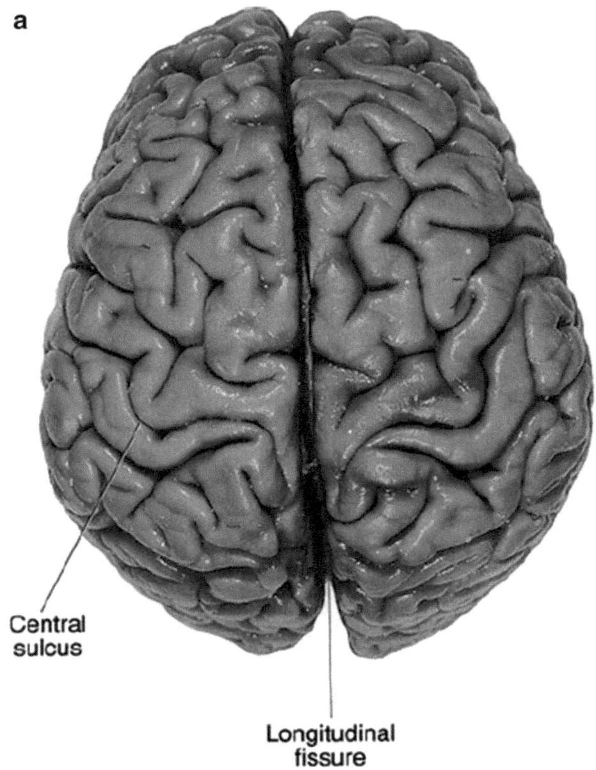

Fig. 2.4 Brain surface. (**a**) Top surface: The cerebrum is separated into two hemispheres by the longitudinal fissure. The central sulcus is the boundary between the frontal lobe and the temporal lobe. (**b**) Lateral surface: Each lateral surface consists of four lobes as shown in the lower-right figure. Numbers refer to Brodmann's areas defined by German neuroanatomist Brodmann. (**c**) Basal surface. A horizontal section through the middle and back parts of the right hemisphere reveals parts of the limbic system such as hippocampus and amygdaloidal body and other structures within the cerebrum. Roman numbers denote different cranial nerves: *I* – Olfactory, *II* – Optic, *III* – Oculomotor, *IV* – Trochlear, *V* – Trigeminal, *VI* – Abducens, *VII* – Facial, *VIII* – Auditory-vestibular, *IX* – Glossopharyngeal, *X* – Vagus, *XI* – Spinal accessory, *XII* – Hypoglossal. Half of them enter the brain via pyramid of medulla oblongata. Just above the pyramid is the pons (Adapted with permission from Strominger NL, Demarest RJ, Laemle LB (2012) *Noback's Human Nervous System Seventh Edition: Structure Function.* Humana Press. Figs. 1.4, 1.2 and 1.9)

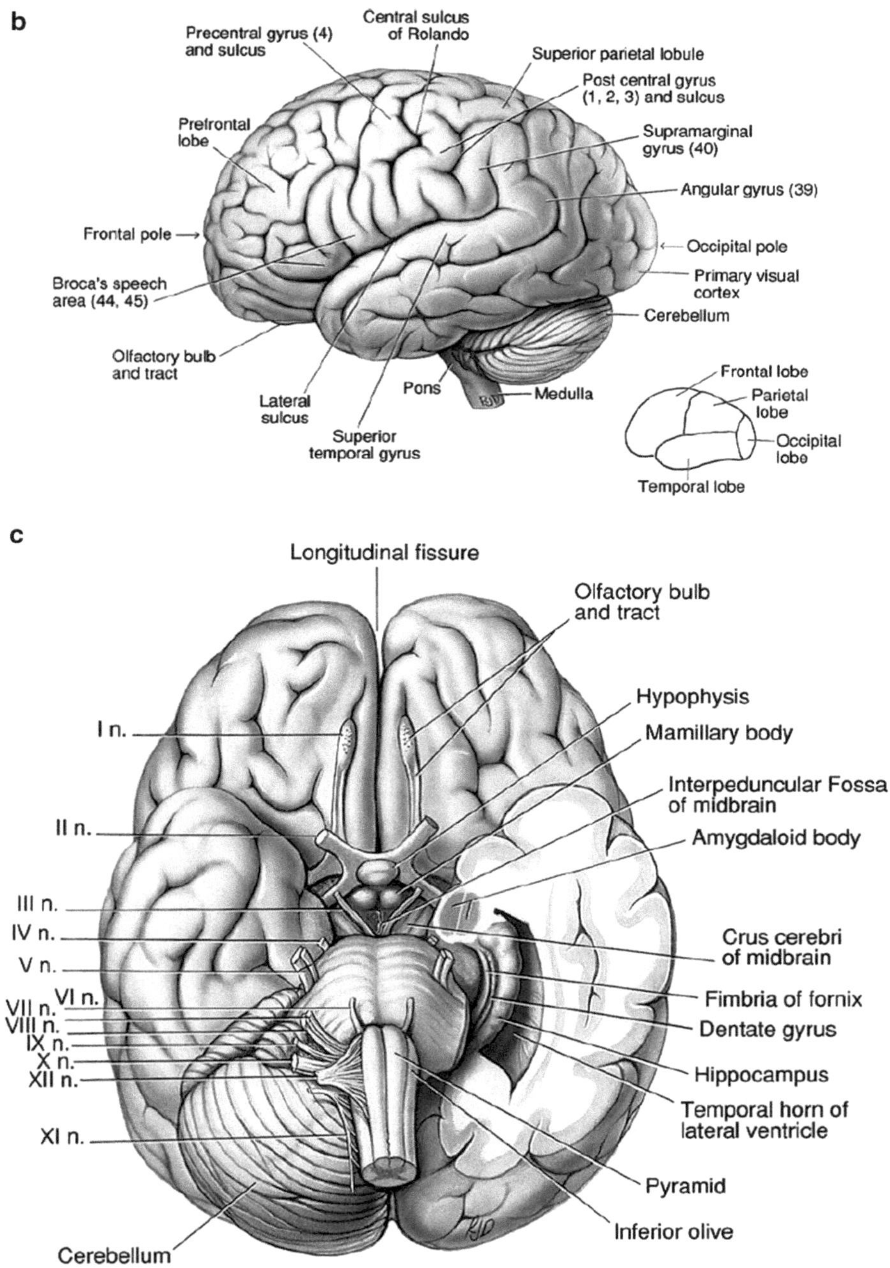

Fig. 2.4 (continued)

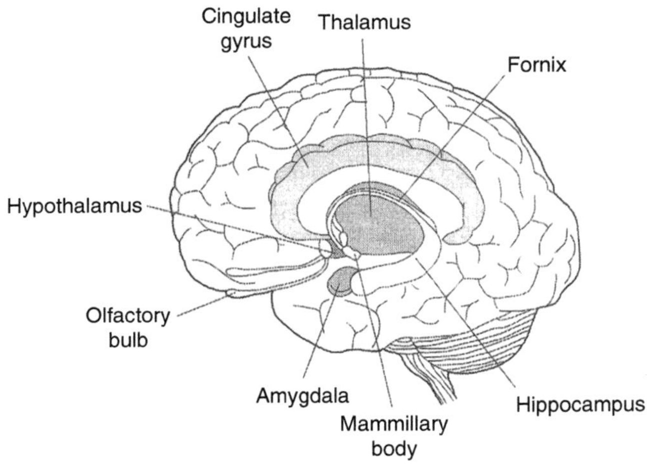

Fig. 2.5 The limbic system, thalamus, olfactory bulb, fornix, and mammillary body (Reproduced with permission from Wagner H and Silber K (2004) *Physiological Psychology*. BIOS Scientific Publisher. P.216 Fig. 1)

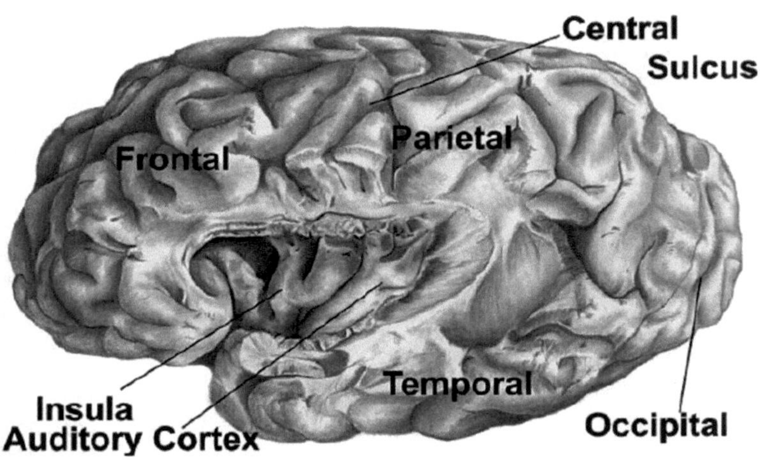

Fig. 2.6 Insula and auditory cortex deep in the lateral sulcus. The figure is taken from a left cerebral hemisphere after the inferior most frontal and parietal cortex, the operculum, has been removed (Reproduced with permission from Jacobson S and Marcus EM (2011) *Neuroanatomy for the Neuroscientist* (2nd Edition). Springer. Fig. 1.8)

There is a pinkish layer of 3–4 mm at the surface of adult cerebrum, cerebral cortex, which is composed of nerve cells and their processes and is often called gray matter,[2] due to its color in anatomical preparations. There are bundles of nerve fibers under this layer, which are covered by white fatty substances and thus are called white matter. The white matter is responsible for the communication between different areas of gray matter and plays an essential role in neural information processing too; its volume occupies almost half of the brain. There are three types of fibers in the white matter: the association fibers connecting gyri in the same hemisphere; commissural fibers connecting the gyri in one hemisphere to the corresponding gyri in the contralateral hemisphere, including corpus callosum and the anterior commissure; and projection fibers connecting the cortex and subcortical structures, such as the basal ganglia, thalamus, brainstem, and spinal cord.

Basal Ganglia Basal ganglia are a collection of nuclei scattered in white matter, including five nuclei above, below, and outside the bilateral thalami: caudate nucleus, putamen, globus pallidus, subthalamic nucleus, and substantia nigra. Caudate nucleus and putamen together is called striatum. Basal ganglia are involved in initiating and commanding movement. Basal ganglia receive their inputs from cerebral cortex via striatum, the outputs of which feedback to the cerebral cortex through globus pallidus and ventral lateral nucleus (Fig. 2.7). Any damage of basal ganglia may lead to movement disorder.

Limbic System Cut the brain along the longitudinal fissure and remove the brainstem and cerebellum away; an area surrounding corpus callosum and the removed brainstem can be shown in Fig. 2.5, which was named limbic system by the great French neurologist Paul Broca. The limbic system is composed of cingulate gyrus, hypothalamus, anterior thalamic nucleus, hippocampus, and amygdale. The function of limbic system is related to olfactory perception, emotion, memory, etc. Especially, the hippocampus plays a key role in transferring short-term memory into long-term memory.

Hippocampus The location of hippocampus is shown in Fig. 2.5. The hippocampus receives information from associative cortex via olfactory cortex and parahippocampal cortex, and its output reaches hypothalamus via fornix.

Thalamus Thalamus is a pair of egg-like structures on the top of brainstem shown in Fig. 2.8. Many nuclei in the thalamus are relays from sensory receptors to cerebral cortex (except for olfactory receptors), for example, the lateral geniculate nucleus is the relay from optic nerves to primary visual cortex. Thus, thalamus is often considered as the gate to the cerebral cortex. In addition, there are complex reciprocal connections between cerebral cortex and thalamus; thus, thalamus and

---

[2] Besides cerebral cortex, there are some nuclei deep in the brain such as basal ganglia, etc., which are also mainly composed of neuronal cell bodies and dendrites; they also belong to gray matter, the same for the outer layer of cerebellum too.

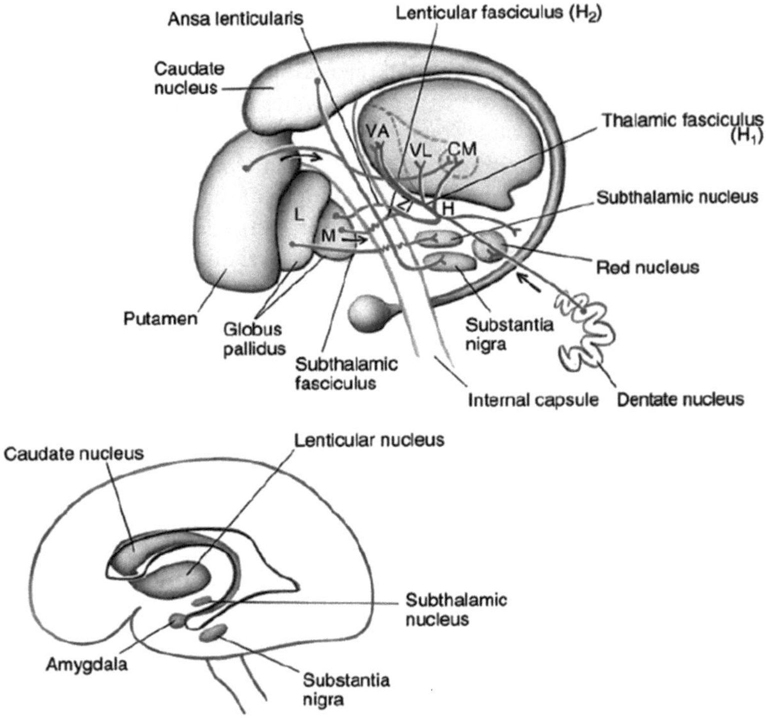

Fig. 2.7 Basal ganglia. *Lower part*: The location of the basal ganglia within the cerebrum (the lateral ventricle, which will be discussed later, is outlined as a reference). *Upper part*: The nuclei in the basal ganglia. *L* lateral segment of the globus pallidus, *M* medial segment of the globus pallidus, *VA* ventral anterior thalamic nucleus, *VL* ventral lateral thalamic nuclei, *CM* centrum medianum, *ZI* zona incerta (Reproduced with permission from Strominger NL, Demarest RJ, Laemle LB (2012) *Noback's Human Nervous System Seventh Edition: Structure Function.* Humana Press. Fig. 24.2)

cerebral cortex together is often considered as one system—the so-called thalamocortical system, which is often considered as the core for consciousness. Edelman and Tononi (2000) called such complex reciprocal connection between cortex and thalamus and the communication via such connections to and fro "reentry" and suggested that such reentry is one of the prerequisites for consciousness and one of the significant differences between brains and computers.

Hypothalamus The location of thalamus and hypothalamus is shown in Fig. 2.5. Nuclei in hypothalamus are centers for keeping body's homeostasis; they regulate many vital functions such as drinking and feeding, sexual activities, body temperature, heart rate and blood pressure regulating, etc. There are many direct connections between hypothalamus and a variety of brain areas. In addition, many nuclei in it release hormone into blood and regulate neuronal activities throughout the nervous system, and its activity is also influenced by hormones in the blood.

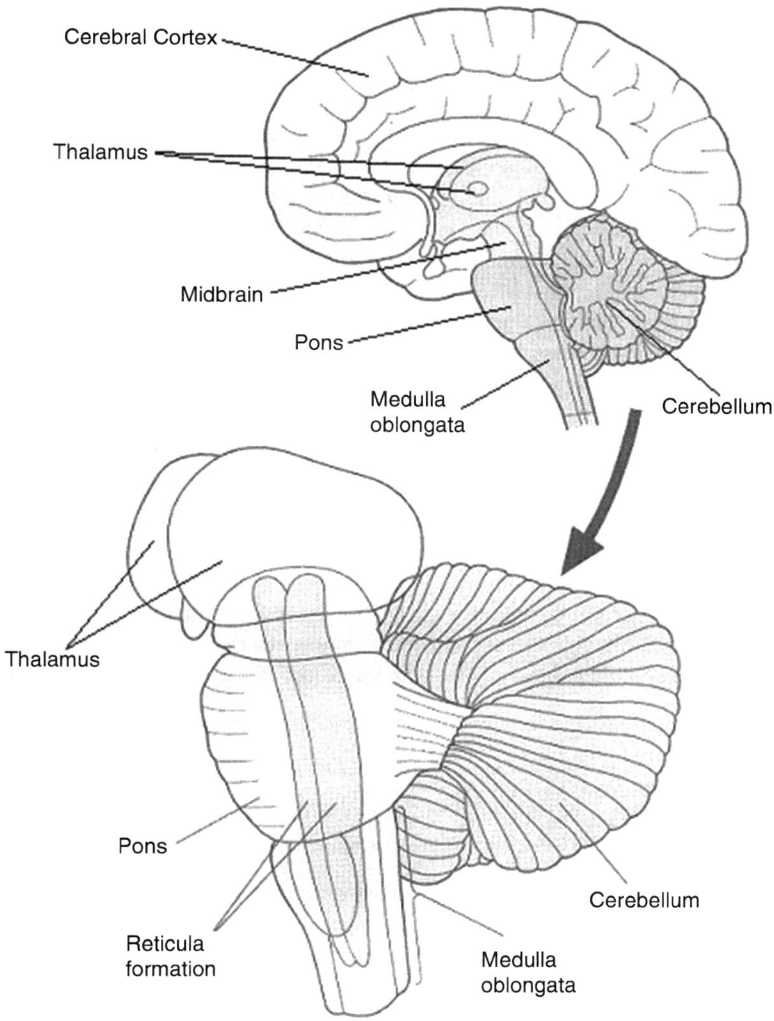

Fig. 2.8 Brainstem, thalamus, and cerebellum. *Upper part*: The location of thalamus, brain stem, and cerebellum in a brain sagittal section. *Lower part*: A 3D diagram of the above structures, reticular formation is also shown (Adapted with permission from Wagner H and Silber K (2004) *Physiological Psychology*. BIOS Scientific Publisher. P.30 Fig. 7)

Therefore, not only neuronal electric activities but also chemical substances play essential roles in neural information processing. Compared with electrical activities, chemical factors effect more broadly and last longer; they may adjust parameters of the nervous system. However, chemical effects have much less been considered in neural modeling than electrical activities.

Brainstem The brainstem and thalamus are shown in Fig. 2.8. The egg-like structure on the top of the figure is the thalamus; the rest is brainstem, including midbrain, pons, and medulla oblongata. In the midbrain, several structures can be seen in this figure: Pineal body is at the middle of the brain, it releases melatonin to modulate circadian rhythm and is related to sexual activities; superior colliculus is related to controlling eye movements, while inferior colliculus is related to hearing. Besides being relay stations among cerebrum, cerebellum, and spinal cord, midbrain is also the center to regulate basic vital functions, such as controlling breath, heart rate, blood pressure, etc. Patient may survive after cerebrum was injured; however, brainstem damage often means death. The health of brainstem may be an enabling factor of consciousness; however, it may not be involved in the neuronal correlates of consciousness.[3]

In brainstem, there is a collection of diffusely organized nuclei and tracts called the reticular formation (Fig. 2.8), which is defined more by its physiological characteristics than by anatomic boundaries. Some of its nuclei release important neurotransmitters spreading to vast regions of the brain, such as serotonin (e.g., from the dorsal raphe nuclei) and norepinephrine (e.g., from the locus coeruleus) (Figs. 2.9 and 2.10), which are involved in sensation, movement, and autonomic function and above all maintaining consciousness. A portion of the reticular formation in the upper pons and midbrain, called the ascending reticular activating system (ARAS), sends projections to the entire cerebrum via thalamus. ARAS is essential for the sleep and wakefulness cycle; damage to the ARAS may lead to coma.

Midbrain The midbrain is a structure between pons and thalamus, which surrounds the cerebral aqueduct connecting the third ventricles and the forth ventricle. Several nuclei in the midbrain are involved in eye movements (such as superior colliculus), auditory relay (e.g., inferior colliculus), and movement coordination (e.g., red nucleus). The location of the midbrain in the brain is shown in Fig. 2.3a.

Pons The pons (Figs. 2.3a and 2.8) is mainly composed of nerve tracts connecting the cerebral cortex and the brainstem, the spinal cord, and the cerebellum, within which some nuclei are scattered. These nuclei are related to audition, balance, sensory and motor functions about the face and the mouth.

Medulla Oblongata All the ascending and descending nerve tracts between spinal cord and higher brain centers pass through medulla oblongata (Figs. 2.3a and 2.8). Half of the cranial nerves enter the brain via the pyramid of medulla oblongata (Fig. 2.4c), in which many nuclei are related to cranial nerves. These nuclei are related to the movements of the tongue, head, and face and also to functions about regulating heart rate, breath, and digestion.

Spinal Cord The spinal cord (Figs. 2.2 and 2.3b) is the most caudle part of the central nervous system, which receives and deals with the information from the skin, joints, muscles from limbs and the trunk, and also controls their movements.

---

[3] The neuronal correlate of consciousness is "the minimal set of neuronal mechanisms or events jointly sufficient for a specific conscious percept or experience" (Koch 2004).

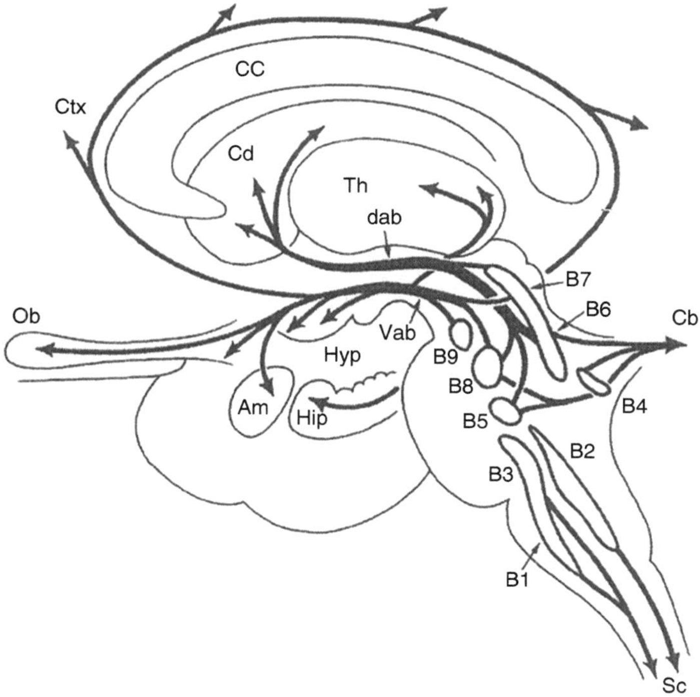

Fig. 2.9 Summary of ascending serotonergic projections. *Am* amygdale, *B1-B9* serotonergic cell groups, *Cb* cerebellum, *cc* corpus callosum, *Cd* caudate nucleus, *Ctx* cortex, *dab* dorsal ascending bundle, *Hip* hippocampus, *Hyp* hypothalamus, *Ob* olfactory bulb, *Sc* spinal cord, *Th* thalamus, *vab* ventral ascending bundle (Reproduced with permission from ten Donkelaar HJ (2011) *Clinical Neuroanatomy: Brain Circuitry and its disorders*. Springer. Fig. 5.10)

Cerebellum The cerebellum (Figs. 2.4b, 2.4c and 2.8) is one of the biggest structures in the brain only next to the cerebrum. Its surface folds like the cerebrum, the area of which is bigger than half of the cerebral surface. The cerebellum is also divided into two hemispheres, concerning with the coordination of limbs, while the bridge connecting the two parts is called the vermis which is devoted to postural adjustment (Fig. 2.11). Mainly, the cerebellum is a movement control center; it has wide connections with cerebrum and spinal cord. It receives both movement instructions from the cerebrum and sensory information from spinal cord and other sensory systems, the integrated information is sent to control movement. Recently, there are evidences showing that the cerebellum not only plays a role of "movement controller" but also plays supporting roles in detection and coordination of sensory inputs and thus may influence memory, attention, and cognition (Bower and Parson 2003). Different from the cerebrum, the left cerebellum hemisphere controls the left part of the body, and the right one controls the right half of the body.

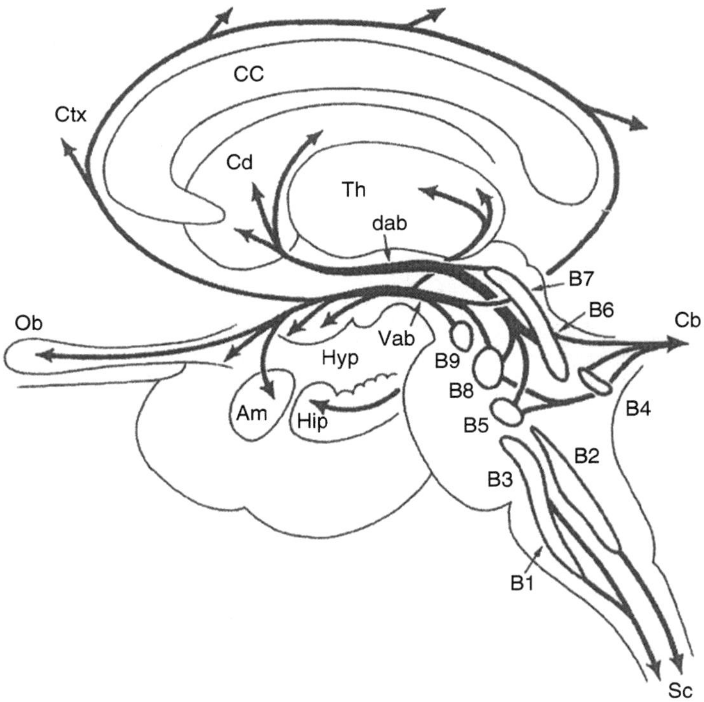

Fig. 2.10 Summary of ascending noradrenergic projections. *Am* amygdale, *A1-A9* noradrenergic cell groups, *Cb* cerebellum, *cc* corpus callosum, *Cd* caudate nucleus, *Ctx* cortex, *dab* dorsal ascending bundle, *dpb* dorsal periventricular bundle, *Hip* hippocampus, *Hyp* hypothalamus, *Ob* olfactory bulb, *Th* thalamus, *vab* ventral ascending bundle (Reproduced with permission from ten Donkelaar HJ (2011) *Clinical Neuroanatomy: Brain Circuitry and its disorders*. Springer. Fig. 5.11)

Ventricle Ventricles are cavities filling with transparent cerebrospinal fluid within the brain. It can be divided into four connected compartments as shown in Fig. 2.12. There is a lateral ventricle within each hemisphere, both of them underlie the frontal, temporal, parietal, and occipital lobes. In the middle, there is the third ventricle, surrounded by thalamus. The cerebrospinal fluid is generated in the lateral ventricles, it flows through the third ventricle, the cerebral aqueduct, and then into the forth ventricle just in front of the cerebellum and dorsal to the brainstem. There is a hole below the cerebellum; the cerebrospinal fluid flows through the hole into the cranial cavity and spinal column. The cerebrospinal fluid is the environment of the nervous system, which exchanges material between them and also plays a role as a buffer to protect the brain. The ventricles can also be treated as a reference for the location of some structures deep into the brain.

In the above, we have given a brief introduction to the brain morphology. Edelman and Tononi (2000) divided the brain into three subsystems from a point

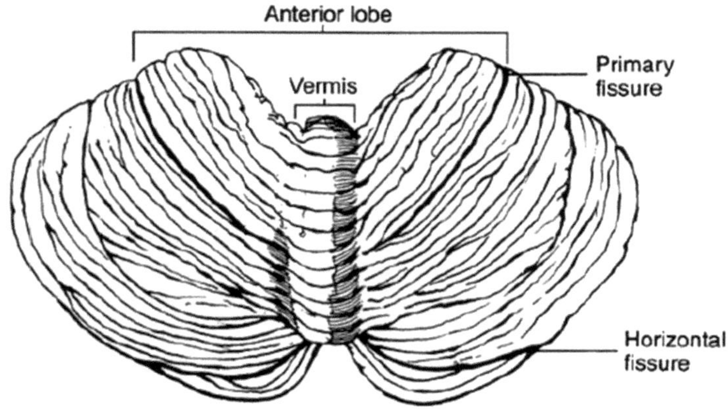

Fig. 2.11 The dorsal surface of the cerebellum (Reproduced with permission from Conn PM (Ed) (2008) *Neuroscience in Medicine (3rd Edition)* Humana Press (in Goodlett CR. The cerebellum. Fig. 1))

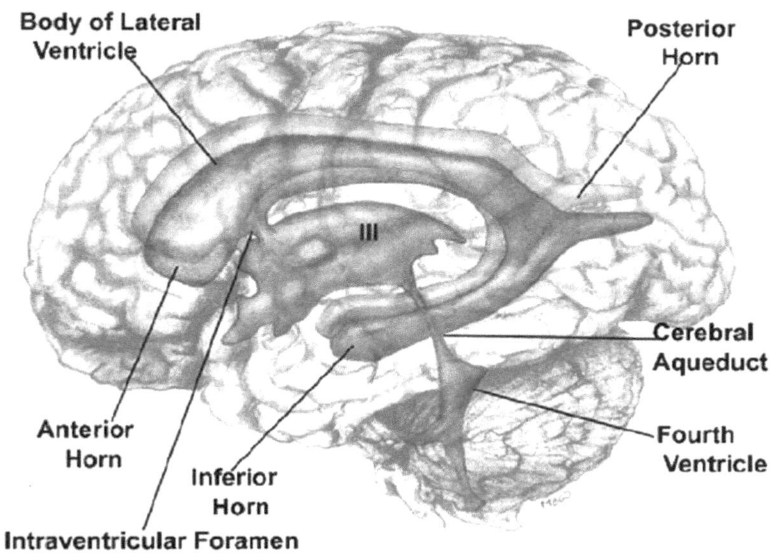

Fig. 2.12 The ventricles (Reproduced with permission from Jacobson S and Marcus EM (2011) *Neuroanatomy for the Neuroscientist (2nd Edition)*. Springer. Fig. 18.3)

of view of functions: (1) the thalamocortical system, (2) cortical appendages, and (3) the diffusely projecting value system.

The thalamocortical system is composed of the cerebral cortex and the thalamus. The sensory information enters the thalamus firstly and then projects to the cerebral cortex via some relays. Now it has been discovered that there are hundreds of thalamocortical areas with specific functions, while each area is composed of ten thousands of neural assemblies. These millions of neural assemblies interact each other with diverging and converging connections; thus, they can keep functional specificity on one hand while form unified and closely coupled networks on the other hand. Disturbances in any part of the network would be sensed by everywhere in the network quickly. Therefore, it seems that the thalamocortical systems are suitable for integrating a large quantity of parts with specific functions to give a unified response. Edelman called such broad interconnections "reentry." He thought that although there is no coordinating center in the brain, it is the reentry which integrates various specific functions of a variety of brain areas.

The cortical appendage includes cerebellum, basal ganglia, and hippocampus. Besides ascending pathway from cerebellum to cortex, the cerebellum also receives descending inputs from the cerebral cortex and projects to thalamus with a large quantity of synaptic exchanges and then back to the cortex again. The basal ganglia receive inputs from many areas of the cortex too and project to thalamus through a series of successive synaptic exchanges and then back to the cortex again. The hippocampus receives inputs from many areas in the cortex and projects back to many places in the same areas after processing. Almost all the cortical appendages have similar organization mode (especially cerebellum and basal ganglia): They have parallel and multi-synaptic pathways from the cerebral cortex, which reach a series of successive relays, and then back to the cortex again, no matter if they pass through the thalamus or not. Such serial multi-synaptic structure is quite different from the one of thalamocortical systems. The former, generally speaking, is monodirectional rather than reciprocal and forms long loops, without horizontal interactions except for mutual inhibition for short distances. These systems seem suitable for a variety of complex movement and cognition procedures; majority of such procedures need functional isolation between different loops, so that such functions can be carried out precisely and quickly.

The diffusely projecting value systems start from a few neurons in some special nuclei in the brainstem and hypothalamus, project diffusely to the vast areas of the brain, and thus influence billions of synapses (Figs. 2.9 and 2.10). These nuclei are named respectively after the chemical substances they release, such as the noradrenergic locus coeruleus, serotonergic raphe nucleus, etc. The firing of these neurons leads to releasing some special chemicals—neuromodulators diffusely. Neuromodulators influence not only neuronal activities but also neuronal plasticity, i.e., they can modify synaptic strengths in neural circuits, leading to adaptive responses. It seems that this system is suitable for telling the whole brain that some important thing has happened and led to modification of the strengths of billions of synapses.

Peripheral Nervous System The main role of the peripheral nervous system (Fig. 2.2) is to provide pathways between the central nervous system and sensory organs, muscles, and glands. It includes all the nerves and ganglia outside the brain and the spinal cord. As what we have mentioned in the beginning of this section, the peripheral nervous system is composed of two parts: the somatic nervous system and the autonomic nervous system. The former is composed of motor nerves connecting the central nervous system and skeletal muscles and somatosensory nerves, while the latter is mainly composed of neurons and nerve fibers innervating smooth muscles, cardiac muscles, and glands. Thus, the former deals with the interaction of the body with the external world, while the latter is involved in the regulation of the internal organs to maintain the relative constancy of the internal environment (homeostasis). In addition, there are also cranial nerves emanating from the brainstem, part of which belongs to the central nervous system, while the rest belong to the peripheral nervous system.

Cranial Nerves Cranial nerves are related to the sensory and motor systems of the head (Fig. 2.4c). There are 12 pairs in total. Some of them are sensory pathways to the brain, such as olfactory nerves, optic nerves, and auditory nerves, while others are motor pathways from the brain, such as oculomotor nerves innervating muscles of eyeball movement, facial nerves innervating facial muscles, and some others are related both to sensation and movement, such as trigeminal nerves, which transfer facial sensation on one hand and control chewing movement on the other hand. All these nerves enter or leave the brain through holes at the skull. The only cranial nerve which can go long away from the head is the vagus nerves, which can extend to the heart and the intestine. The vagus nerves are also part of the parasympathetic nervous systems. The two most anterior nerves, the olfactory and the optic tract, are considered as part of the central nervous system.

Spinal Nerves Thirty-one pairs of spinal nerves are arranged along the spinal cord; a branch of each pair emanates from each side, leaving through holes at spinal column. There are both sensory nerves and motor nerves in these branches. Its sensory nerves transmit somatosensory information, including touching, temperature, and pain sensation; they enter the spinal cord from its dorsal side, while motor nerves leave spinal cord from its ventral side. The spinal cord and the spinal nerves emanating from it are shown in Fig. 2.13; the spinal cord surrounded by a layer of hard membrane is at the center of the spinal column; a section of the spinal cord is shown on the right; there is a butterfly-like structure at the center of the spinal cord, which is its part of the gray matter composed of cell bodies of interneurons and motor neurons, while the rest is the white matter composed of nerve fibers entering or leaving the spinal cord. The motor fibers leave from its ventral part, ventral root, and sensory nerves enter the spinal cord from its dorsal part, dorsal root. The cell bodies of motor nerves are within the spinal cord, while the cell bodies of sensory nerves gather together into the so-called dorsal root ganglia outside the spinal cord. After entering the spinal cord, sensory nerves may contact motor neurons directly, or contact an interneuron, or go up to higher part of the spinal cord or even the brain itself. Nerves transmitting different sensory information or motor information gather into specific ascending or descending tracts in the spinal cord.

Autonomic Nervous System There are two lines of sympathetic chains along the bilateral sides of the spinal cord, which innervate smooth muscles of guts and vessel walls. The sympathetic nervous system is composed of sympathetic chains and the

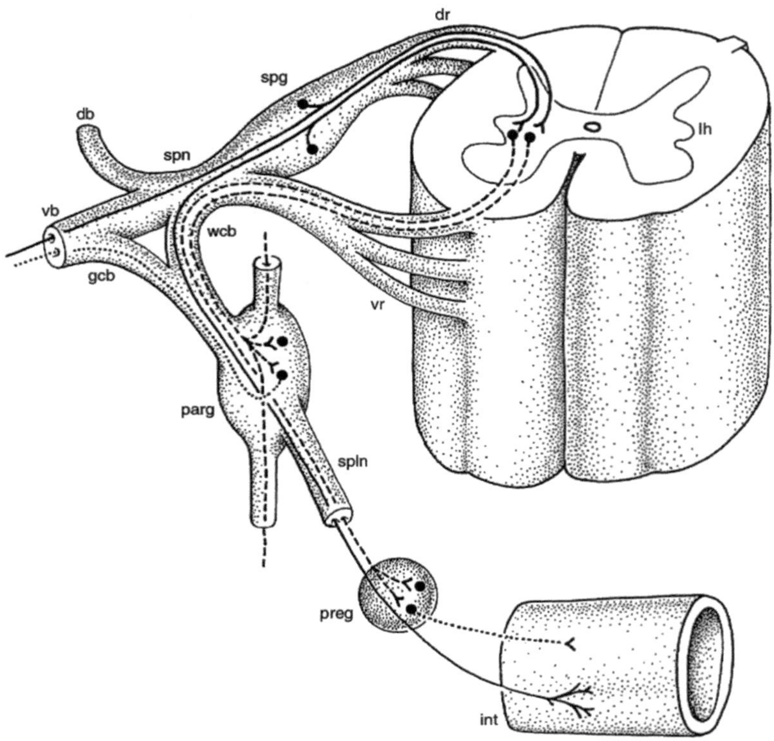

Fig. 2.13 The spinal cord and spinal nerves. *db* dorsal branch, *dr* dorsal root, *gcb* gray communicating branch, *int* intestine, *lh* lateral horn, *parg* paravertebral ganglion, *spg* spinal ganglion, *spn* spinal nerve, *spln* splanchnic nerve, *vb* ventral branch, *vr* ventral root, *wcb* white communicating branch (Reproduced with permission from ten Donkelaar HJ (2011) *Clinical Neuroanatomy: Brain Circuitry and its disorders*. Springer. Fig. 12.1 (After Lohman et al. 2007))

nerves run out from them. The parasympathetic nervous system comes from the top and sacrales of the spinal cord, in which the vagus nerves coming from the top of the spinal cord are part of the cranial nerves, as shown in Fig. 2.14. It is astonishing that the body is controlled by both sympathetic nerves and parasympathetic nerves, while the effects of them are just opposite, so that the movements can be controlled precisely. For example, excitation of the sympathetic nerves fastens the heart beat, while the excitation of parasympathetic ones slows it down; the former dilates pupils, while the latter contracts them. In majority of cases, both sympathetic system and parasympathetic system are active to reach some balance, thus controlling activities of their target precisely. In nineteen century, people thought the autonomic nervous system dominated majority of involuntary movements and misunderstood that its activity was independent of the brain; this is why they called it "autonomic." However, now we know that activities of the so-called autonomic nervous system are not independent absolutely, which will be controlled in some way by the cerebrum too.

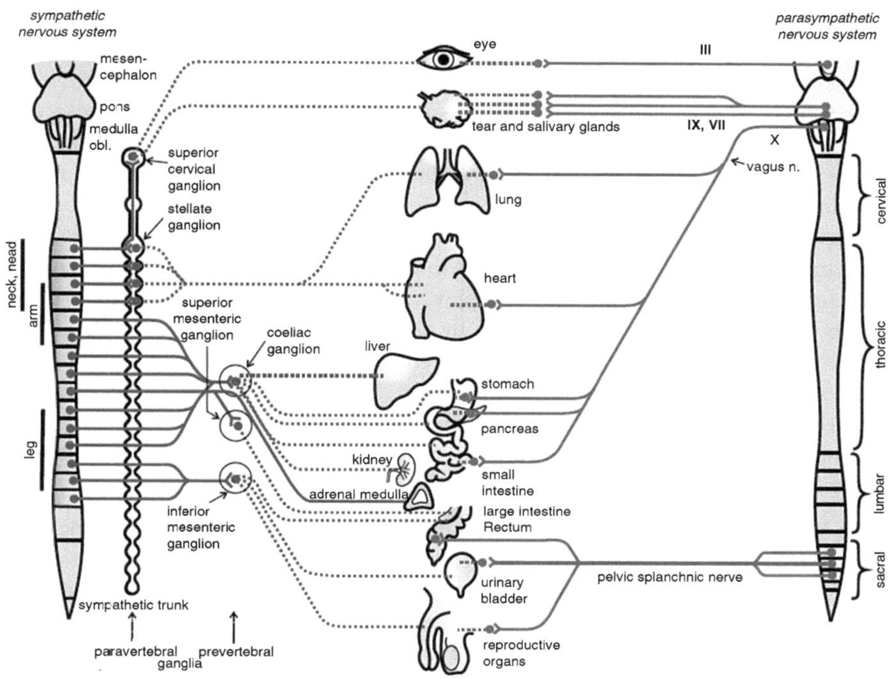

Fig. 2.14 The autonomic nervous system. *Bold lines*: preganglionic axons, *dotted lines*: postganglionic axons. The sympathetic innervation of blood vessels, sweat glands, and arrector pili muscles originates in all thoracolumbar spinal segments. The origin of the sympathetic innervation for the extremities and the head are indicated on the *left side* (Reproduced with permission from Galizia CG and Lledo PM (Eds) (2013) *Neuroscience – From Molecule to Behavior: A University Textbook*. Springer Spectrum. Fig. 10.2)

2.2 Nerve Cells

What is the basic building block of the nervous system? There are still debates: someone thinks that it is ionic channels, while others propose synapses, neurons, or neuronal assemblies. At the present moment, the majority suppose that the nerve cell or neuron is both the functional and structural unit of the nervous system. However, the nervous system is never only a simple collection of neurons. The complicated connections of neurons, hierarchical structure, and diversity of neurons themselves play key roles in sophisticated functions of the nervous system, not mentioning the role of glial cells, the quantity of which is almost ten times of the one of neurons.

Generally speaking, the quantity of neurons in the human brain is about 10^{11}–10^{12}, and every neuron can connect several thousands of other neurons on average, "ranging from a minimum of very few (exceptionally perhaps only one) to several hundreds of thousands (synapses between parallel fibers and a single

Purkinje cell in the cerebellum)" (Braitenberg 2007). As what we have emphasized before, the nervous system is a hierarchical system with many levels, every level has its own emergent properties; thus, it seems to the author that it is difficult, if not impossible, to elucidate higher function such as cognition and consciousness with properties of single neuron, or even lower level elements such as ionic channels as some scientists suggested. However, cellular level is an important level, neural assemblies are composed of neurons, and the neuron itself is a system composed of synapses, ionic channels, etc. Thus, to realize the structure and function of neuron is still important, and it will be elucidated in this section. We will also emphasize in this section that neurons are very complex, not simple elements similar to switches or so described in some public media or popular science publications. Of course, some simplification is unavoidable and even necessary for modeling, but oversimplification may mislead readers and hinder the development of new models and new theories. Thus, one should keep in mind that most hypotheses for modeling may just be simplified and should always prepare to give up some such hypothesis, when it is necessary to make the model more realistic.

The Morphological Structure of Neurons A "standard picture" of motor neurons is shown in Fig. 2.15, the most significant feature of such neuron is that there are two different kinds of processes sent out from the cell body: many branched tiny processes called dendrites, which receive input from other neurons, and a long process called axon, which sends output to other target cells. Similar diagrams are widely used in popular science publications; such pictures are so impressive that many readers thought that all the neurons should look like that. As a matter of fact, the motor neuron is only a special type of neurons. The morphology of neurons varies wildly; no two neurons share the same shape exactly. Neurons could be roughly classified into more than 50 types at least. Several of them are shown in

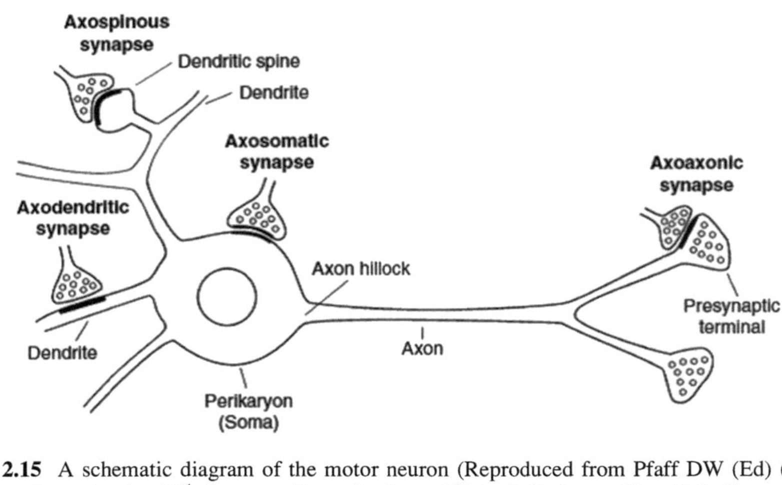

Fig. 2.15 A schematic diagram of the motor neuron (Reproduced from Pfaff DW (Ed) (2013) *Neuroscience in the 21st Century: From Basic to Clinical*. Springer. Fig. 11.2 (from Cohen RS. Cell biology of the synapse))

Fig. 2.16 as examples. In spite of the diversity of their forms, there are still some common features in all neurons, say, there are many processes emanating from the cell body, it is also possible that some process is quite long, and it is these processes that make complicated connections between neurons. For many neurons, the volume of their processes occupies more than 90 % of its volume; thus, a great part of the brain is composed of neuronal processes. Only a small part of interconnections between neurons takes place at the cell body; the majority takes place between different processes.

Generally speaking, neurons could be roughly classified into two categories, according to the nomenclature given by the great Spanish anatomist Ramon y Cajal, the founder of neuron doctrine: Golgi type I cells and Golgi type II cells. The pyramidal cell and the Purkinje cell in Fig. 2.16 are examples of Golgi type I cells, and the interneurons within the cerebral cortex are examples of Golgi type II cell.

Compared with Golgi type II neurons, the main features of Golgi type I neurons are their bigger cell bodies and long axons. The motor neuron (Fig. 2.15) is a typical example of Golgi type I neuron; its long axon can transmit the information from one neuron to another neuron or an effector such as muscles far away. As an extreme example, the axon of a motor neuron in the spinal cord which innervates muscles in the toe can be as long as more than 1 m, while the diameter of its cell body is less than 100 μm, i.e., its axon length is longer than the diameter of its cell body 10,000 times. For most readers with non-neuroscience background, they would almost always think of Golgi type I neuron, when the term neuron is mentioned. The majority of the available neuron models are models of Golgi type I neurons.

Although the motor neuron is only a special case of Golgi type I neurons, it has some features quite typical for this type of neurons. As shown in Fig. 2.15, there are many processes emanating from its cell body. Such processes can be classified into two sorts: the dendrites are tree-like process branching thinner and thinner; there are stems called "spines" emanating from such branches, which have larger and rounder heads and thinner necks arising from the dendritic shaft; on the head, there is only one excitatory synapse, and the synapse at the neck is inhibitory or modulatory. The function of spines is increasing the surface area not only for synaptic contact but also for synaptic plasticity and isolating individual synapses. Neurons that have large numbers of spines on their dendrites are called spiny neurons, while those have smooth dendrites or few dendritic spines aspiny. The second type of process is called "axon"; generally speaking, there is only one axon for one Golgi type I neuron; it is relatively long and rarely branches except at its end, where it branches into a lot of ends with some ball-like structure—bouton, which interacts with the next neuron or effector. The diameter of the axon almost remains the same for its whole length, only except for its terminals. The locations for its input and output are clear for Golgi type I neuron: it receives its inputs at its dendrites or soma (cell body) and gives out its outputs at the ends of axon. The axon leaves the cell body at axon hillock—a cone-like structure, which plays a key role in integrating information from its various inputs.

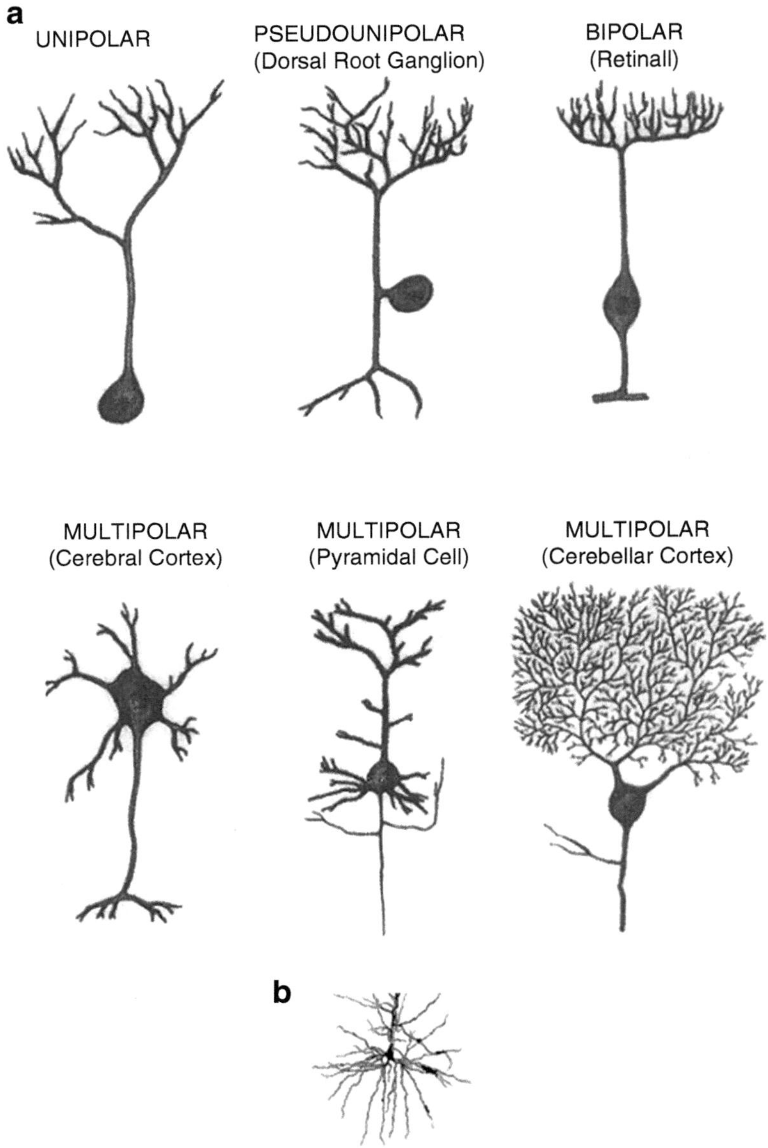

Fig. 2.16 A few schematic diagram of neurons with different shapes. (**a**) Unipolar, bipolar, and multipolar cells (Reproduced with permission from Jacobson S and Marcus EM (2011) *Neuro-anatomy for the Neuroscientist (2nd Edition)*. Springer. Fig. 2.1). (**b**) Stellate cell – an example of Golgi type II neuron

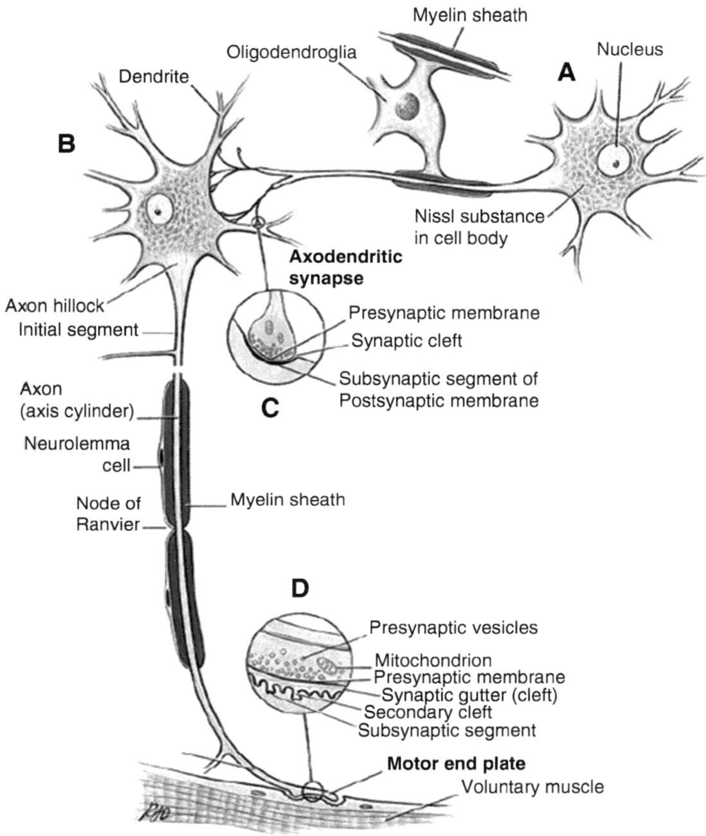

Fig. 2.17 A schematic diagram of a neuron with myelinated axon (Reproduced with permission from Strominger NL, Demarest RJ, Laemle LB (2012) *Noback's Human Nervous System Seventh Edition: Structure Function.* Humana Press. Fig. 2.1)

All the axons are wrapped by glial cells. There are two different wrapping types. Small axons can be wrapped by one glial cell, and one glial cell may wrap several such axons. For large axons, one glial cell can wrap around an axon again and again and forms so-called myelin sheath, which is some kind of electrical insulator. In the peripheral nervous system, a Schwann cell (a kind of glial cell) wraps only one axon and forms one segment of its myelin sheath with a length of 1–2 mm. There are many such segments if the axon is long, and there are gaps of 1–4 μm between neighboring segments, which are called nodes of Ranvier (Fig. 2.17). The axon with myelin sheath is called the myelinated axon, while the axon without myelin sheath is called the unmyelinated axon.

For Golgi type II neuron, their axons are much shorter or even no axon at all. In addition, its cell body is smaller. Generally speaking, they only connect their neighboring neurons and do not contact neurons far away in the nervous system;

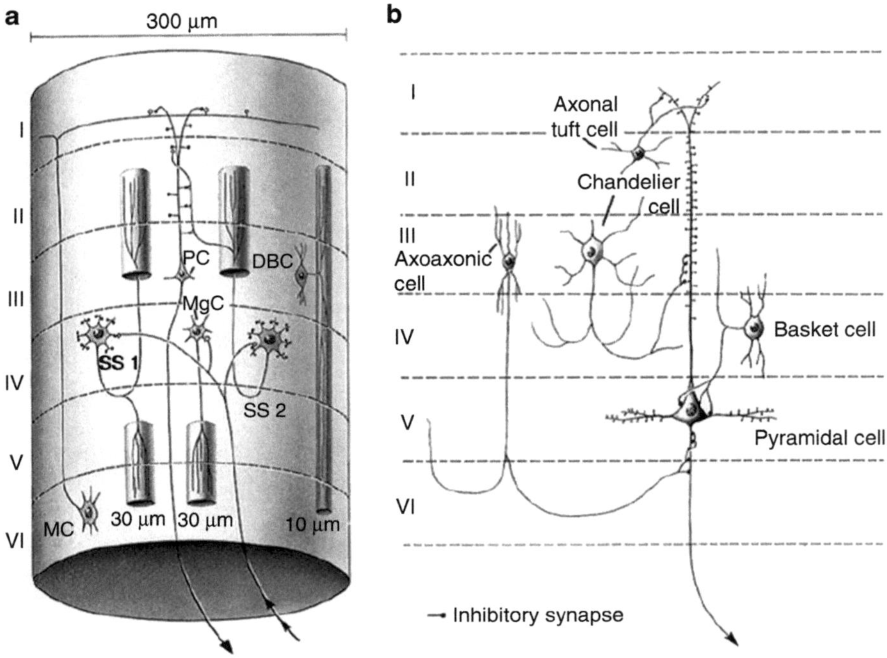

Fig. 2.18 Golgi type II neurons within cerebral cortex. (**a**) Modular organization of the neocortex as formed by the afferent inputs from specific thalamic nuclei, commissural, and association fibers. (**b**) In this figure, only pyramidal cell is Golgi type I cell, which transmits output to other region or subcortical structures. The input to this neural circuit is omitted (Reproduced with permission from Strominger NL, Demarest RJ, Laemle LB (2012) *Noback's Human Nervous System Seventh Edition: Structure Function*. Humana Press. Fig. 25.2 (adapted from Szentagothai, 1978))

their processes are limited within individual nucleus or cortical layers (Fig. 2.18b). Therefore, their role is to interact with other neurons locally, and they are also called association neuron. Another distinct feature of such neuron is that its dendrites and axon ends can play the roles of both input and output. Thus, there could be interactions between such neurons within local areas around dendrites and axons. In a typical nucleus, there are both types of neurons: the inputs to and outputs out of the nucleus are via Golgi type I neurons, while Golgi type II neurons form complicated internal interconnections, which may play important roles in information processing, that are rarely considered in the previous neural network models.

Synapse The synapse is the place where two neurons interact. At synapse, in most cases the two interacting neurons do not merge each other; there is a gap called synaptic cleft between them. There may be thousands of synapses on a neuron surface; there are even more than 100,000 synapses at the dendrites of Purkinje cells in the cerebellum.

There are two kinds of synapses: chemical synapses and electrical synapses. In the human brain, most of the synapses are chemical. At the chemical synapse, there are many vesicles containing specific chemicals in the presynaptic membrane, which can be released into the synaptic cleft, when it is activated. These chemicals can diffuse across the synaptic cleft and reach the postsynaptic membrane of the postsynaptic neuron, binding with specific proteins, receptors there, and changing its configuration to make specific ions enter or leave the cell selectively, depending on the property of the receptor (Fig. 2.19). Excess neurotransmitters are either broken down by enzymes located in the cleft, diffused away from the synaptic site,

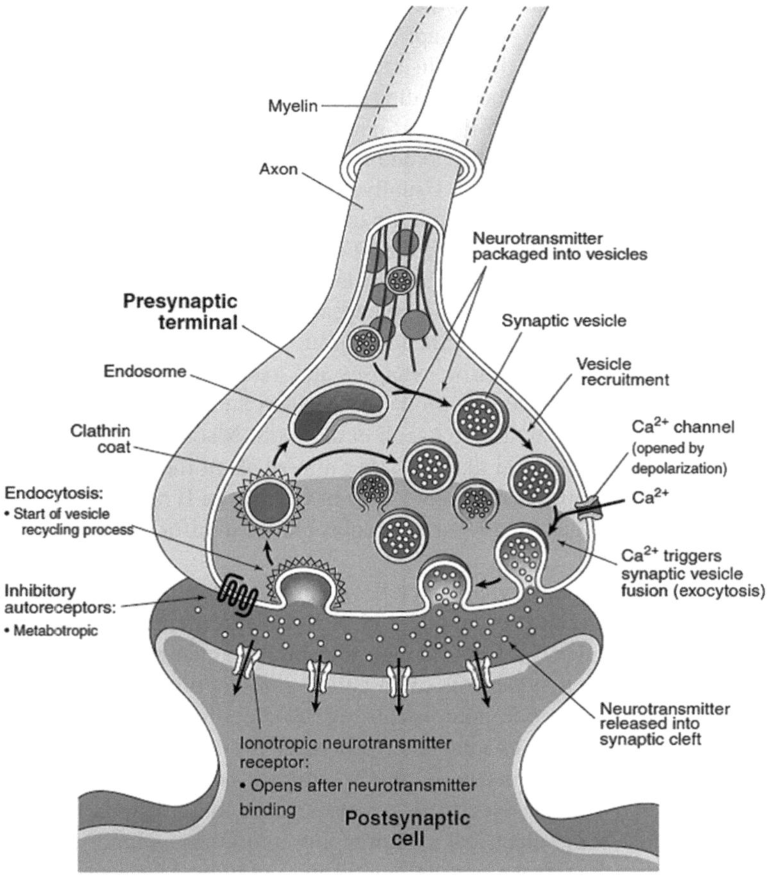

Fig. 2.19 The chemical synapses. The presynaptic terminal (or bouton) is filled with neurotransmitter containing synaptic vesicles. A presynaptic spike depolarizes the nerve terminal and opens Ca^{++} channels. Ca^{++} ions trigger synaptic vesicle fusion and release the contained neurotransmitter into the synaptic cleft. The neurotransmitter binds to postsynaptic receptors after diffusion across the cleft. The binding makes the specific ion channels on the postsynaptic membrane open and let the corresponding ions flow through the channels (Reproduced with permission from Conn PM (Ed) (2008) *Neuroscience in Medicine* (3rd Edition) Humana Press (in von Gersdorff H. Synaptic transmission. Fig. 3))

or retaken up by the presynaptic cells again, and thus its effect cannot be kept for a long time. For the chemical synapse, the width of the synaptic cleft is about 20–40 nm. There are two subtypes of chemical synapses: excitatory and inhibitory. Most of the excitatory synapses are at dendrites, their synaptic clefts are wider, their vesicles in the presynaptic membrane are ball like, and their effects to the postsynaptic membrane are excitatory, i.e., making the postsynaptic neuron more likely to fire, while most of the inhibitory synapses are at cell bodies, their synaptic vesicles are flat, their synaptic clefts are narrower, and their effects to the postsynaptic membrane are inhibitory, i.e., making the postsynaptic neuron more difficult to fire. There are also some synapses; the chemicals which are released are neither excitatory nor inhibitory; they modulate the postsynaptic membrane and trigger a series of chemical reactions in the postsynaptic neuron.

Most of the synapses in the brain are between the end of the axon and the dendrite of the next neuron, the so-called axodendritic synapse, and between the end of the axon and the soma of the next neuron, the so-called axosomatic synapse (Fig. 2.15). Generally speaking, axodendritic synapses are excitatory, while axosomatic synapses are inhibitory. Usually, there is a structure called the neuropile at the axodendritic synapse, where a large quantity of dendrites, axonal terminals, and processes of glial cells tangle together. For Golgi type II neuron, there are also synapses between dendrites and dendrites (dendrodendritic synapse) and synapses between axons and axons (axoaxonic synapse). Usually axoaxonic synapses are inhibitory and are called presynaptic inhibition. The dendrodendritic synapse is more complicated. Dendrodendritic synapses and axoaxonic synapses were rarely considered in the available neuron models. Recently, studies on interneurons have increased (Druckmann et al. 2012; Ascoli et al. 2008; Maciokas et al. 2005; Richardson et al. 2005; Markram et al. 2004), but almost all of them are about neurons with short axons, as for the studies on models of Golgi type II neurons without axon are still rare, if there is any, and what their roles played in information processing is still an open problem.

The width of electrical synaptic cleft is only 1–2.5 nm. And some protein channels even across both presynaptic and postsynaptic membrane, so that some ions and small molecules can flow through the common channel directly from one neuron to the next. Such synapse is often called gap junction channel (Fig. 2.20). The electrical changes in one neuron can lead to electrical changes in the other neuron through gap junction channels; the process is faster than the one of chemical channels; however, the process is much simpler and lack of complicated processing. For most of the electrical synapses, such ionic flow is bidirectional. However, there are some electrical synapses, the ionic flow is unidirectional, just like a diode, and such synapse is called rectifying junction.

Glial Cells As what we have mentioned above, Schwann cells are special cases of glial cells, which are also called neuroglia. As a matter of fact, the nervous system is composed of glial cells and neurons. Although people think that the neuron is both the structural and the functional unit of the nervous system, there are much more glial cells in the nervous system (ten times of the neurons or even more). Half of the cerebrum volume is occupied by glial cells. However, it is not clear what the roles the glial cell plays.

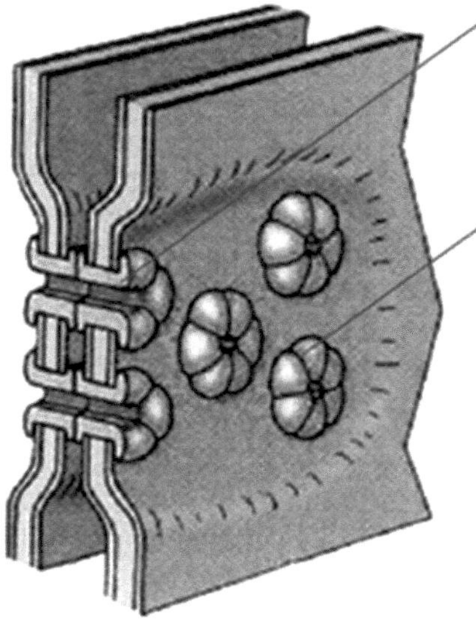

Fig. 2.20 Electrical synapse (Adapted with permission from Galizia CG and Lledo PM (Eds) (2013) *Neuroscience – From Molecule to Behavior: A University Textbook.* Springer Spectrum (in Lüscher C and Petersen C. The synapse. Fig. 8.8b))

At the present moment, most of the people think that glial cells play roles of support and nutrition for neurons and fill the space between neurons; they may regulate the compositions of the extracellular space outside neurons and thus may modify their functions indirectly. Schwann cells—a specific kind of neuroglia—form insulating layers around nerve fibers to make nerve impulse propagate more quickly. Some of neuroglia contact both neurons and vesicles to make a barrier between tissues in the central nervous system and the blood and to protect the cerebrum from the damage by harmful substances in the blood. Recent studies have hinted that the neuroglia may also play a role in learning and memory. There are intercommunications not only between glial cells and neurons but also between different glial cells themselves. Glial cells may also regulate the information transmission at synapses and even have some effect in synaptic formation (Fields 2004). These kinds of information exchanges may mainly depend on chemical signals; however, their detailed underlying mechanisms and the exact roles the glial cells play in neural information processing are still unclear.

2.3 Electrical Characteristics of Neurons

In ancient times, people thought there were some spirit flows in hollow nerves to make muscles contracting. Such ideas were given up only after the great Italian physician Luigi Galvani (1737–1798) found that it was electrical signals rather than the mysterious spirit that propagated along the nerve fibers and controlled muscle contract. It has been well acknowledged that electrical signals play a key role in neural information processing. It is the electrical potential changes across the cell membrane of the neuron that encode and transmit information. As a matter of fact, there is some change of the potential difference across the cell membrane, when a neuron receives some input; and the information transmission depends also on potential changes. Therefore, it is important for understanding neural information processing to comprehend the electrical characteristics of neurons.

There are three categories of membrane potentials: resting potentials, graded potentials, and action potentials. The resting potential is the constant potential difference across the neuronal membrane without any stimulation. The graded potential is a continuously changeable membrane potential biased from resting potential when a neuron is stimulated. And the action potential is a propagatable electrical impulse with fixed shape and amplitude when some graded potential reaches a fixed value—the threshold. A kind of graded potential—receptor potential—is generated when a sensory receptor is stimulated, which may lead to an action potential along its axon or a synaptic potential, a graded potential at the postsynaptic membrane of its neighboring neurons. Both the synaptic potential and receptor potential are often called generator potential. A synaptic potential can lead to an action potential or a synaptic potential of other neuron. Another kind of graded potential is the subthreshold potential at the axon hillock, which is formed by the sum of all the postsynaptic currents flowing through the axon hillock.

A graded potential can only spread with short distance and its amplitude will decay with distance, while an action potential can propagate with long distance and keep its amplitude unchanged. From a point of view of information processing, action potentials are suitable for long distance signal transmission, while graded potentials play essential role in signal integration. A brief introduction of the three kinds of potentials will be given in the following. As all these events happen at neuronal membrane, thus a brief description of the membrane is necessary:

Neuronal Cell Membrane All neuronal membranes are composed of lipid bilayer with some specific proteins embedded in it (Fig. 2.21). Neither ions nor other solutes can penetrate the lipid bilayer; thus it plays a role as a capacitor; ions and small polar molecules may pass through the channel composed of membrane proteins embedded within the lipid bilayer, depending on the configuration change of such proteins, which plays a role as some variable conductance. There are five different types of ionic channels:

Leaky channels: They are responsible for basic permeability and can be treated as fixed resistors.

Voltage-gated channels: The permeability of which depends on the membrane potential; these channels mainly scatter on the axon hillock and the axon.

Ligand-sensitive channels: They respond to specific chemical agents—the so-called neurotransmitters—and open or close depending on the presence of that agent; they are found mostly in dendritic and soma membrane.

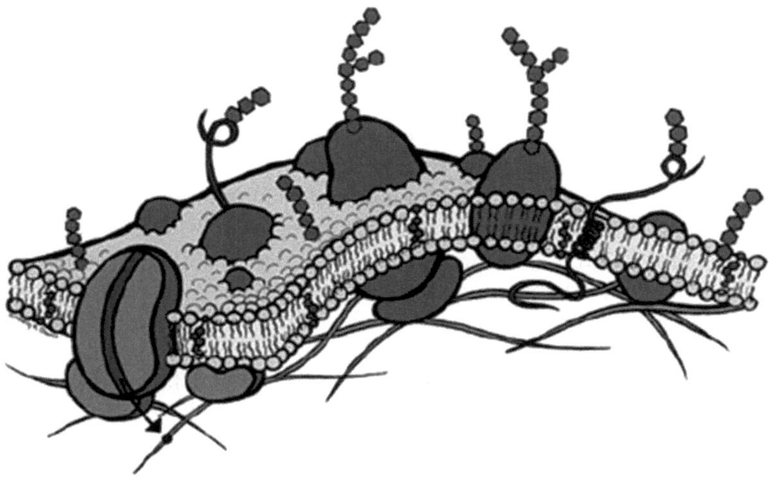

Fig. 2.21 The structure of neuronal membrane. Amphipathic phospholipids make up approximately 50 % of cell membranes. They form double layers (bilayers), with the charged phosphate head groups (represented by the *circles*) facing the water on both sides and the oil-like lipids in the middle. A channel is shown on the *left* of the diagram, which extends through the bilayers and has both extracellular domains that can bind ions, drugs, hormones, or other factors in the extracellular solution and intracellular domains that can interact with proteins on the inside of the cell (Reproduced with permission from Pfaff DW (Ed) (2013) *Neuroscience in the 21st Century: From Basic to Clinical*. Springer (in Papke RL. Water, ions, membranes, pumps, and transporters. Fig. 1.2))

Mechanosensitive channels: They respond to deformation of the channel or the membrane surrounding the channel; they are found in some receptor cells.

Gap junctional channels: They allow ions to move between cells, which form electrical synapses.

Most channels, especially voltage-gated channels, show a high degree of ion specificity; only certain specific ions can pass through such channels. However, a few ligand-sensitive channels do not show significant ionic specificity (Dowling 2001). Thus, the final four types of ionic channels can be considered as variable conductors. In summary, the structure of neuronal membrane can be modeled by a circuit composed of parallel branches with capacitor and resistors with batteries driving different ions to enter or leave the neuron.

Resting Potential Usually, there is a potential difference of about 60–70 mV between the two sides of the neuronal membrane without any stimulation, the inside potential is negative compared with the outside one, which is often taken as the reference or the base line of the membrane potential. This potential is often called resting potential. This potential difference resulted from the different ionic composition between the intracellular and extracellular fluid and also the different permeability of the membrane to different ions. Thus, there are more anions inside and more cations outside. The main intracellular cations are potassium ions, and anions are small organic molecular ions; their concentrations are much higher than the ones in extracellular fluid, while the main extracellular cations are sodium ions, and anions are chlorine ions, and their concentrations are also much higher than the one in intracellular fluid. Such concentration difference between both sides of the neuronal membrane is maintained by a specific protein called ionic pump at the membrane, which drives the potassium ions and sodium ions moving against their concentration gradient. The membrane permeability for potassium ions is the biggest, next for chlorine ions, and small for sodium ions, when the neuron is at resting state. There is almost no permeability to organic molecular ions. Thus, at resting state, there are only potassium ions and chlorine ions that pass through the membrane in opposite directions driven by their concentration gradients. Considering these two ions with opposite charges, the result of such flows makes the inside be negative to the outside. On the other hand, the accumulated electric field will drive these ions moving in opposite directions, and at last some equilibrium will be reached to make the net flow be zero. The potential difference between both sides of the membrane is the resting potential. According to Goldman equation in physical chemistry, the equilibrium, i.e., the resting potential, can be estimated, if the concentrations of various intracellular and extracellular ions and their permeability are known.

Action Potentials Changes of membrane potential can be observed when current is injected into the giant axon of squids. Such changes would follow the law of paralleled passive circuit with capacitor and resistors, if it makes the membrane potential even more negative than the resting potential, such process is called "hyperpolarization." If the change makes the membrane potential less negative than the resting potential, it is called "depolarization." Such change would also be passive, if its membrane potential does not reach a certain level (the threshold). However, once the depolarization is above the threshold (an increase about 15 mV), i.e., the membrane potential increases from −70 mV to about −55 mV, it will rise abruptly and reach as high as +50 mV within 1 ms and then decrease quickly until −90 mV in another 1 ms; after that the potential rises gradually again to restore to its resting potential. Such impulse-like potential changes are called action potentials or spikes.

The change processes of membrane potentials with different injected currents are shown in Fig. 2.22. The lower-right diagram shows the hyperpolarization currents (with negative charge) injected into the axon with successively growing amplitudes and the corresponding hyperpolarization potentials. The upper-right diagram shows the depolarization currents (with positive charge) with growing amplitudes and the corresponding depolarization potentials. As what was shown in the diagrams, there will be an action potential only when the injected positive current is strong enough so that the membrane potential is depolarized to a certain degree, i.e., reaches its threshold. On the left is a schematic diagram of the experiment setup. There is a current generator shown on the left to inject current

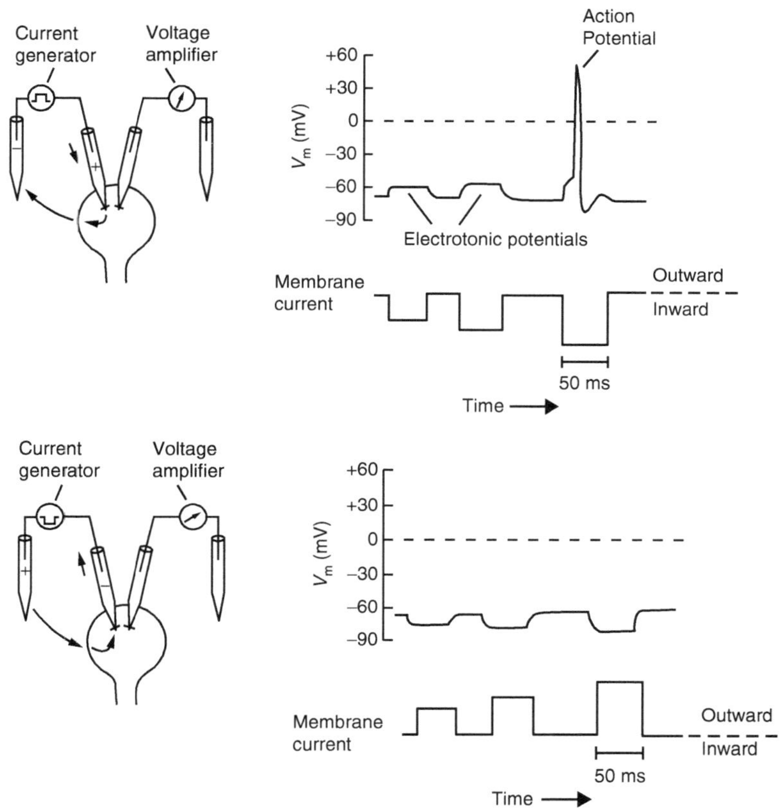

Fig. 2.22 Changes of the membrane potential by injecting currents with different intensities and polarities (Reproduced with permission from Pfaff DW (Ed) (2013) *Neuroscience in the 21st Century: From Basic to Clinical.* Springer (in Hendry S. Membrane potentials. Fig. 5.9))

into the cell and a voltage amplifier on the right to pick up and amplify the potential change.

What will happen if the injected depolarization current increases further? In Fig. 2.23, the upper diagram shows three successively increased depolarization currents and the lower their corresponding changes of the membrane potential. It can be seen that there is no action potential emerging when the inject current is weak; action potentials emerge only when the inject current is strong enough. However, the action potentials induced have the same shape and the same amplitude, independent of the intensities of the stimuli, only if a threshold is reached. Neuroscientists call such phenomenon the "all-or-none" property of the action potential. Increasing stimulus intensity does not change the shape and amplitude of the action potential; however, if a step stimulus is used, then the larger the stimulus intensity, the shorter the interval between the successive action potentials. The reciprocal value of the interval between neighboring action potentials is called

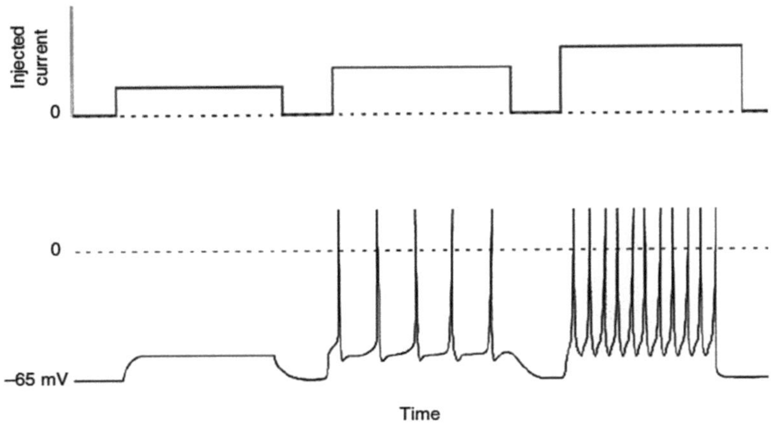

Fig. 2.23 The shapes and the amplitudes of the spikes triggered by three positive current stimuli with different amplitudes. Once action potentials are induced, their shape and amplitude are all the same, no matter how strong the stimulus is; this property is often called "all or none." However, the stronger the stimulus, the shorter the interval between neighboring spikes (Reproduced with permission from Pfaff DW (Ed) (2013) *Neuroscience in the 21st Century: From Basic to Clinical.* Springer (in Hendry S. Membrane potentials. Fig. 5.11))

the transient frequency or discharging rate. Nevertheless, there is a upper limit for such frequencies—about 1200 spikes/s, i.e., the shortest interval between two neighboring spike is about 1 ms. Thus, it is impossible for two spikes within 1 ms to occur. This is also manifested as the fact that if the interval between two successive stimuli is shortened beyond some value, further shortening will make the second stimulus more difficult to induce a spike, i.e., only stronger stimulus may trigger the second spike. Decreasing the interval further blow about 1 ms; no matter how strong the second stimulus is, there will be no second spike which can be induced. This time interval is called absolute refractory period, while the former is called relative refractory period as shown in Fig. 2.24. This means that the threshold is not a constant, it will rise to infinity during absolute refractory period and then decrease gradually in the relative refractory period until restoring to its resting level.

Observing the waveform of an action potential, it is found that it does not return to its resting value monotonically, but with some undershoot, which is called the afterpotential. It means that the excitability of the neuron varies with time after its firing.

The key factor in producing an action potential is that there are voltage-dependent ionic channels on the membrane. The permeability of potassium ions across the membrane is 30 times bigger than the one of sodium ions, when the membrane is at its resting state. However, the permeability of the sodium ions across the membrane increases with depolarization, and more sodium ions come into the intracellular fluid. Once the inward flow of sodium ions exceeds the outward flow of potassium, a depolarization of about 15 mV is reached, and the

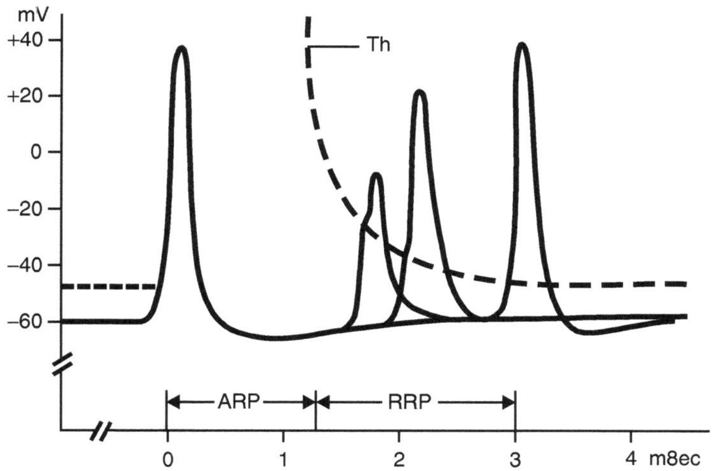

Fig. 2.24 Threshold variations immediately after spiking. Suppose there is no spiking before the moment (0), a proper depolarization can trigger an action potential; after that in a period marked by "ARP (Absolute refractory period)", no matter how strong a stimulus is, no spike can be triggered, i.e., in this period, the threshold value is infinity. Only after this absolute refractory period, the threshold decreases gradually as shown by the *dashed line*. In the period marked by "RRP (relative refractory period)," spikes can be triggered with stronger stimulus, until the threshold (Th) restores to its resting value at last

membrane potential at that moment is the threshold for producing an action potential. The continuing entrance of net cation enhances depolarization further, and the more depolarization makes more sodium ions enter into the intracellular fluid; such a positive feedback makes the leading edge of the action potential rise abruptly. Nevertheless, such high permeability of sodium ions can only keep for a short period. After 1 ms or so, an inactivation process for sodium channel starts, the entering sodium ionic flow becomes less and less, and the permeability of potassium ions increases gradually; thus, a process of repolarization and even hyperpolarization starts, i.e., the membrane potential not only returns to its resting potential but also becomes even more negative. After that, the permeability of potassium ions across the membrane decreases, and the membrane potential rises again and return to its resting value at last (Fig. 2.25).

The key factors here are that both conductance of sodium channel and potassium channel depend on both voltage and time. Both conductance of sodium channel and potassium channel change with time, when there is any change of membrane potential. Once the membrane potential is depolarized to a threshold, the sodium channel conductance increases quickly, but returns to its original value swiftly after 1 ms. There is a delay of the increase of the conductance of the potassium channel after the conductance of the sodium channels decreases, and its speed is also slower. It decreases only when the membrane potential begins to decrease. And the chlorine conductance almost remains constant during the whole process, thus has little contribution to the production of the action potential.

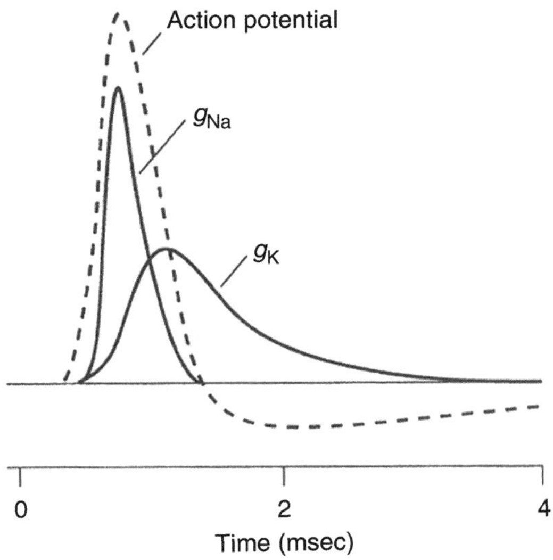

Fig. 2.25 The changing process of membrane potential (V), the permeability of sodium ions across the membrane (manifested as the conductance of sodium channel g_{Na}), and the one of potassium ions (manifested as the conductance of potassium channel g_K) (Reproduced with permission from Conn PM (Ed) (2008) *Neuroscience in Medicine (3rd Edition)* Humana Press (in Stojilkovic SS. Ion channels, transporters, and electrical signaling. Fig. 8b, adapted from Hodgkin and Huxley AF, J. Physiol 1952; 117:500))

Sodium ions rush into the axon during the rising phase of an action potential; the large quantity of rushing positive charges also neutralize the negative charges in its neighborhood and make the membrane depolarized. The permeability of sodium ions across the membrane increases with the rising of depolarization; this increase makes the membrane even further depolarized until an action potential is triggered. Thus, an action potential occurred at one place will trigger another action potential at its neighboring membrane segment with voltage-sensitive ionic channels, which is the very reason why an action potential can propagate along an axon. Such propagating speed varies with different types of axons. The slowest speed may be less than 1 m/s, while the fastest may be greater than 100 m/s. The bigger the diameter of the axon, the faster the speed. In addition, wrapping with myelin sheath may greatly enhance the speed. Owing to the high resistance of the myelin sheath and the low resistance of intracellular and extracellular fluid, it is easier to form a loop between neighboring nodes of Ranvier. Only if the distance between them is not too long so that the current at the neighboring node is still strong enough to exceed the threshold, a new action potential can be triggered; thus, it seems that there is a jump of the action potential from one node to the next node; such conduction is called saltatory conduction. Most of the projection neurons in vertebrate animals have myelin sheath. If it were not the case, and to get the same speed by increasing axon diameters, then the brain would be 10 times bigger than what we have now. In addition, the propagation of the action potential along the axon is unidirectional, owing to the abruptly dropping of the permeability of sodium ions to a very low value at the location where an action potential has just been triggered, while the permeability of potassium ions is very high there; thus, the threshold there is very high. This explains why a refractory period will follow the firing. This is the very reason why an action can only propagate forward, but not backward.

One of the most important features of the action potential is that it obeys the "all-or-none" law, i.e., besides the fact that the amplitude and waveform of the action potential does not depend on the stimulus strength which we have mentioned above, it also means that once an action potential is triggered, it will propagate along the whole axon without any decay in normal cases. The action potential recorded at the end of the axon terminals has just the same amplitude and waveform as the one at the very beginning. This guarantees that there will be no distortion of the action potential during its propagation, a necessary condition for the reliability of the information transmission.

Graded Potentials For convenience, we will discuss the synaptic potential at first. A diagram of the motor neuron in the spinal cord is shown in Fig. 2.26. Synaptic vesicles within the presynaptic terminals fuse with the membrane and release neurotransmitters within them into the synaptic cleft, when a spike reaches there. The neurotransmitter diffuses across the synaptic cleft and binds to a specified protein, receptor at the postsynaptic membrane, thus promoting specific ions to enter or leave the postsynaptic neuron directly or indirectly and make the postsynaptic membrane depolarized or hyperpolarized, depending on the nature of both neurotransmitters and receptors.

Depolarization of the postsynaptic membrane (there is inflow of cations or outflow of anions) facilitates spiking of the postsynaptic neuron, and the postsynaptic potential then is called excitatory postsynaptic potential (EPSP) (lower leftist curve in Fig. 2.26). The effect of hyperpolarization of postsynaptic membrane is just the opposite; thus it is inhibitory, and the corresponding postsynaptic potential is called inhibitory postsynaptic potential (IPSP) (the next lower leftist curve in Fig. 2.26). All the following processes need time: transmitter releasing from the presynaptic membrane, diffusing across the synaptic cleft to the postsynaptic membrane, binding with receptors on it to open ionic channels, and triggering membrane potential changes at last. The total period for the above processes is about 0.5 ms, which is called the synaptic delay. It is often misunderstood that the synaptic delay of 0.5 ms or so is mainly due to the diffusion of neurotransmitters across the synaptic cleft. However, as a matter of fact, only 50 μs is needed for such diffusion. Releasing transmitters needs 300 μs, activating and opening ionic channels at the postsynaptic membrane needs 150 μs. The postsynaptic potential induced by only one spike is small, and its effect is quite weak. However, if a series of impulses reach presynaptic membrane at a synapse successively, before a postsynaptic potential triggered by one spike goes down completely, a second spike is arriving; then its effect will be superimposed on the remain of the previous one; thus, a bigger postsynaptic potential will be formed. This process is called

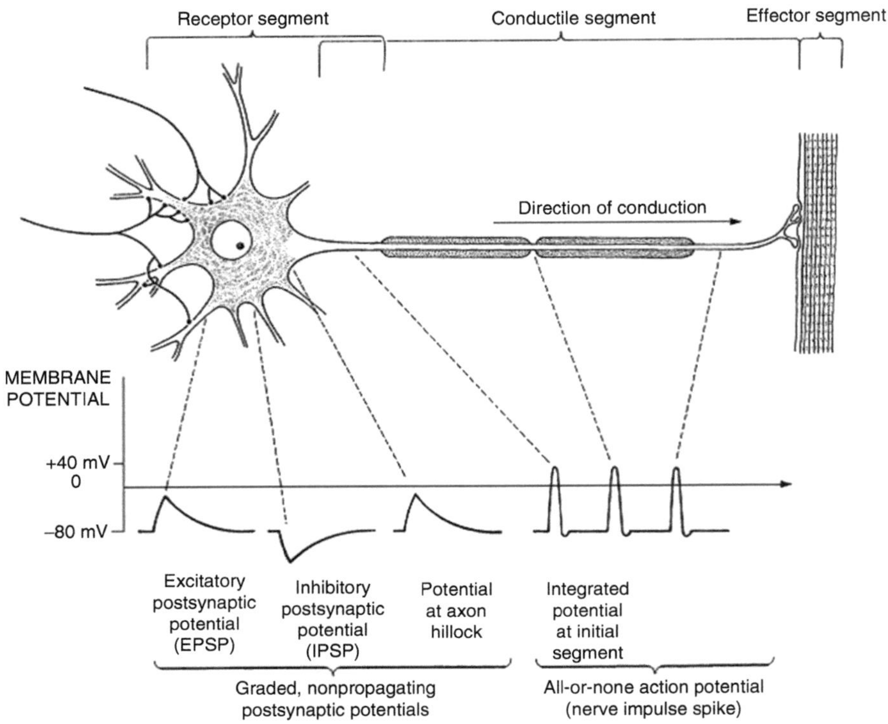

Fig. 2.26 Postsynaptic potentials, their summation and triggering action potentials. *Upper*: a schematic diagram of a motor neuron; *lower*: action potentials (the *right three curves*) will be triggered, when a summated potential at the axon hillock reaches a threshold (the *middle curve*). The *left two curves* show excitatory postsynaptic potential (EPSP) and inhibitory postsynaptic potential (IPSP) (Reproduced with permission from Strominger NL, Demarest RJ, Laemle LB (2012) *Noback's Human Nervous System Seventh Edition: Structure Function*. Humana Press. Fig. 3.1)

temporary summation. If several spikes arrive at different synapses at approximately the same time, their effects can also be summed up, and such effect is called spatial summation. In case the summation exceeds a threshold, then a spike is triggered. The effects of excitatory postsynaptic potentials and inhibitory postsynaptic potentials are just contrary to each other (Fig. 2.26). It is the effect of such summation that makes a neuron integrate various inputs which are received by the neuron and makes information processing possible.

Finally, it should be pointed out that, generally speaking, an action potential can only be triggered at axon hillock. There are two reasons for that:

Firstly, there are abundant voltage-sensitive sodium and potassium channels there; therefore, the threshold for spiking is lowest, while there are much less such channels either at soma or at dendrites.

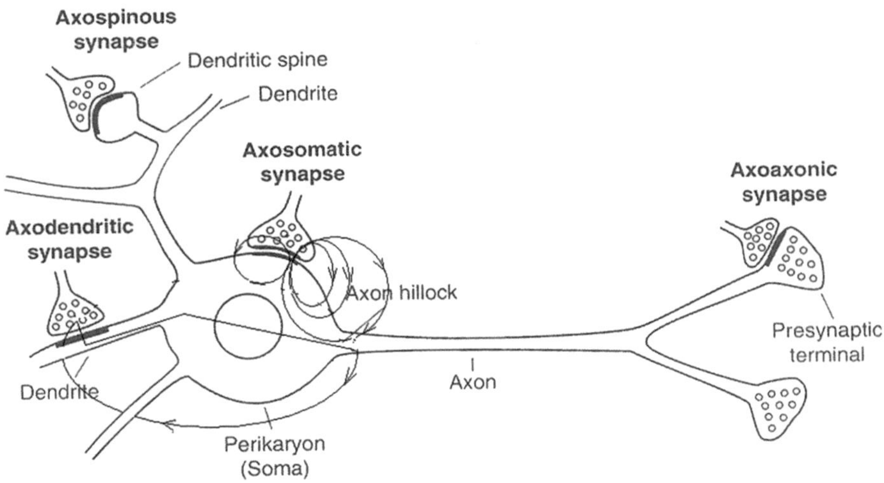

Fig. 2.27 Integration of synaptic currents distributed at different locations on dendrites and soma at the axon hillock. Only some of the local postsynaptic currents are shown as examples. The picture of the neuron is a copy of the neuron from Fig. 2.1. Note the direction of the excitatory postsynaptic current and the inhibitory postsynaptic current at axon hillock are just opposite to each other, and only part of the postsynaptic current can reach that location

Secondly, generally speaking, the surface of dendrites or soma is covered with a large quantity of boutons of axonal terminals from other neurons and also surrounded by glial cells. Only the axon hillock is exposed to extracellular fluids, thus the resistance here is minimal, and majority of currents from synapses will flow through it to form loops; thus, the axon hillock is like a current node to superimpose all the inflow and outflow ions from all the synapses with different weights depending on their locations and natures, just as shown in Fig. 2.27.

In short, the total ionic flow through the axonal hillock is the weighted summation of all synaptic ionic currents. When the potential difference between the inside and outside of axon hillock membrane is less than the threshold for spiking, it varies gradually with varied amplitudes and waveforms; it decays with distance and thus is a graded potential and called subthreshold potential. Only when such potential accumulates to a threshold value can an action potential be triggered here and propagated along the axon.

There are also abundant voltage-sensitive channels at the sensory terminals of some sensory unipolar neurons; action potentials can also be triggered at these places.

In this section, as far as graded potentials are concerned, we have mainly discussed synaptic potentials and subthreshold potentials, and almost nothing has been mentioned about receptor potentials, owing to the fact that the underlying mechanisms to transfer a variety of stimuli (light, mechanical forces, sounds, chemical agents, electricity, heat, etc.) are quite different, while the nature of the potentials themselves is quite similar to other graded potentials. Therefore, the basic properties of graded potentials discussed in this section can also be used for receptor potentials on the one hand, while the mechanisms to transfer different stimulus to the corresponding receptor potential are different on the other hand, which might be better to be explained in later sections when the corresponding sensory systems are discussed. Nevertheless, there is a common feature for all the receptor potentials, adaptation, which is manifested as a decay of the potentials with time when a stimulus is on and then keeps constant, and should be mentioned here. Such decay could be either very fast or slow; in addition, it can either return to its resting state or maintain at a lower constant level. It is this feature that makes the sensory system be able to detect the temporal change of stimuli. Only the changes in their environment are important to animals' survival. Besides, detection of changes can significantly reduce the redundancy of the information, which must be transmitted to the brain.

As a summary, the action potential can be compared with graded potentials in the following list:

Graded potentials	Action potentials
1. Its amplitude is variable, and can vary continually	1. Both amplitude and waveform are fixed, obey "all-or-none" law
2. Its duration is variable	2. Its duration is fixed
3. No threshold	3. There is some threshold
4. It can be superimposed	4. It cannot be superimposed
5. No refractory period	5. There is some refractory period
6. Significant decay with propagation	6. No decay with propagation
7. Either excitatory, or inhibitory	7. Only excitatory

It should be emphasized again that the all-or-none property of the action potential indicates that its waveform and amplitude are fixed, and thus they carry no information at any level above neuron from a point of view of information processing. It is the spike train that carries information! The all-or-none law guarantees that no decay or distortion would happen during the propagation process; thus, it guarantees the reliability of information transmission. Therefore, information processing depends on graded potentials rather than action potentials. On the contrary, it is not suitable for graded potentials to transmit information for long distance owing to its decay and apt to be distorted by disturbances.

In summary, for Golgi type I neurons and part of Golgi type II neurons, the information transmission and processing processes can be summarized approximately as follows: Synaptic vesicles within presynaptic membrane release neurotransmitters into the synaptic cleft when spikes arrive at presynaptic axonal terminals, these transmitters diffuse across the synaptic cleft and bind with receptors on the postsynaptic membrane, and open the ionic channels to let corresponding ions flow into or out of the postsynaptic neuron to form local currents directly or trigger complicated biochemical reactions to open the channels indirectly. Part of the local currents from different synapses flows through axon hillock and is summarized there. If the summarized potential there is above some threshold, an action potential is triggered and propagates along the axon until its terminals.

At last, it should be noted that the neuron displays much more diversity than any other cell; the above description has only given a rough picture of the structure and function of the neuron. Many facts, even rather important, have been omitted, e.g., besides the ionic channels listed above, there are also calcium channel and calcium-dependent potassium channel at the membrane, although the ion flow, generally speaking, is weak and plays a little role in the generation and propagation of action potentials; however, the small calcium inflow may change the intracellular calcium concentration and trigger the calcium-dependent potassium channel to open when its concentration is high enough and thus to modulate the spiking pattern which carry important neural information. We will discuss this problem with more details in the next chapter.

2.4 Cerebral Cortex

For a variety of brain functions, is only certain localized small area responsible for one of such functions, or is the whole brain needed for carrying out such function? This is a significant problem which has been disputed for centuries in neuroscience history. In the nineteenth century, an Austrian doctor Franz Joseph Gall proposed a theory, according to which the cerebral cortex could be divided into 35 different areas responsible for 35 different mental functions. If some function is outstanding, the corresponding area would be expanded, and there would be a bump on the skull just above that area; thus, an examination of the bumps on a skull could tell the personality of the subject. His theory was called phrenology in history. Although his theory lacks sound scientific basis, has not been verified by experiments, and is wrong on the whole, it is him who raised the idea of functional localization at first, which should not be denied. Gall's theory was opposed by experiment physiologist Pierre Flourens strongly, who destroyed different locations in bird brains and found no specific deficit in their behaviors; thus he thought that the behavior of animals depended on the whole brain. Unfortunately, his subjects mainly limited to lower vertebrates or very young animals, and the functions he examined usually were those nonspecific ones such as sleep, wake, feeding, and so on and so forth. Thus he used the correct methods with wrong ideas and got wrong conclusions.

In 1861, French neurologist Paul Broca examined a patient who could say nothing other than the sound "tan" (thus he was named as Mr. Tan in scientific literatures afterward), he died six days after Broca examined him, and Broca performed an autopsy of his brain and found that there was a significant lesion in his left front lobe neighboring to temporal lobe. Later, Broca found six other patients with similar symptoms, i.e., they can understood other's words but could not express themselves with words. Therefore, Broca announced that the area he found was the center for expressing with words and the area was later named as Broca's area in honor of him (Fig. 2.30). For distinguishing from phrenology, Broca emphasized that his area was different from what Gall had said. A few years later, a German doctor Carl Wernicke found a patient who could speak nonsense but could not understand the words. After the patient's death, he also performed an autopsy and found a lesion around the boundary between parietal lobe and temporal lobe—an area which is named after him as Wernicke's area (Fig. 2.30), which he thought is responsible for understanding words instead of saying. These classical works strongly support the functional localization theory. Almost immediately after these discoveries, German physiologist Gustav Fritsch found that stimulating different brain areas could induce movements of different muscles. Thus, the functional localization theory was revived again.

Just before the First World War, German neuroanatomist Korbinian Brodmann divided the cerebral cortex into 52 different areas based on their neuronal morphology, density, and layer organization (Fig. 2.28); further studies show that these different areas are really related to different functions. Since the beginning of the twentieth century, most of the people have agreed that there is some degree of functional localization in the cerebral cortex; however, the debate has not gone to an end. It was suggested that higher brain functions such as thinking and memory need the whole cerebral cortex to be involved in and no functional localization exists for such functions. The most influential hypothesis belonging to this category is American psychologist Karl Lashley's theory of memory. In his experiments, rats tried to learn finding foods in a maze. He compared the healthy rats and those with brain damage. He found that for those rats with cortex damage, they had to learn more times to avoid blind ends and get the food. In another series of experiments, the rats had learned how to avoid blind ends and get the food, then some part of their cerebral cortex was destroyed, and the animal was apt to make mistakes. His most impressive results are that the degree of memory deficit is proportional to the area of the damaged cortex and not related to where the damage was made. His results were considered as a strong support to the holistic theory of higher brain function. However, further studies indicated that his conclusion was problematic and the key point is that the job the animal did in his experiments involved many sensory modalities. Thus, lesion in one modality alone does not prevent the animal to find the reward, as other intact modalities can still guide the animal to its goal. Only very large area involved in many sensory modalities is damaged, the deficit in learning becomes obvious. Lashley gave wrong explanation to his results! However, there is one of his points which is still correct, i.e., learning and memory are not only related to a small local area in the cortex. Since the 1930s, Clinton Woolsey, Philip Bard, and others have studied sensory and motor maps in the brain in detail; they found that there were not only one such map for every sensory modality. Now it is found that there are more than 30 maps for visual information processing.

In summary, only very simple function is localized in some brain region accurately; more complicated behavior or perception is produced by many areas, located in various parts of the brain. Thus, "complex functions such as perception,

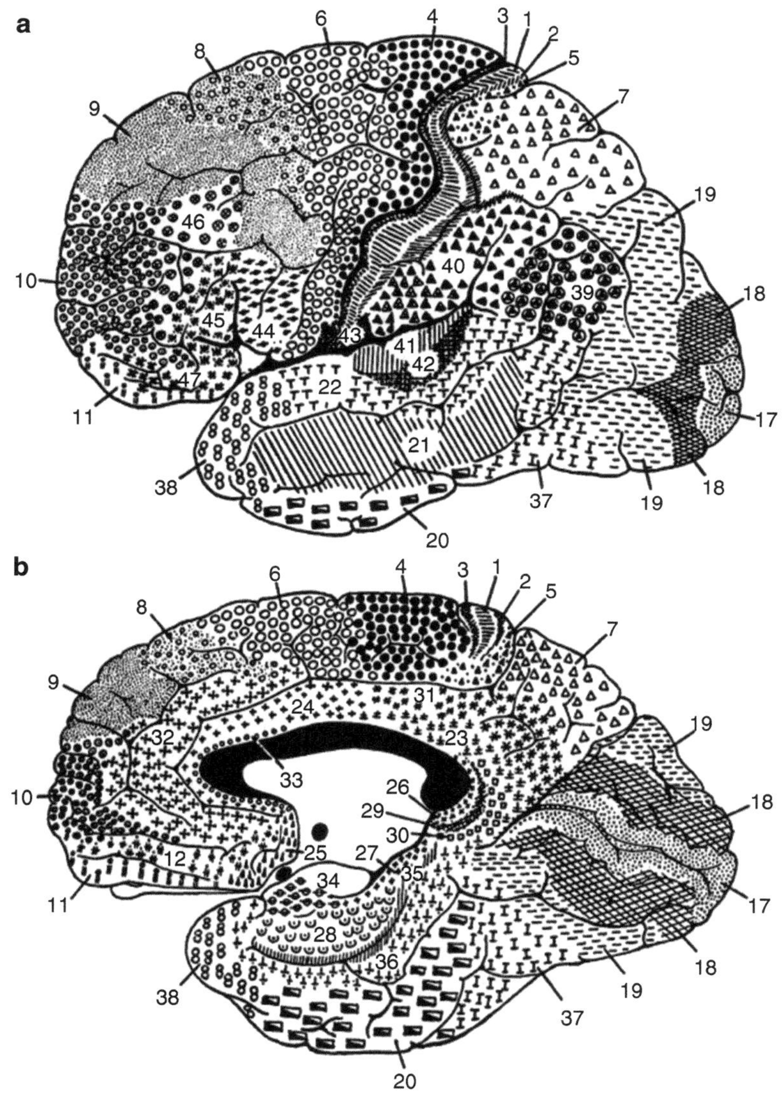

Fig. 2.28 Brodmann's maps on the lateral (**a**) and medial (**b**) surfaces of the cerebral cortex (Reproduced with permission from Conn PM (Ed) (2008) *Neuroscience in Medicine (3rd Edition)* Humana Press (in Miller MW and Vogt BA. The cerebral cortex. Fig. 6))

memory, reasoning, and movement are accomplished by a host of underlying processes that are carried out in a single region of the brain.. . .Any given complex ability, then, is not accomplished by a single part of the brain. . ... However, simple processes that are recruited to exercise such abilities are localized" (Kosslyn and

Andersen 1992). In short, there have to be a trade-off between the holistic view and functional localization view.

The debate between the localizationism and the holism is concerned with the main principle of neural information processing. From the debate, we understand that there is no "homunculus" in the brain who is in charge of a complex brain function and there is no such a commander to command. It is more probable that there is a network which is in charge of the function. This is important for understanding the underlying mechanism of cerebral cortex. Thus, when we say that some cerebral region is in charge of some function, it does not mean that the region is the only one which relates to that function. On the contrary, almost every complex function is concerned with a network, elements in which may locate at different lobes, or even in subcortical structures.

The precentral gyrus just in front of the central sulcus is often called primary motor cortex, as weak electrical stimulus to a location within it can lead to corresponding muscle movement, while the front lobe is involved in planning and initiating a movement. The postcentral sulcus just behind the central sulcus is often called primary somatic sensory cortex. Experiments show that the different location within it corresponds to sensations from different points on the body surface. If the location at which a movement or a sensation would be induced is noted along the central sulcus, or the corresponding part of the body is drawn along it intuitively, then an inverse homunculus will be shown both in front of and behind the central sulcus as shown in Fig. 2.29. The lower diagram is a representation of the sensory mapping in primary somatic sensory cortex, while the top is the one of the motor mapping in primary motor cortex. It should be noted that the scale of such representation is not proportional to the size of its corresponding part of the body; the finer the control of muscle is needed or the sensation is, the bigger the corresponding mapping area in the cortex is.

The superior temporal gyrus below the lateral sulcus is the location for primary auditory cortex, while the primary visual cortex at the pole of occipital cortex is around the calcarine fissure. However, not all the regions on cerebral cortex are related to sensory or motor functions directly; the prefrontal cortex, posterior parietal cortex, and inferior temporal cortex are such examples; these regions are called association areas, which are considered as regions related to higher functions in general (Fig. 2.30). As what we have mentioned above, Brodmann divided cerebral cortex into 52 different regions based on their cellular architectures and labeled them with numbers according to the orders he investigated them. For example, the back pole of the occipital lobe was labeled as area 17, while the region just in front of the central sulcus area 4 (Fig. 2.28). Although Brodmann's nomenclature was based on cellular architectures, generally speaking, his division is also related to their functions in some way. However, it should be noted that the same cortical region may have several names, say, Brodmann's area 17 is also called striate cortex based on its morphological feature, or primary visual cortex or V1 owing to its function. In this example, all the four different names denote the same region; however, in other cases, there is no such one to one correspondence,

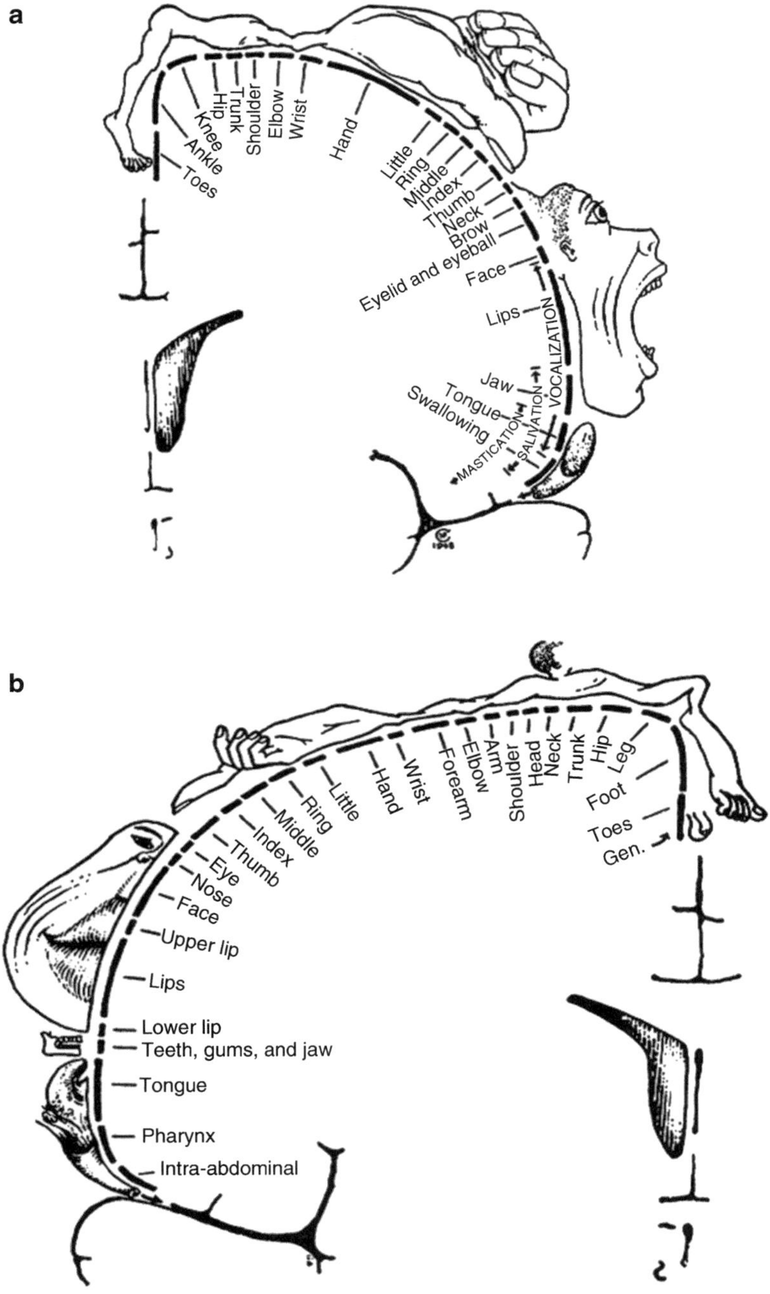

Fig. 2.29 Representations of the body surface in the primary motor cortex (**a**) and the primary somatic sensory cortex (**b**) (Adapted with permission from Jacobson S and Marcus EM (2011) *Neuroanatomy for the Neuroscientist* (2nd Edition). Springer. Fig. 10.11 and Fig. 14.1 (from Penfield W and Rasmusen T. The Cerebral Cortex of Man. MacMillan, 1955))

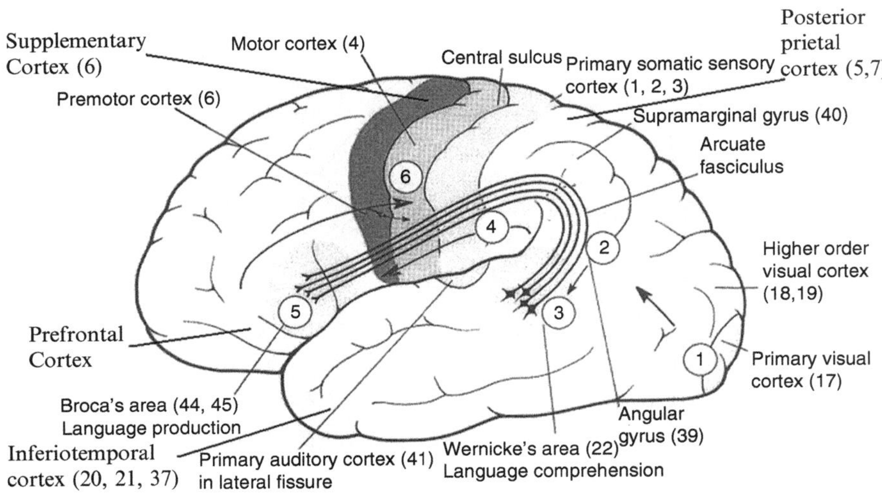

Fig. 2.30 Diagram of functional division of cerebral cortex. Here the number within a circle denotes (*1*) the visual cortex (Brodmann's areas 17–19); (*2*) the angular gyrus (area 39), which is related to face recognition; (*3*) Wernicke's area (area 22); (*4*) the arcuate fasciculus; (*5*) Broca's speech areas 43 and 44; and (*6*) motor area 4 (Adapted with permission from Strominger NL, Demarest RJ, Laemle LB (2012) *Noback's Human Nervous System Seventh Edition: Structure Function*. Humana Press. Fig. 25.7)

for example, although Brodmann's areas 18 and V2 denote regions approximately very close, but they are not identical.

From a point of view of information processing, there are three features in many regions of the cerebral cortex, especially in sensory and motor cortex: topological mapping and reentry, layer structure, and functional columns.

In visual and somatic sensory cortex, neurons having neighboring receptive fields, i.e., neurons responding to stimuli from neighboring spatial locations, aggregate together, and such topological relation keeps while they map to other brain regions. The auditory has similar features, although such relationship is about the stimulus frequency rather than spatial location. It was considered that there is rather rigorous hierarchical structure in cerebral cortex; the closer to the periphery the location is, the simpler and the more specific the neuronal function is; now we know such idea is not accurate. Although sensory information processing does manifest some degree of hierarchical structure, such mapping is not unidirectional; there are large quantities of reciprocal connections within a region and between different regions, in which information goes to and fro, i.e., the so-called reentry by Edelman (Edelman and Tononi 2000). Thus, as what we have mentioned above, there is no single region which is in charge of the complicated function as a top commander; a broad network is needed.

Majority of human cerebral cortex, which is called neocortex, can be divided into six layers (Fig. 2.31). The most prominent neurons in neocortex are pyramidal cells (Fig. 2.16 lower middle); their cell bodies scatter in layers 2–6. In most cases, there is a dendrite of such neuron extending up to the cortical surface, which is

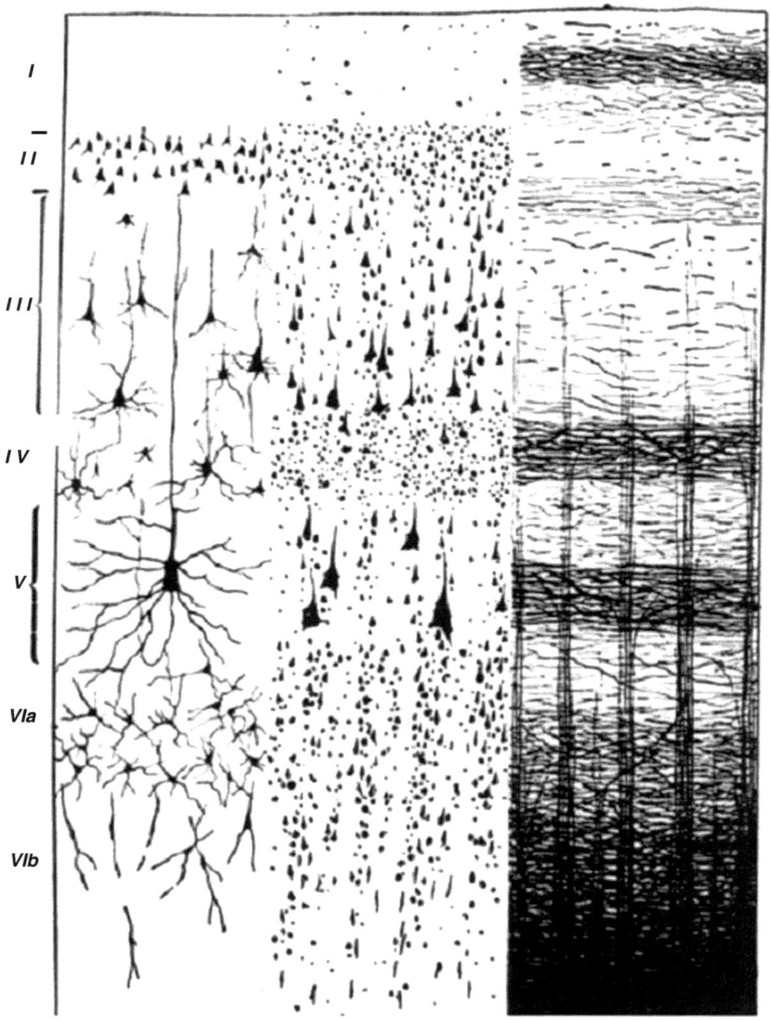

Fig. 2.31 Layer structure of the cerebral cortex. (*left*) A Golgi stain; (*middle*) a Nissl stain, which only shows cell bodies; (*right*) a myelin stain, which shows nerve fibers. *I* – molecular layer, *II* – external granular layer, *III* – external pyramidal layer, *IV* – internal granular layer, *V* – large or giant pyramidal layer, *VI* – fusiform layer (Reproduced with permission from Jacobson S and Marcus EM (2011) *Neuroanatomy for the Neuroscientist (2nd Edition)*. Springer. Fig. 10.1)

named apical dendrite. There are also a lot of dendrites extending in every directions at the base of its soma, which are called basal dendrites. Layer 2 contains mainly small pyramidal cells. Layer 3 contains medium-size pyramidal cells, forming connections with other cortical regions. Layer 4 contains numerous small cells including both stellate cells and small pyramidal cells. Layer 5 contains both medium-sized and large pyramidal cells. Layer 6 contains small cells including stellate cells, pyramidal cells, and fusiform cells. Beginning in layer 3 through 6, there are pyramidal cells sending axons internally. The pyramidal cells in layers 5–6 send axons to subcortical structures. The effect of pyramidal cells is mainly excitatory. Another main type of neurons in neocortex is the stellate cell (Fig. 2.16b); there is a large quantity of dendrites extending to a variety of directions. The effect of the spiny stellate cell is also excitatory, while the smooth stellate cell is inhibitory. The stellate cell belongs to Golgi type II neurons (Abeles 1991). The most outstanding feature of neurons in the neocortex is their variability both from their morphology and electrical physiological properties. The classification of neocortex neurons into pyramidal cells and stellate cells is only a rough division (Toledo-Rodriguez et al. 2002). The input fibers from the thalamus generally terminate at the intrinsic stellate cells in the 4th layer and to some extent in layer 6, while inputs from other cortical regions and other subcortical structures terminate at pyramidal neuron dendrites in layers 1–3. The stellate cells in layer 4 terminate at apical dendrites of pyramidal cells in layers 2 and 3, which project back onto pyramidal cells in layer 5 and back further onto pyramidal cells in layer 6, which in turn project back to layer 4 to form local circuits. In addition, the pyramidal cells in layer 4 excite the aspiny inhibitory neurons to inhibit the neighboring regions. The output fibers to other cortical neurons, even in the opposite hemisphere, generally start from pyramidal cells in the 3rd layer. There are also some neurons deep in the cortex, axons of which extend up to the cortical surface and branch a lot at the surface layer to make synaptic connection with apical dendrites of pyramidal cells. In addition, there are also some neurons having horizontally extended axons in the surface layer—layer 1, which is a fiber- and synapse-rich layer with few cell bodies, containing many axons running parallel to the brain surface. A simplified diagram is shown in Fig. 2.18a. The thickness of different layers is different, e.g., the 4th layer is thicker in sensory cortex. Especially, the 4th layer in primary visual cortex is so thick that it can be recognized by naked eyes and thus has the name of striate cortex.

On contrast, some brain structures such as primary olfactory cortex, which have been evolved earlier, have less layers and are called allocortex.

Besides horizontal layer structure, cortex may also have vertical column-like structure, in which every layer is connected by the axons that run up and down, making synapses along the way. All the neurons at the line perpendicular to the surface have nearly identical receptive fields. Neurons within such minicolumn seem to encode similar features, e.g., neurons in the primary visual cortex at such line have the same preferred orientation. Vernon Mountcastle (1978, p. 37) defined the minicolumn as follows: "I define the basic modular unit of the neocortex as a minicolumn. It is a vertically oriented cord of cells formed by the migration of neurons from the germinal epithelium of the neural tube along the radial glial cells to their destined location in the cortex, as described by Rakic." Mountcastle (1997, p. 701) declared that a column is "formed by many minicolumns bound together by short-range horizontal connections." He proposed that the neocortex was rather uniform both in appearance and structure; perhaps there were some basic operation rules underlying their different functions. Inspired by Mountcastle's ideas, Hawkins (Hawkins and Blakeslee 2004) thought that "There is a single powerful algorithm implemented by every region of cortex. If you connect regions of cortex together in a suitable hierarchy and provide a stream of input, it will learn its environment." And he suggested that "a column is the basic unit of prediction" for his memory-prediction framework of intelligence.

Hubel and Wiesel (1974) used the term "hypercolumn" to denote "a unit containing a full set of values for any given receptive field parameter," i.e., a hypercolumn contains cells tuned to all values of every receptive field variable. Various estimates suggest there are 50–100 cortical minicolumns, each comprising around 80 neurons, in a hypercolumn. However, the theory that the cortical column is a distinct unit composed of an aggregate of minicolumns still lacks solid experimental evidence, and there are no anatomical boundaries between different columns, and a canonical microcircuit that corresponds to the cortical column has not been found. Horton and Adams (2005) pointed out: "No one has demonstrated a repeating, canonical cellular circuit within the cerebral cortex that has a one-to-one relationship with the minicolumn." Walter Freeman said: "The 'Cortical column' as defined by Mountcastle is not an anatomical feature. It is a dynamic collection of cortical neurons in an area defined by coherent firing of the neurons in phase-locked synchrony, not necessarily in fact not usually at the same phase, but with fixed-phase differences" (Freeman, personal communication).

Many people suggested that the minicolumn is "the most basic and consistent template by which the neocortex organizes its neurons, pathways, and intrinsic circuits" (Buxhoeveden and Casanova 2002). Henry Markram has simulated a rat cortical column based on a variety of realistic neuron models on a supercomputer "Blue Gene"; he wishes that connecting all such column models could simulate a whole brain (Markram 2012). Besides the fact that general principle would be inundated within too many details, not everybody agreed that the column is a functional unit of the cortex. In a review for summarizing half a century of research after Mountcastle's discovery of the cortical column in 1955, Jonathan C Horton and Daniel L Adams (2005) even concluded that the column may have no function at all! They said that: "Although the column is an attractive concept, it has failed as a unifying principle for understanding cortical function......Species variation in columnar structure is hard to reconcile with ideas about the functional importance of columns......The ultimate blow comes with the realization that some members of a single species have ocular dominance columns in only part of the visual cortex, leaving other regions of binocular cortex bereft of columns (Adams and Horton 2003). At some point, one must abandon the idea that columns are the basic functional entity of the cortex. It now seems doubtful that any single, transcendent principle endows the cerebral cortex with a modular structure. Each individual area is constructed differently, and each will need to be taken apart cell by cell, layer by layer, circuit by circuit and projection by projection to describe fully the architecture of the cortex."

A conclusion may not be given at the present moment; anyway, the above arguments may have to be kept in mind when people wish to elucidate the brain mechanism underlying neural information processing. Is the cerebral cortex a modular system? And is the minicolumn a basic universal module of such system? These are still in hot debate.

2.5 Visual System

The visual system endows us with the ability of "seeing" the environmental objects around us. Visual processes include various aspects, such as allocation of visual targets, perception of depth, motion detection for moving targets, as well as identification of visual features such as shape and color, etc. These processes involve complex neural mechanisms. In human brain, nearly half of the cortical areas participate the analyzing and processing of visual information.

In mammalians, visual processes start with eyes. The eye is an organ that specified for light detection, allocation, and analysis (Fig. 2.32). The anterior half of the eye is mainly an optical system, the function of which is to adjust the intensity of light flux and focus the visual images on the retina. In the meantime, moving target can be traced via eye movement, so as to allow for the visual target to form clear image on the retina. The retina is located at the back of the eye and is a part of the central nervous system. Visual information is preliminarily processed in the retina before it reaches other parts of the central visual system.

The nature of light is electromagnetic waves. The wavelengths of visible light are between 400 and 700 nm. Light is absorbed, reflected, and refracted at the surface of objects. Most of what we see is the light reflected by object surfaces. An object normally absorbs light within a certain range of wavelengths, while reflexes light of other wavelengths. The transparent media in the eye mediate refraction of the light in the eye and focus it onto the retina to form a clear image.

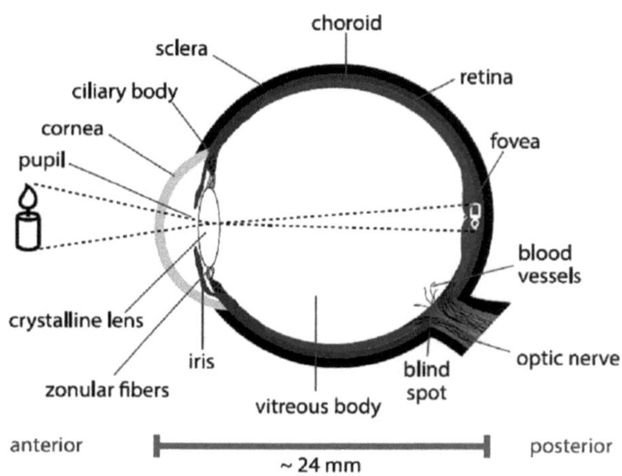

Fig. 2.32 Gross anatomy of the human eye and the position of retina in the eye (From Galizia and Lledo 2013, Springer, Neurosciences – From Molecule to Behavior, Fig. 18.1)

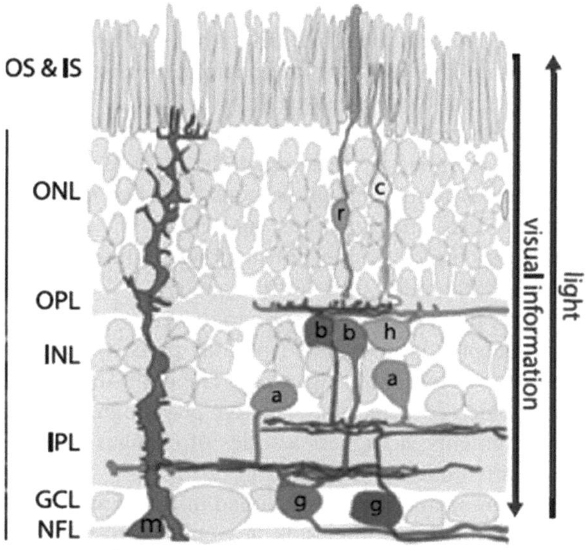

Fig. 2.33 The retinal structure. *NFL* nerve fiber layer, *GCL* ganglion cell layer, *IPL* inner plexiform layer, *INL* inner nuclear layer, *OPL* outer plexiform layer, *ONL* outer nuclear layer, *OS & IS* outer and inner segments of photoreceptors, *g* ganglion cells, *a* amacrine cells, *h* horizontal cell, *b* bipolar cells, *r* rod photoreceptor, *c* cone photoreceptor, *m* Müller cell) (From Galizia and Lledo 2013, Springer, Neuroscience – From Molecule to Behavior, Fig. 18.4)

In the retina, the neurons which are responsible for light detection are called photoreceptors; these cells are rich in photopigments which absorb light in a wavelength-dependent way and result in membrane potential changes via certain transduction mechanisms.

A schematic illustration of retinal structure is given in Fig. 2.33. In the retina, the vertical pathway consists of signal transmission from photoreceptors to ganglion cells via bipolar cells. In the meantime, retinal visual information processing is also subject to the lateral adjustments mediated by other two types of neurons: horizontal cells and amacrine cells. Horizontal cells receive input from photoreceptors and send signal back to photoreceptors via feedback pathways, so as to adjust the activities of photoreceptors and surrounding bipolar cells which are postsynaptic to photoreceptors; amacrine cells receive input from bipolar cells and send signals to nearby bipolar cells, ganglion cells, as well as other amacrine cells via lateral projection. Ganglion cells are output neurons of retina; their photo-responses are characterized by action potentials, which are sent to the central visual part via optic nerve.

One of the fundamental concepts related to retinal function is that photoreceptors are the only retinal neurons that are sensitive to light, while all the other neurons respond to light via their direct or indirect synaptic connections with photoreceptors; another important concept is that ganglion cells are the only output neurons of the retina; no neurons other than ganglion cells can convey the retinal output to the central visual part via the optic nerve.

From the schematic retinal structure illustrated in Fig. 2.33, one can tell that various types of neurons are nicely layered in the retina. The innermost layer is ganglion cell layer, which is formed by the cell body of ganglion cells; its adjacent layer is the inner nuclear layer, which contains cell bodies of horizontal cells, bipolar cells, and amacrine cells; the next layer is the outer nuclear layer, which contains cell bodies of photoreceptors; the distal layer is the outer segment layer of photoreceptors, which is sensitive to light. In the eye, light must penetrate all the transparent layers, including ganglion cells and other cells located in the inner nuclear layer, before it can reach the photoreceptors.

Synapses among retinal neurons form two reticular layers. The synapses among bipolar cells, amacrine cells, and ganglion cells form the inner plexiform layer, while the synaptic connections among photoreceptors, horizontal cells, and bipolar cells form the outer plexiform layer.

Light input is converted into neural signal in retinal photoreceptors. Each photoreceptor is constituted of four parts: outer segment, inner segment, soma, and synaptic terminal (Fig. 2.34). The outer segment contains a stack of membrane disks, which contain light-sensitive photopigments. When photons are absorbed by photopigments, it triggers a series of changes in photoreceptor and induces membrane potential changes.

Photoreceptors can be classified into rods and cones following the shape of their outer segments. While rods are characterized by their long and slender outer segments, cones are with short and tapered outer segments. The structural difference results in differentiation in function of these two types of photoreceptors. Rods contain abundant membrane disks with photopigments of higher density; the rods are thus endowed with higher light sensitivity, which is about a thousand times higher than that of the cones. Therefore, while in dark vision, the rod system is the main one which is responsible for phototransduction. However, in light vision, the rod system is susceptible to be saturated, so that the phototransduction during light vision is mainly conducted by the cone system. This is why the retina is normally considered as containing two overlapping systems, with the rod system mediating dark vision and the cone system mediating light vision.

The differences between rods and cones are also exhibited by other aspects. While mammalian retinas typically possess only one kind of rods with a certain kind of photopigments, animals with color vision normally possess three different kinds of cones, with each containing a particular type of photopigments. The structure of photopigments determines their light-absorbing characters and thus determines the spectral sensitivity of photoreceptors. This forms the structural basis of color vision (Fig. 2.35).

Rods and cones are not equally distributed across the whole retina. In the peripheral part, rods are far more numerous than cones. On the other hand, the central part contains cones with higher density. Such structural property endows the peripheral retina with better light sensitivity, while the central part of the retina is with higher visual acuity. In the retina, the fovea located in the center of macular is the area with the best visual acuity. This part of retina is rod-free and with most densely packed cones (Fig. 2.36). This fovea area with a diameter of about half a millimeter is thinner than any other parts of the retina. The formation of the fovea is due to the displacement of ganglion cells toward the peripheral part, so that light

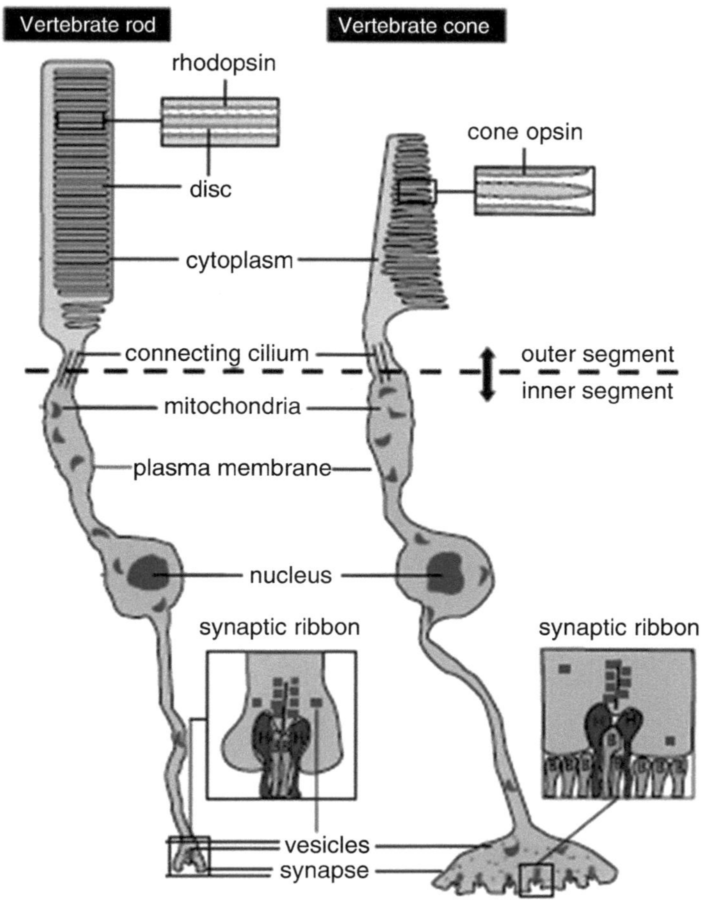

Fig. 2.34 Structure of rod and cone photoreceptors (Redrawn from Galizia and Lledo 2013, Springer, Neurosciences – From Molecule to Behavoir, Fig. 18.2)

can arrive the photoreceptors directly within this subregion, and therefore to enhancement of the visual acuity of this region.

The function of photoreceptors is to transfer the light signal into the membrane potential changes. This process somehow analogues the transferring of chemical signal to electrical signal in synaptic transmission.

In photoreceptors, the membrane potential of rod outer segment is around −30 mV in the dark. This is different from the typical resting membrane potential of −65 mV in other neurons. Such depolarizing effect of rod in the dark is due to sustained Na^+ influx via specific channel on the membrane in the outer segment. This positive transmembrane current is called dark current. The relevant sodium channel is gated by intracellular second messenger cyclic guanosine monophosphate (cGMP). When photons are absorbed by photopigments

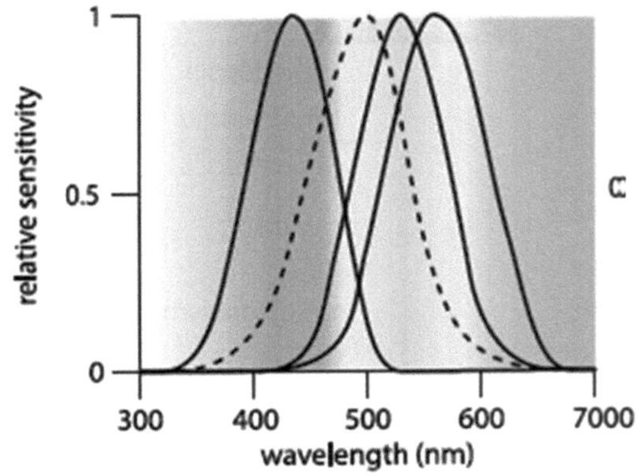

Fig. 2.35 Spectral sensitivity curves for rod (*dashed curve*) and three cone subtypes (*solid curves*) (From Galizia and Lledo 2013, Springer, Neurosciences – From Molecule to Behavior, box 18.1)

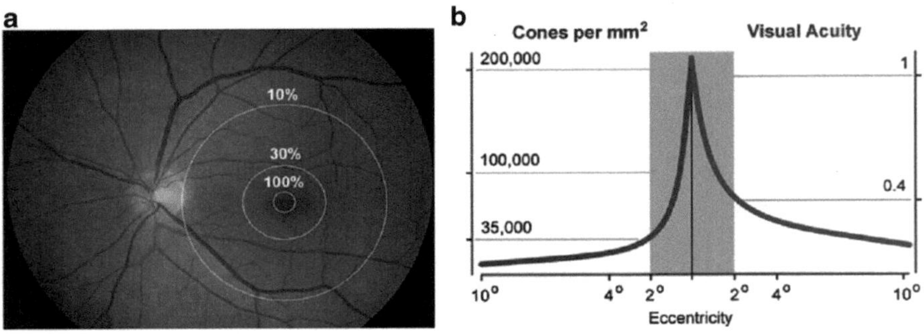

Fig. 2.36 The location of fovea on the retina and the visual acuity. (**a**) In the human retina, the fovea has the highest visual acuity. (**b**) Cone density (*left*) and visual acuity (*right*) of the human retina (From Pfaff 2013, Springer, Neuroscience in the 21st century, Fig. 18.29)

(rhodopsin) located on the membrane disks in rod outer segment, the phototransduction response is trigged. Rhodopsin is a receptor protein (opsin) prebound to a particular chemical agonist (retinal, a derivative of vitamin A). Retinal absorbs light and then leads to a conformation change and activates opsin, which is the only light-dependent step in vision. This process results in the activation of a G-protein which is called transducin on the membrane disk, which in turn activates an effector enzyme phosphodiesterase (PDE). PDE dissociates cGMP in the cytoplasm, which results in closure of the cGMP-gated Na^+ channel and reduces Na^+ influx. The light response of rod is therefore a hyperpolarization of membrane potential (Fig. 2.37).

Sustained light illumination keeps cGMP at a low level in rod and saturates its light response; more light stimulation cannot cause any further hyperpolarization. Hence, visual process in light condition is mainly dependent on the cone system.

The phototransduction process in cones resembles that in rod, with the only difference being the subtype of opsins located on the membrane disks of cone outer

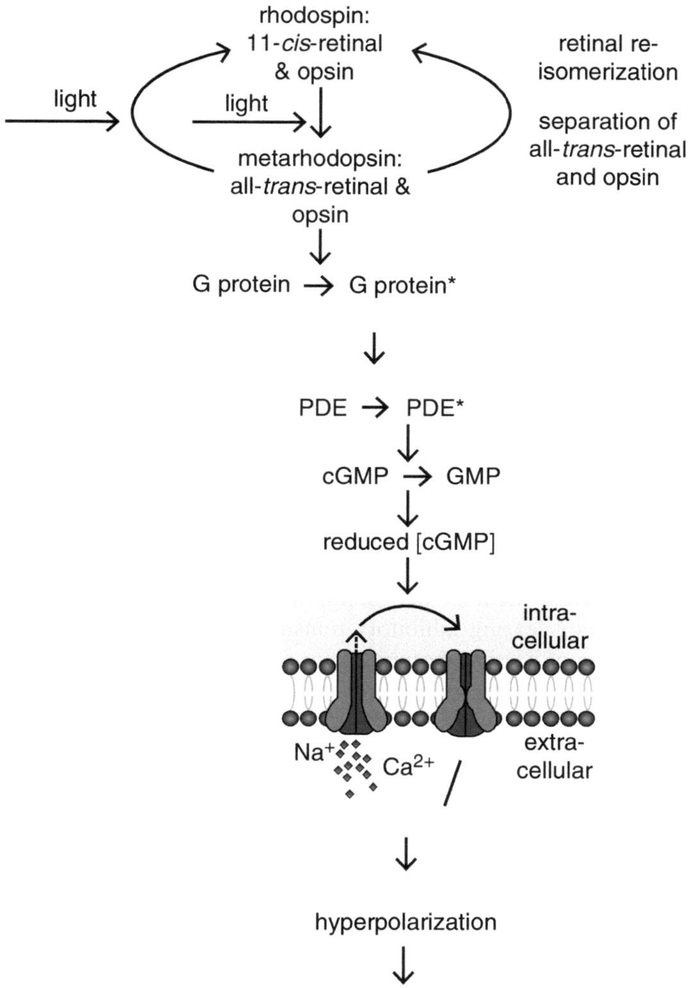

Fig. 2.37 Main steps of phototransduction cascade (From Galizia and Lledo 2013, Springer, Neurosciences: From Molecule to Behavior, Fig. 18.3)

segments. In humans, there are three different kinds of opsins in cones. According to the spectral sensitivity of the opsins, cones can be classified as "blue sensitive" (most sensitive to light of 430 nm), "green sensitive" (most sensitive to light of 530 nm), and "red sensitive" (most sensitive to light of 560 nm).

The identification of color signals in the visual system is largely dependent on the relative contribution of these three kinds of cone photoreceptors. British physicist Thomas Young indicated in 1802 that all the colors shown in rainbow, including white, can be represented by mixing red, green, and blue lights with proper ratios. He suggested that retinal photoreceptors can be classified into three different kinds, with each being most sensitive to blue, green, or red lights. This hypothesis was later supported by German physiologist Hermann von Helmholtz in the nineteenth century, with the theory being termed as "Young–Helmholtz trichromacy theory." Based on this theory, brain can identify color signals according to the ratio of output signals from three kinds of cones. When all the cones are equally activated, "white light" can be perceived. Loss of one or more opsin subtypes will cause various kinds of color blind. This also explains why cones are not active in the dark environment, and no change of color can be perceived.

In the retina, the most direct signal pathway is the one from photoreceptors to ganglion cells via bipolar cells. However, in each plexiform layer of synaptic connections, light response is also subject to lateral integration mediated by horizontal cells and amacrine cells.

Photoreceptors release neurotransmitter while they are depolarized, which is similar to other types of neurons. Photoreceptors are depolarized in the dark and are hyperpolarized in response to light stimulation. Thus photoreceptors release more excitatory transmitter glutamate in the dark. In the outer plexiform layer, photoreceptors are connected to both bipolar cells and horizontal cells. Bipolar cells are the interneurons in the direct signal pathway from photoreceptors to ganglion cells, while horizontal cells mediate the lateral information adjustment in the outer plexiform layer via releasing inhibitory transmitter γ-aminobutyric acid, so as to adjust the activities of nearby photoreceptors and bipolar cells.

According to their light response properties, bipolar cells can be classified into two subtypes. Off-center bipolar cells express ionotropic glutamate receptors. Glutamate-gated cation channels are closed by light stimulation when glutamate release is reduced, so that light projected onto their central parts causes hyperpolarization. On the other hand, on-center bipolar cells express metabotropic glutamate receptors (G-protein-coupled receptors). Glutamate binding to metabotropic glutamate receptor causes closure of cation channels, so that light projected onto the central parts of these cells results in hyperpolarization.

Each bipolar cell receives direct synaptic inputs from a number of photoreceptors and in the meantime makes indirect connections with photoreceptors located in its surrounding area via horizontal cells. The receptive field (the light stimulation within which the neuron's response will elicit) of a bipolar cell is therefore composed of two parts: a round-shaped central area, which receives direct inputs from photoreceptors, and a circular peripheral region, which receives signals modulated by horizontal cells. In bipolar cells, the membrane potential changes are antagonistic in responses to light stimulation located in its receptive field center and that located in the peripheral part. If a light stimulation projected in its central area causes depolarization of a bipolar cell, light illumination on its peripheral part will hyperpolarize it due to the inhibitory signal from horizontal cells. Similarly, when a dark stimulation of the central part depolarizes a bipolar cell, dark shaded in the periphery will hyperpolarize it.

Such antagonist center-surround receptive field property is not limited to bipolar cells; it is passed onto ganglion cells via synaptic connections in the inner plexiform layer. In the meantime, the lateral adjustments via amacrine cells also contribute to the formation of ganglion cell's receptive field.

In the retina, photoreceptors and other neurons in the inner nuclear layer all respond to light stimulation with graded potentials. Ganglion cells are the only type of cells that responds to light with action potentials, which form the output signals from the retina to the central visual part, so as to encode the visual stimulation.

Similar to that of the bipolar cells, ganglion cells also have eccentric center-surround receptive field structure. Ganglion cells receive glutamate inputs from bipolar cells. Those ganglion cells which receive inputs from on-center bipolar cells respond to light stimulation in their receptive field center with increased action potential firings and are called "on-center ganglion cell." For these cells, a light spot projected onto their receptive field center will elicit firings with high frequency, while for the peripheral area of their receptive field, only the withdrawal of light will increase the cells' firing activities. Accordingly, "off-center ganglion cell" receives input from off-center bipolar cells. In this ganglion cell subtype, dark shaded onto their receptive field center will enhance their firing activity, and light illuminated in the peripheral subregion will also enhance their firing activities. For these two subtypes of ganglion cells, light projected in their central part and peripheral parts will antagonize each other; therefore, ganglion cells respond optimally to contrast in their receptive field (Fig. 2.38).

Ganglion cells can also be classified according to other properties such as morphological and electrophysiological properties. In primate retinas, most ganglion cells fall into two main classes. While 5 % of the ganglion cells are identified as magnocellular cells (M-cells), nearly 90 % other ganglion cells are of parvocellular subtype (P-cells). Both M and P classes include on-center and off-center cells. There are still other 5 % of the ganglion cells being none-M-none-P cells, the properties of which are still not clear. M-cells differ from P-cells in several aspects. Firstly, M-cells are with bigger dendritic arbors and larger receptive field and faster propagating speed and being more sensitive to stimulus with low contrast; secondly, M-cells fire relatively transiently in response to sustained

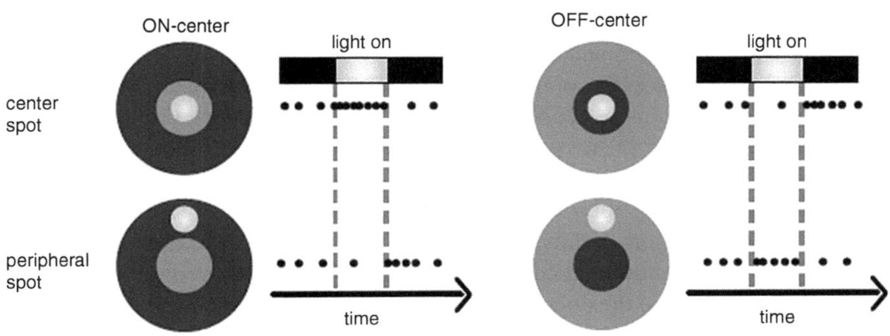

Fig. 2.38 The center-surround antagonist receptive field property (From Galizia and Lledo 2013, Springer, Neurosciences: From Molecule to Behavior, Box 18.2)

stimulation, whereas P-cells respond with sustained firings; in addition, M-cells are more sensitive to moving targets, while P-cells are more sensitive to shape and finer details. Another important difference among ganglion cells is that P-cells and some of the non-M-non-P cells are spectral sensitive, while M-cells are insensitive to color. The response property of spectral-sensitive neurons is that central and peripheral parts of the receptive field show color-antagonistic property, e.g., with a red on-center and a green off-surround. Yellow and blue are another pair of antagonist colors. Our color perception is based on the relative activities of these color-sensitive ganglion cells.

One important feature of retinal information processing is parallel processing of signals. First of all, the two retinas generate two parallel signal flows. Secondly, in each retina, light information is processed separately by on- and off-pathways. Thirdly, M-cells, P-cells, and non-M-non-P cells detect different visual aspects. Information is processed by different ganglion cell subtypes independently in the retina and transferred to the central visual part via segregated pathways.

Having known these, one can tell that although the optical part of the eye resembles the structure of a camera, the visual neural system starting from retina is delicate and is far more complicated than a camera. Retina does not passively transmit the visual information about the image projected onto it, but retrieves and processes various aspects of information about the image. Visual perception about the environment therefore depends on the information provided by the retina, as well as the analyses of such information by the central visual part.

The axons of all retinal ganglion cells converge to form the optic nerves, which leave the eyes from the optical disks. The optical disk is the region containing no photoreceptors and therefore is a blind spot in the retina. After leaving the optic disks, the optic nerves exchange their fibers at the optic chiasm anterior to the hypothalamus. Axons from the nasal half of each retina are crossed in the optic chiasm and go to the contralateral side of the brain, while axons from the temporal half of each retina go straight to the ipsilateral side of the brain. Because the objects in the left visual field are imaged on the nasal part of the left retina and the temporal part of the right retina, and in the meantime the nasal axons from the left retina and the temporal axons from the right retina merge to form the right optic tract at the optic chiasm, therefore, all the information about the left visual field is directed into the right half of the brain and *vice versa* (Fig. 2.39).

Most of the axons in the optic tract terminate at the lateral geniculate nucleus (LGN), while the axons of LGN neurons project to the primary visual cortex via optic radiation. Injury at any part of the neural projection from the eye to visual cortex via LGN will cause blindness. In the meantime, there are a minority of ganglion cells sending their axons to the central areas other than LGN. The projection to the hypothalamus is related to biorhythms, while the projection to the pretectum of the midbrain controls pupillary reflexes and eye movements.

LGNs are located in the dorsal part of the thalamus, with each including six cell layers numbered from 1 to 6, ventral to dorsal (Fig. 2.40). M-cells, P-cells, and non-M-non-P cells from the two retinas make synaptic contacts with LGN neurons at separated LGN layers. Besides, signals from the two retinas are processed by

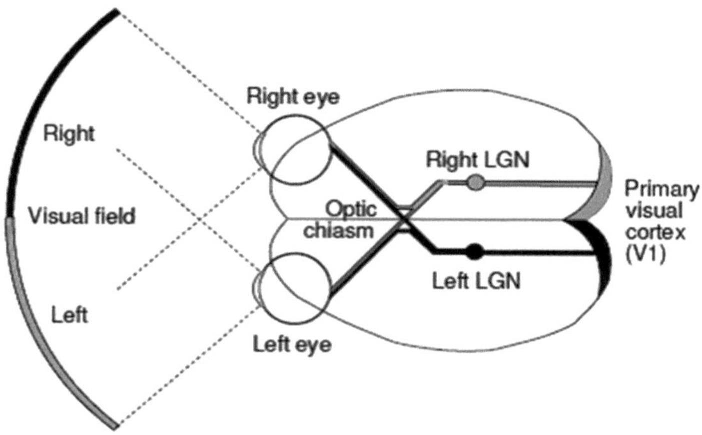

Fig. 2.39 Human visual pathways (From Miikkulainen et al. 2005, Springer, Computational maps in the visual cortex, Fig. 2.1)

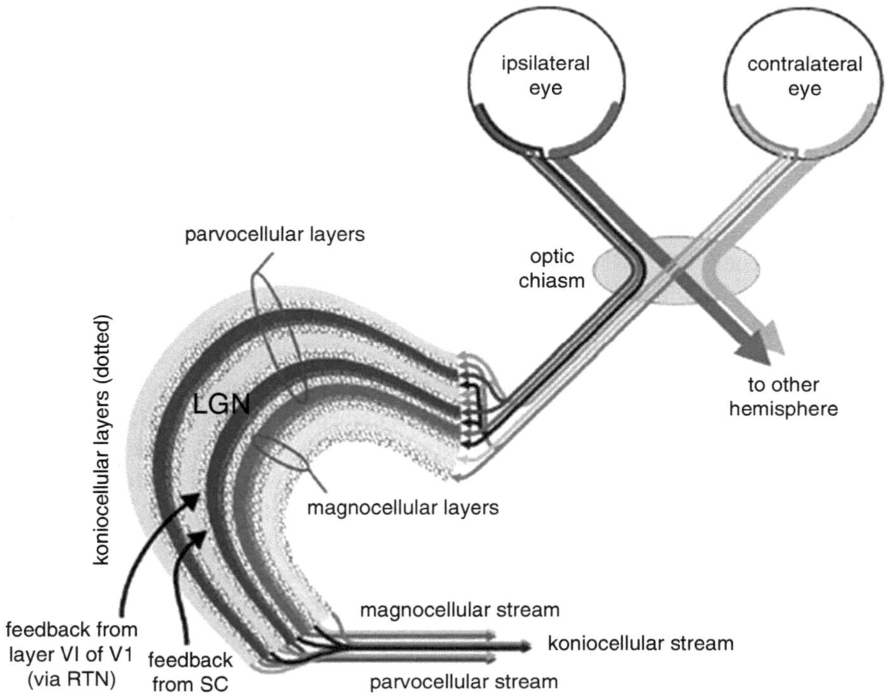

Fig. 2.40 LGN structure and information flow (From Galizia and Lledo 2013, Springer, Neurosciences: from molecule to behavior, Fig. 18.5)

LGN neurons located in different layers. Axons from ipsilateral retina form synaptic connections with LGN cells in the 2nd, 3rd, and 5th layers, while axons from contralateral retina form synaptic connections with LGN cells in the 1st, 4th, and 6th layers. The axon fibers from the two eyes are segregated; in the meantime, the retinal projection forms retinotopic map on each LGN layer.

In LGN, the two ventral layers (layer 1 and layer 2) contain M-cells and are called M-cell layers; the four layers (layers 3 through 6) on the dorsal part contain P-cells and are called P-cell layers. They receive projections from retinal M-cells and P-cells, respectively. In addition to the cells located in these six main cell layers, there are abundant tiny neurons located in the ventral part of each cell layers, which form koniocellular layers. Cells in these koniocellular layers receive input from non-M-non-P retinal ganglion cells. In LGN, information from these three subtypes of ganglion cells from two retinas is still segregated.

LGN cells keep the same receptive field properties as the retinal ganglion cells providing inputs to them. Similar to retinal M-type ganglion cells, LGN M-cells have relatively large center-surround receptive field and have transient action potential firings in response to visual stimulation, but these cells are insensitive to spectral difference. On the contrary, the properties of LGN P-cells resemble that of retinal P-cells, having small center-surround receptive field, firing sustained action potentials, and with color opponency. Some of K-cells also have color-opponent property.

Albeit most retinal ganglion cells' axons are terminated at LGN, up to 80 % excitatory input LGN neurons received are from primary cortex. Although the functions of these inputs are not entirely clear, we still have reasons to believe that such cortical-fugal feedback pathway is important for modulating the LGN neurons' activity. Apart from this, LGN neurons also receive synaptic inputs from brain stem. Therefore, LGN is not only a relay from the retina to the visual cortex; it should be involved in some important visual information regulation.

The main target of LGN neurons is the primary visual cortex (Fig. 2.41). In primate brain, it is located in the occipital lobe and is also named as Brodmann area 17 (or V1, striate cortex). The same as neurons in other neocortical areas, cell bodies in this region are arranged in six layers, named as layers I to VI, with layer I underneath the pial surface and layer VI close to the white matter in the center. But the striate cortex differs from other cortical area in a way that layer IV can be further subdivided into IVA, IVB, and IVC (including IVC_{α} and IVC_{β}) sub-layers. Synapses from LGN neurons mainly terminate at IVC sub-layer.

The striate cortex contains neurons with various shapes. Stellate cells are with small cell body and with abundant spines on the dendrites that radiate out from the cell body. These cells are distributed in the two sub-layers of IVC; their axons are confined to the primary visual cortex to form local connections. Pyramidal cells are distributed in other layers out of IVC. These cells have one or more thick and branched apical dendrites and laterally stretching basal dendrites. Pyramidal cells are the only ones whose axons project to cortical areas out of the striate cortex and form connections with neurons located in other part of the brain (Fig. 2.42).

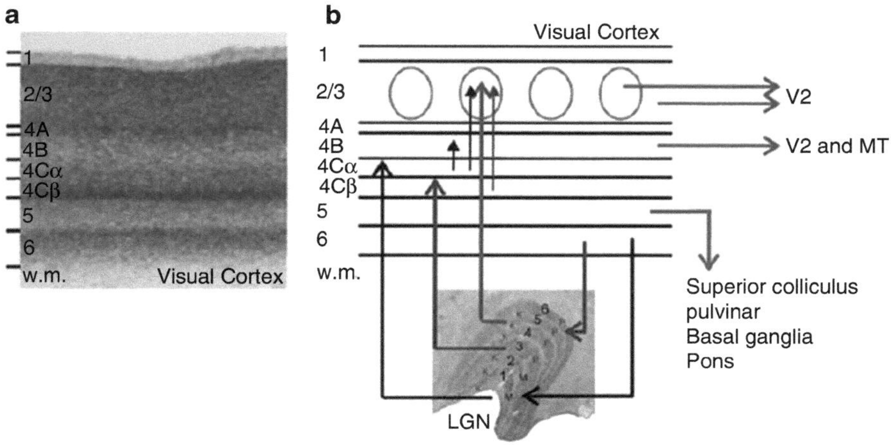

Fig. 2.41 Primary visual cortex (Area 17) in the macaque monkey. (**a**) The six layers of cells in V1 (*w.m.* white matter). (**b**) Signal flow among LGN, VI, and other visual cortices (From Pfaff 2013, Springer, Neuroscience in the 21st century, Fig. 20.3)

Fig. 2.42 Distribution of neurons within the striate cortex. Stellate cell is located in IVC; pyramidal cells are in III, IVB, V, and VI layers (From Bear et al. 2007, Lippincott Williams & Wilkins, Neuroscience – Exploring the Brain, Fig. 10.13)

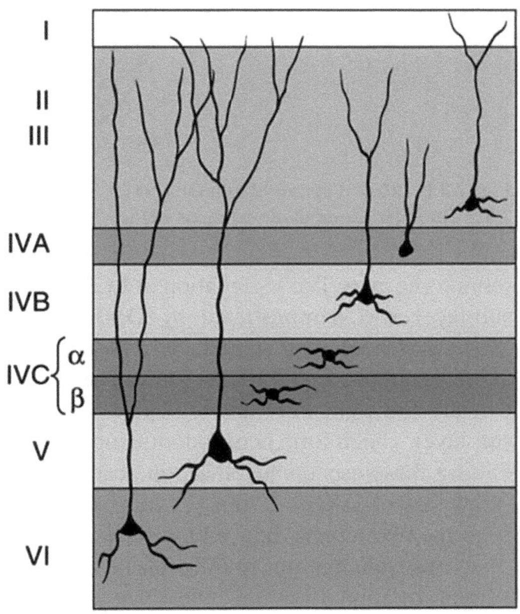

The neuronal projections from LGN neurons to the IV layer of area 17 form parallel signal pathways. Information flows from the two eyes, and various cell types are segregated in IVC sub-layer. Generally speaking, LGN M-cells and P-cells project to IVC_α and IVC_β sub-layers, respectively, with each containing a

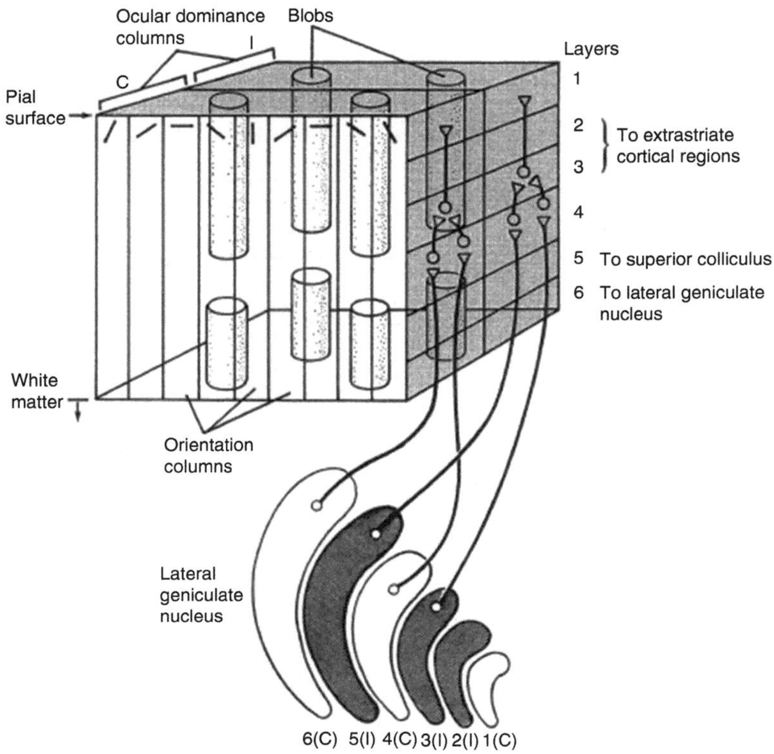

Fig. 2.43 Block diagram of primary cortex with the LGN afferents (From Miikkulainen et al. 2005, Springer, Computational maps in the visual cortex, Fig. 1.1)

retinotopic map. The segregation of LGN signals from the left and right eyes in IVC sub-layer was demonstrated by David Hubel and Torsten Wiesel of Harvard Medical School who shared 1981 Nobel Prize for Physiology or Medicine, via autoradiography in the 1970s. They observed that the axons conveying signals from the left and right eyes are segregated into alternatively appearing strips in IVC sub-layer, which form ocular-dominance columns with width of 0.5 mm (Fig. 2.43). Putting these aspects together, one can tell that one small patch in IVC (0.5 mm (thickness of IVC) $\times$ 1 mm $\times$ 1 mm) contains inputs from various kinds of cells from the two retinas, relayed by LGN.

Axons from neurons in IVC sub-layer make vertical projection to IVB and layer III; information from the two eyes starts to converge at this stage. However, information processing is still segregated for M-cell and P-cell pathways. IVC_α neurons receiving LGN M-cell inputs project to IVB, while IVC_β cells receiving LGN P-cell inputs project to layer III.

Apart from inputs from IVC_β P-cells, layer III cells also receive direct inputs from LGN neurons. Margaret Wong-Riley of University of California discovered in 1978 that cytochrome oxidase is not evenly distributed in the striate cortex. Cells with high cytochrome oxidase density are clustered in columns through layers II–V in the striate cortex (Fig. 2.43). In layer III, these areas with high cytochrome oxidase density are scattered, appearing like blobs on leopard. The areas between the blobs are called "interblob area." Later experiments discovered that cells in both blob and interblob areas receive LGN P-cell inputs via IVC_β, while in the meantime, blob cells also receive direct inputs from LGN K-cells, i.e., P-cell and K-cell inputs merge in the blobs. On the other hand, each blob has its center located in an ocular dominance column in layer IV. Thus, blob cells still receive input from single eye. The input from two eyes merges at interblob area in layer IVB and layer III.

Taken together, there are three independent pathways for visual information processing in the striate cortex: M-cell pathway, P-cell-interblob pathway, and blob pathway. They are responsible for different aspects of visual information. We will discuss the physiological properties of these information pathways separately.

LGN M-cells are driven by signals from a single eye; they respond to visual stimulation with transient responses and are spectral insensitive. In IVC_α sub-layer, many neurons receiving LGN M-cell inputs have narrow elongated excitatory and inhibitory subareas responding to light-on and light-off transients, respectively (Fig. 2.44). Hubel and Wiesel called these cells "simple cells." The elongated receptive field is considered as due to convergent excitatory inputs from several LGN cells whose receptive fields are aligned along a certain direction (Fig. 2.45). A simple cell is most sensitive to stimulation along the long axis of its receptive field, but insensitive to stimulation orthogonal to the long axis. Therefore, such IVC_α cells are orientation selective. Cells in IVB layer that receive IVC_α cell inputs also have such simple receptive field structure and therefore have orientation selectivity. But one important difference between cells in these two layers is that IVC_α cells

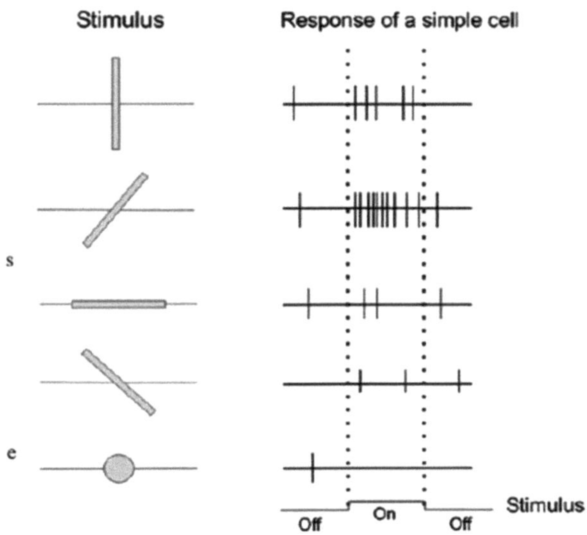

Fig. 2.44 Response property of simple cell (From Pfaff 2013, Springer, Neuroscience in the 21st century, Fig. 20.4)

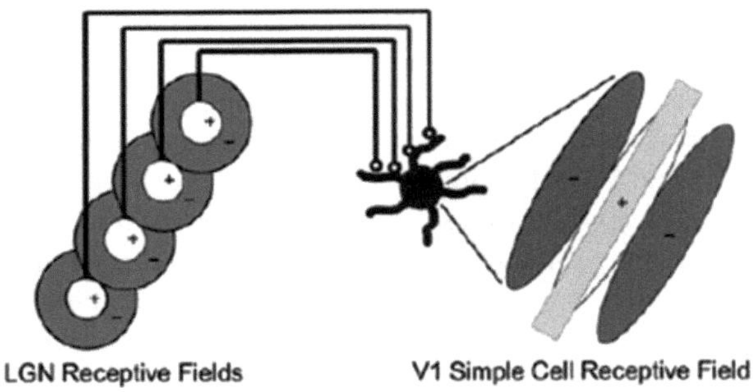

Fig. 2.45 Receptive field properties of simple cell and possible mechanism (From Pfaff 2013, Springer, Neuroscience in the 21st century, Fig. 20.5)

receive inputs from single eye and form ocular dominance columns, while signals from two eyes are merged in IVB.

Another important physiological feature of the IVB cells is motion direction selectivity. Such cell can be optimally activated by an elongated object moving with a certain direction across its receptive field. Being selective to moving direction of stimulation is a character of neurons in M-cell pathway; the physiological function of M-pathway is therefore thought to be specified to process motion information of visual targets.

The interblob cells in layer III receive input from IVC_β sub-layer. These cells do not have obvious light-on or light-off subareas in their receptive fields, but respond to light–dark contrast appearing in their receptive fields with certain orientation. Hubel and Wiesel called these cells "complex cells." Complex cell is considered as receiving converging input from a group of simple cells with similar receptive field property, and the pattern of input determines the receptive field property of the complex cell. Most of the complex cells receive binocular inputs and are highly selective to stimulation orientation, but are insensitive to color. The function of P-IB pathway is therefore thought to analyze the shape of visual objects.

Now we have the impression that many cortical neurons are sensitive to orientation, i.e., having orientation selectivity (Fig. 2.43). Although these neurons are distributed in various layers in area 17, according to Hubel and Wiesel's initial work, if an electrode vertically penetrates different cell layers, the neurons being recorded will have the same orientation selectivity. Hubel and Wiesel thus called such cell column vertical to the cortical surface "orientation column." However, if an electrode penetrates the cortex tangentially, the cells being recorded should have gradually changing orientation selectivity. Hubel and Wiesel found that an entire changing of 180° of optimal orientation occupied a width of about 1 mm in layer III.

Neurons in the blob areas are different from that in the interblob areas. Most blob cells have round receptive fields, and some of them have color-opponent center-surround property. The blobs contain most spectral-sensitive neuron outside the IVC layer. But these cells are not orientation selective and only receive inputs from single eye. Therefore, the specified function of the blob pathway is to analyze the color information of visual targets.

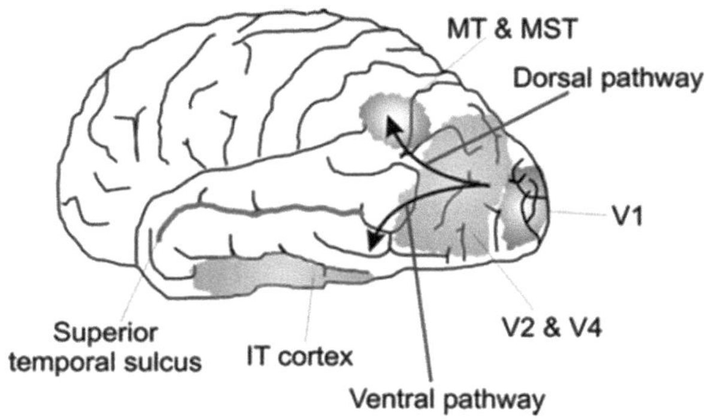

Fig. 2.46 Dorsal and ventral pathway of visual information flow (From Pfaff 2013, Springer, Neuroscience in the 21st century, Fig. 20.12)

Taken together, each of the visual aspects can be analyzed by a specific type of cells located within a limited area in the visual cortex. Actually, a small cortical region of 2×2 mm contains two complete ocular dominance columns, 16 blobs, and an entire group of orientation columns covering 180° of orientation which are formed by interblob cells. Therefore, Hubel and Wiesel proposed that the 2×2 mm region in area 17 formed a basic unit for visual information processing, which they defined as a cortical module (Fig. 2.43). Area 17 might contain up to a thousand of such functional modules.

V1 is the cortical area receiving direct LGN inputs; its projection targets are V2 and V3 accordingly, which further project to MT area in the temporal lobe and MST area in the parietal lobe. This forms the dorsal pathway of visual information flow, the function of which is to analyze the moving information of visual targets. Another projection is to the ventral pathway including V4 and IT areas, the function of which is object identification (Fig. 2.46).

MT area receives retinotopic projection from cortical areas such as V2 and V3; it also receives neuronal input from IVB layer of V1. Because IVB cells are involved in the M-cell pathway which is characterized by big receptive fields and with transient light response and direction selectivity, MT neurons therefore also have big receptive fields and are with motion direction selectivity. While being sensitive to motion of visual targets, MT neurons are insensitive to shape of visual targets. This means that the specified function of MT area is to process the object motion; the activities of neurons in this area are crucial for motion perception. Similar with orientation columns in V1, the ordered arrangement of cells in MT area builds up motion direction selective columns. The perception of object motion is dependent on the comparison of activities across the whole 360°.

In addition to MT area, there are also other specified motion-sensitive areas in the parietal lobe. For example, some neurons in MST area are sensitive to rectilinear motion, radial motion, or circular motion.

Being parallel to the dorsal pathway, the ventral pathway is originated from areas V1, V2, and V3. It builds up a pathway leading to the temporal lobe, with the function being analyzing the visual information apart from object motion. Among these temporal areas, the area being understood the best is area V4, which receives blob and interblob inputs from V1 via V2. As compared to V1 neurons, V4 neurons are with larger receptive fields, and many V4 cells are both orientation selective and spectral sensitive, so that V4 area is important for cognition of both shape and color. In the ventral pathway, areas outside V4 all contain neurons with complex spatial receptive fields. Area IT in inferior temporal lobe receives inputs from V4; colors and abstract shapes are both good stimulations for IT cells. This area is significant for visual cognition. One interesting aspect of IT is that a small portion of neurons can be activated by facial images, and some particular faces may induce more severe activity of these neurons. At the same time, these cells also contribute to other information processing.

Comparing the receptive field properties of neurons of different stage in the visual pathways, it leads to some interesting findings. In photoreceptors, the receptive fields are small spots, while in retinal bipolar cells and ganglion cells, the receptive fields are with center-surround structure. For the cortical neurons, the response features are more specified, being highly selective to some of the visual aspects and even being responsible to special facial stimulation. It was proposed that our perception to particular objects is based on the activity of a few highly specified neurons in some brain areas, which are not yet well identified. However, this hypothesis is not commonly accepted, because it is not compatible with the basic principle that neurons in the nervous system are widely tuned.

On the other hand, according to the fact that parallel information processing mechanisms are throughout the whole visual system, another hypothesis is proposed such that the cognition tasks are "distributed" in the brain, i.e., many cells with broad tuning curves in certain cortical areas are activated coordinately for the analyses of particular visual properties, while in a larger scale, the coordination among neurons involved in processing different visual information (including color, shape or movement) contributes to the overall visual perception.

2.6 Auditory System

The audition is one of the main sensations for animals, also a kind of information processing. However, the progress in audition studies is far behind vision research owing to the fact that the importance of audition is less than vision for animal's survival in general. In addition, there are the following specific difficulties for audition research:

Firstly, the frequency band of auditable sound is rather wide; for human beings, its auditable range is between 20 Hz and 20,000 Hz, i.e., almost 10 octaves. In contrast, the spectrum of visible light is within 1 octave.

Secondly, all the problems about audition are always dynamic in their essence; temporal factor plays a key role in these problems.

Thirdly, the auditory pathways are very complex; there are many levels, and the classification of levels is rather ambiguous. Besides the indirect connections via intermediate levels, there may also be direct connections between different levels. There are also cross connections between the left pathway and the right one. In addition, besides ascending pathways, there are also descending pathways from the central nuclei down to the nuclei in periphery to make feedback connections.

Fourthly, there is significant nonlinearity in their functions, e.g., there are characteristics such as rectification effect to sound waves, saturation effect to stimulus intensity, intensity-dependent filtering, asymmetry of two-tone inhibition, etc. Thus, it is impossible to predict the response to complex signals, such as a variety of natural sounds and speech, based on the responses to "simple signal" such as pure tones or clicks, which were extensively and intensively studied before the end of last century.

Fifthly, the nervous structures of auditory system are deeply in the head; thus it is difficult to do experiment on them.

Finally, the wavelength of sound waves is quite long so that the problems due to reflex and diffraction are significant.

It is the problems listed above that make the research on auditory information processing focused on auditory periphery, especially on coding of the firing pattern in single fiber or single neuron to pure tones or clicks before the nucleus olivaris superior (Hawkins and McMullen 1996), and the studies on the central structures in the auditory system and responses to complex sounds mainly started late last century, so that it is difficult to model auditory systems. Maybe an exception is the model of bat echo-localization (Shamma 2002).

Anatomy of Auditory Systems The auditory system is composed of ear and auditory nervous system. The ear can be divided into three parts: outer ear, middle, and inner ear (Fig. 2.47a). The main role of the outer ear, including pinnae and external auditory meatus, and the middle ear is to collect and transmit sound effectively from the air into the lymph within the inner ear. The middle ear also plays a key role in impedance match between gas and fluid. The sound stimulus makes the tympanic membrane at the end of the external auditory meatus vibrate. The vibration of tympanic membrane makes the stapes vibrate via a lever system consisting of the malleus, the head of which is attached at the tympanic membrane, and the incus. The stapes plate at the end of the stapes attaches the oval window membrane at the base of cochlea in the inner ear; thus, the movement of stapes will lead to the movement of the lymph within the cochlea (Fig. 2.47a). In case there is no such a structure and the sound is transmitted from air to the lymph directly, then most of the energy would be reflected at the interface between gas and fluid, and the transduction efficiency would be very low. As the area of tympanic membrane is much larger than the one of oval window and there is a lever system composed of malleus, incus, and stapes to amplify the force, the conduction efficiency is much higher. In its essence, the outer ear and middle ear is a kind of passive and linear mechanical system; the main function of which is to collect and conduct sounds but not information processing; therefore, we will not discuss these apparatus in detail. It is only pointed out here that the outer ear and middle ear could be considered as a

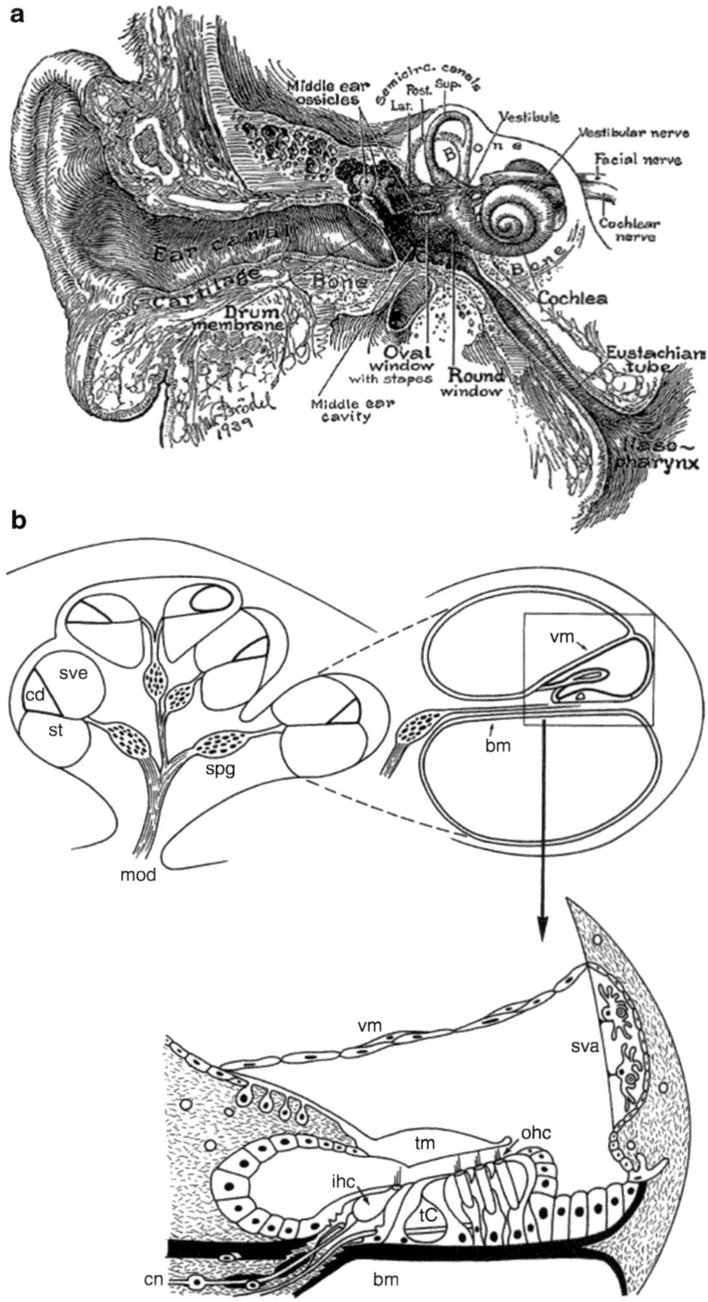

Fig. 2.47 Auditory system. (**a**) Outer ear, middle ear, and inner ear (Reproduced with permission from ten Donkelaar HJ (2011) *Clinical Neuroanatomy: Brain Circuitry and its disorders*. Springer. Fig. 7.1). (**b**) structure of the cochlea (*upper*) and a section diagram of the cochlea and the organ of Corti (*bottom*), *cn* the cochlear nerve, *mod* the modiolus, *spg* the spiral ganglion, *bm* basilar membrane, *cd* cochlear duct, *ihc* inner hair cells, *ohc* outer hair cells, *st* scala tympani, *sva* stria vascularis, *sve* scala vestibuli, *tC* tunnel of Corti, *tm* tectorial membrane, *vm* vestibular (Reissner) membrane (Reproduced with permission from ten Donkelaar HJ (2011) *Clinical Neuroanatomy: Brain Circuitry and its disorders*. Springer. Fig. 7.2)

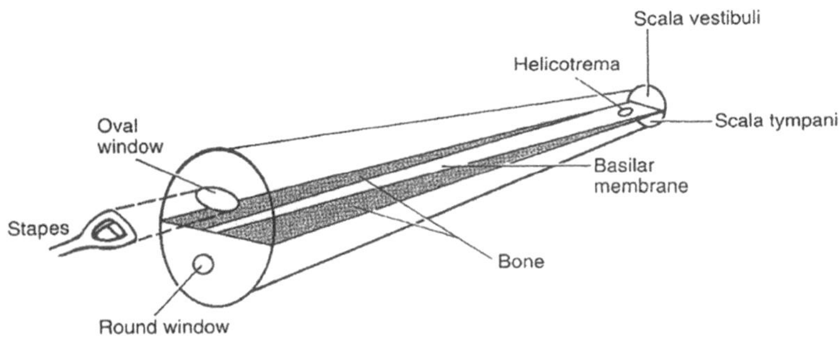

Fig. 2.48 A schematic diagram of cochlea. The cochlea is shown uncoiled. The basilar membrane is narrow near the base and wider near the helicotrema, where the scala tympani and the scala vestibule meet (Reproduced with permission from Conn PM (Ed) (2008) *Neuroscience in Medicine (3rd Edition)* Humana Press (in Yin TCT. Audition. Fig. 4B))

kind of band-pass filters. The inner ear can be considered as both a transducer which transforms sound signal into nervous electrical signal and an analyzer which transforms frequency variables into spatial variable in some degree and plays the role of some band-pass filter bank. Every auditory nerve plays a role of some band-pass filter with different passing band. Thus the original sound stimulus with very wide spectrum could be transformed into a series of signals with frequency components within much narrower band.

The inner ear is composed of two parts: the cochlea and the semicircular canals, while the latter is related to the balance sense. The cochlea is a snail-shell-like canal, which is divided into three parts: the scala vestibule, scala media, and scala tympani (left upper diagram in Fig. 2.47b). The scala vestibule and the scala tympani interconnect with each other at the top of the cochlea—the helicotrema (Fig. 2.48). Perilymph, a kind of fluid similar to cerebrospinal fluid, is full in these canals, which contains low potassium ion concentration and high sodium ion concentration, while in the scala media, it is full of endolymph—a kind of fluid similar to intracellular fluid, in which there is high concentration of potassium ions and low concentration of sodium ions. Although the scala vestibule and the scala media are separated by a thin membrane, they could be considered as one compartment from the point of view of hydromechanics. The scala media and the scala tympani are separated by a thick membrane—the basilar membrane with the organ of Corti on it. Hair cells as transducers are arranged along the basilar membrane. There are two kinds of hair cells: inner hair cells form one column and outer hair cells are arranged in three columns. There are elastic hairs on the top of every hair cell, a tectorial membrane with one free end, and another end fixed is just above these cells. The tectorial membrane contacts outer hair cells directly, but not so with the inner hair cells (lower diagram in Fig. 2.47b).

There is another membrane covering the round window at the end of the scala tympani. Owing to the incompressibility of the fluid, the vibration of the oval window membrane will lead the round window membrane to vibrating correspondingly via the basilar membrane. The vibration of the basilar membrane makes a relative movement between the tectorial membrane and the hairs of outer hair cells, which bend to and fro. The hairs of the inner hair cells are also bent to and fro by the vibration of the lymph. Bending of the hairs opens the deformation-sensitive channels, producing receptor potentials in the hair cells. While the hairs bend toward one direction, the hair cell is depolarized, the hairs bend toward the opposite direction, and the cell is hyperpolarized, thus only one way bending may trigger firing of the ganglion cell next to the hair cell, leading to a half-wave rectification for the dilatational wave. The base of the basilar membrane is narrower, while the top is wider (Fig. 2.48). In addition, the stiffness at the base is stronger, while at top is weaker. Thus the different location at the basilar membrane is sensitive to different frequency components in the sound stimulus.

There is synaptic connection between the base of the hair cell and the dendrite of the spiral ganglion cell, which is a kind of bipolar cell and soma of which is located in spiral ganglion at the cochlea axis. There are about 30,000 ganglion cells for human beings. And the axon of the spiral ganglion cell is the auditory nerve. There are about 3500 inner hair cells and 12,000 outer hair cells in the human cochlea. About 95 % of the spiral ganglion cells connect with inner hair cells with only one bouton, while every inner hair cell contacts the terminals of about 8–16 spiral ganglion cells; only 5 % of spiral ganglion cells have synaptic connection with outer hair cells; however, every such spiral ganglion cell contacts many outer hair cells. Recent studies have shown that the length of outer hair cells changes when they are stimulated; thus, the interrelationship between such cells and the tectorial membrane also changes, and the responses of inner cells are amplified at last.

The auditory nerve terminates at the dorsal cochlear nucleus and ventral cochlear nucleus in the ipsilateral medulla oblongata; from these nuclei, there are three nerve tracts ascending to lateral lemniscus via bilateral superior olive nuclei and then to the inferior colliculus in the midbrain. These connections are rather complex; there are both bilateral cross connections and shortcut between levels. From inferior colliculus, there are bilateral projections to the medial geniculate nucleus (MGN) and finally to the primary auditory cortex (Fig. 2.49). The primary auditory cortex (A1) corresponds to Brodmann's area 41, and the secondary auditory cortex is around A1; both of them are located in the temporal lobe (Fig. 2.28). There are about 1000 efferent fibers descending to the inner ear from the brainstem; there are also nerves descending to MGN and inferior colliculus from the auditory cortex.

Frequency Selectivity of Auditory Systems The beauty of Fourier analysis in mathematics and the success of its application to many approximately linear systems make people consider pure tones as the simplest sound and the composing elements of complex sounds. Based on the same linear thinking, a click is also considered as an approximation of the impulse and a "basic element" of various complex sounds. Thus, it is expected to predict responses to a variety of complex sounds based on the responses to these "elementary" sounds. In addition, the frequency selectivity of different location at the basilar membrane in some degree

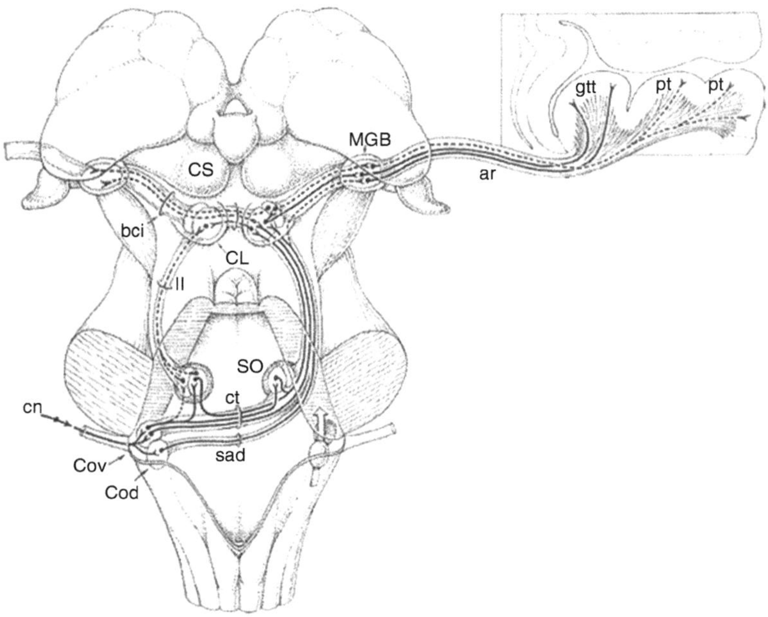

Fig. 2.49 The diagram of auditory pathways. *ar* acoustic radiation, *bci* brachium of colliculus inferior, *CI* colliculus inferior, *cn* cochlear nerve, *Cod* dorsal cochlear nuclei, *Cov* ventral cochlear nuclei, *CS* colliculus superior, *ct* corpus trapezoideum, *gtt* gyrus temporalis transversus (Heschl's gyrus), *ll* lateral lemniscus, *MGB* medial geniculate body, *pt* planum temporale, *sad* stria acoustica dorsalis, *SO* superior olive (Reproduced with permission from ten Donkelaar HJ (2011) *Clinical Neuroanatomy: Brain Circuitry and its disorders*. Springer. Fig. 7.8a)

convinced people that the ear is some kind of Fourier analyzer. Auditory researches focused on researches using clicks or pure tones as stimuli for a long time. Before the 1980s, many hearing studies did like to study the frequency selectivity of auditory systems to pure tones with different frequencies.

The first experiment evidence of frequency selectivity of auditory systems was from the pioneering works of Nobel Laurent Georg von Békésy, who observed that the movement of basilar membrane in corpse cochlea induced by pure tones is a kind of traveling wave, starting from the cochlear base and propagating to the top. Although the vibration frequencies at different locations on the basilar membrane are all the same, their phases are different. Their amplitudes increase at the beginning during the propagation, reach a maximum value at some point, and then decrease rapidly again (Fig. 2.50). The location of the maximum amplitude depends on the frequency of the pure tone: the higher the frequency, the lesser the distance from the base; thus, there is a spatial representation of the stimulus frequency along the basilar membrane. The distribution depends on the physical properties of the basilar membrane itself and its interaction with its surrounding

Fig. 2.50 A diagram of the propagation of the traveling wave along the basilar membrane. In this diagram, the cochlea is flared out, and the amplitude of the membrane is exaggerated greatly (Reproduced with permission from Conn PM (Ed) (2008) *Neuroscience in Medicine* (3rd Edition) Humana Press (in Yin TCT. Audition. Fig. 6))

fluid. This is the so-called place theory of the frequency selectivity of auditory systems.

The frequency selectivity of basilar membranes to pure tone stimuli may be the basis of the similar frequency selectivity of auditory neurons. The so-called tuning curves are used to characterize the frequency selectivity of auditory cells to pure tone stimuli. The tuning curves of hair cells, auditory nerve, neurons in cochlear nuclei, superior olive complex, lateral lemniscus nuclei, inferior colliculus, medial geniculate nuclei, and auditory cortex have been studied.

The so-called tuning curves are essentially a set of curves related to sound stimulus frequencies, stimulus intensities, and firing rates of neurons. Thus, the tuning curves include three classes about the relationship between two variables, while the third is fixed as a parameter: frequency-threshold curves, iso-intensity curves, and iso-rate curves. For frequency-threshold curves, the frequency is the abscissa, while the threshold above which a neuron or a nerve fiber can be excited by the stimulus with that frequency is the ordinate. Such curves form a valley at some frequency, a stimulus which has the lowest threshold to excite the cell, and this frequency is called the characteristic frequency or the best frequency. All the pure tones with the parameters above this curve can excite the cell; this area is called its response area. The frequency-threshold curve of the auditory nerve fiber with lower characteristic frequency is almost symmetry, while the one with higher characteristic frequency has long tail on its lower frequency side and is steep on its higher frequency side (Fig. 2.51). The tuning curves of the single neuron in cochlear nuclei are similar to those of auditory nerve fibers, except for the widths of their frequency-threshold curves which are wider than the ones of the auditory nerves, and the shapes are not as consistent as the ones of nerve fibers and even have multiple peaks in some cases. Although many frequency-threshold curves of neurons in the superior olive complex are similar to the ones in cochlear nuclei, there are also several ones quite different. Generally speaking, neurons in the nuclei above the superior olive complex do not respond to pure tones and respond only to short pure tones when they are on or off. The widths of their frequency-threshold

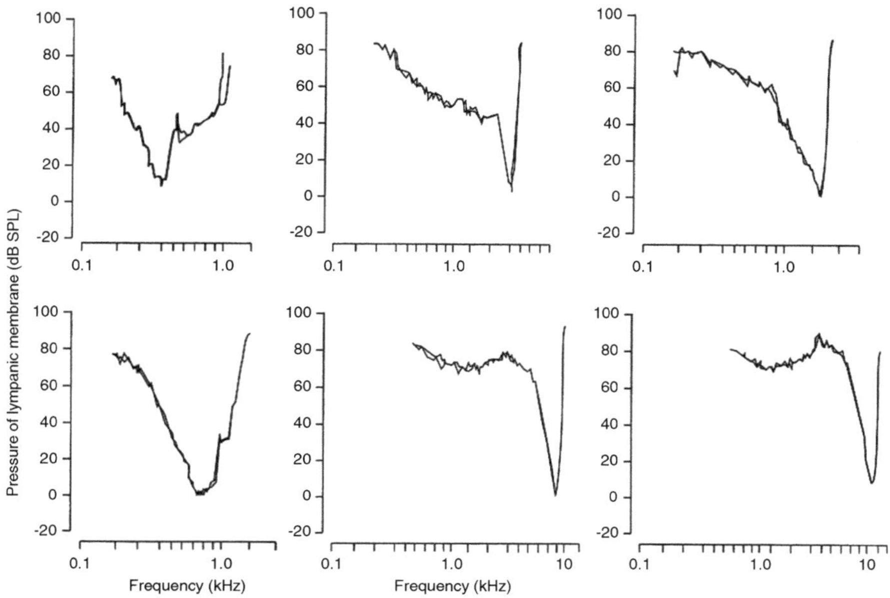

Fig. 2.51 The typical frequency-threshold curves recorded from single auditory nerves in a cat. In each panel, two fibers from the same animal of similar characteristic frequency and threshold are shown (Reproduced with permission from Conn PM (Ed) (2008) *Neuroscience in Medicine (3rd Edition)* Humana Press (in Yin TCT. Audition. Fig. 9))

curves vary a lot, and they may have multiple peaks. As a matter of fact, these neurons are more sensitive to natural sounds and transient sounds.

For all the auditory ascending pathways, there are ordered distribution of the neurons and nerve fibers anatomically according to their characteristic frequencies, which is named as "tonotopic organization." For example, the auditory nerves with higher characteristic frequencies are arranged around the periphery of the nerve tract, while the ones with lower characteristic frequencies are at its center.

The tonotopic organization and frequency-threshold curves are the basis of the place theory which suggests that the auditory system analyzes frequencies based on their exciting locations. However, it should be remembered that all the above results are from experiments using pure tones as stimuli, and the stimulus intensities are quite low.

The long tail of the frequency-threshold curve at its low-frequency side means that the nerve fiber or neuron with higher characteristic frequency will respond to low-frequency pure tones when its intensity is 40–60 dB higher than its threshold value, although such intensity is just the one met in normal physiological conditions. Besides, as the tail is quite flat, the responses to the stimuli with parameters above the tail are almost independent of the stimulus frequency. Such phenomenon, in which the auditory fibers show no significant frequency specificity to normal stimulus intensity, is more significant in the other two tuning curves, especially in

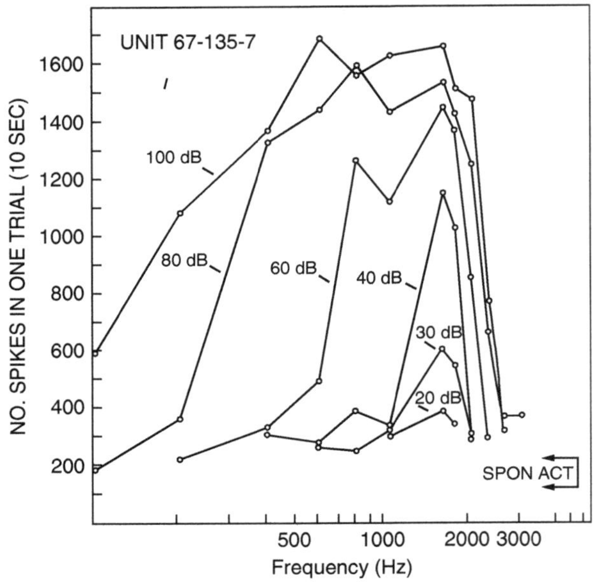

Fig. 2.52 Iso-intensity curves recorded from a single auditory nerve in a squirrel monkey. The ordinate is the total number of spikes induced by a 10 s pure tone stimulus, the abscissa is the frequency of the pure tone stimulus, and the number near the curve is the stimulus intensity (Reproduced with permission from Aage R. Møller (1983) *Auditory Physiology*. Academic Press. Fig. 2.18 (from Rose et al. 1971))

the iso-intensity curves (Fig. 2.52). In Fig. 2.52, the top of the iso-intensity curve is almost flat within a rather wide frequency band for the intensity of 80 dB SPL.[4] Thus, with such intensity, this fiber does not select only one frequency, but responds to pure tones with frequencies in a rather wide band almost identically. This means the filtering characteristics of the auditory periphery vary with the sound intensity; this is an obvious nonlinear phenomenon. As a matter of fact, the frequency selectivity of auditory neurons is quite poor even with the intensity of 70 dB SPL.

Another problem for place theory is a contradiction between the temporal resolution and the frequency resolution. If the frequency selectivity totally depends on the location, to get a high Q value for the recorded frequency resolution (e.g., 3 Hz for low-frequency stimuli), the damp of the filtering characteristics must be very low; thus, an oscillation would last for a long period for its corresponding temporal signal. Thus the recorded temporal resolution could not be reached. On the contrary, if it is required to reach the temporal resolution, the damp has to be quite big, and the recorded frequency resolution could not be reached.

---

[4] dB SPL is an abbreviation of unit (decibels sound pressure level) to measure the sound intensity, which is defined by the following formula:

$$L = 20 \times \log_{10}(P/P_{ref})$$

where P is the magnitude of the stimulus, which is given by the root mean square of the sound pressure (in units of pascals or Pa), and P_{ref} is a reference level. In this scale, 0 dB SPL is defined as the sound pressure whose root mean square value is 20 μPa, which corresponds to the approximate threshold of human hearing at 4 kHz, the frequency at which our ear is most sensitive.

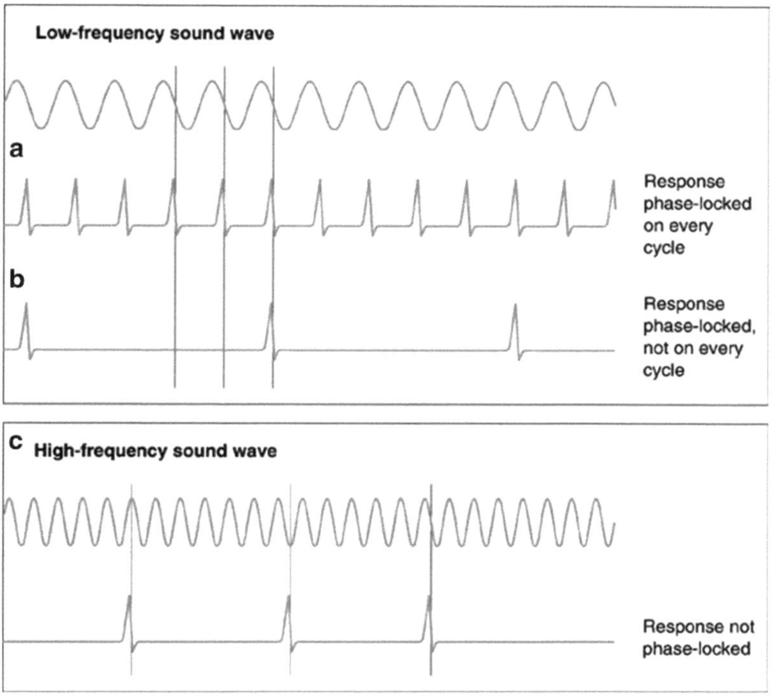

Fig. 2.53 Phase locking in the response of auditory nerve fibers. Low-frequency sound can elicit a phase-locked response, either (**a**) on every cycle of the stimulus or (**b**) on some fraction of the cycle. (**c**) At high frequency, the response does not have a fixed-phase relationship to the stimulus (Reproduced with permission from Bear MF, Connors BW, Paradiso MA (2007) *Neuroscience: Exploring the Brain (3rd Edition)*. Lippincott Williams & Wilkins. Fig. 11.21)

The above contradiction makes it doubtful if the frequency selectivity of the auditory system could be explained only by place theory. In addition, for very low frequencies, say, below 50 Hz, the tuning curves are almost the same.

An alternative hypothesis is the so-called temporal principle, i.e., it is suggested that the frequency information is carried about by the temporal pattern of the spiking in the auditory nerves. The experiment evidence is the so-called phase-locking phenomenon, i.e., the probability of firing varies in the stimulus period systematically; it always reaches a maximal value at a certain phase (Fig. 2.53). This can also be shown in the period histogram, in which a histogram of firing is drawn in a stimulus period starting from some fixed phase, even for complicated periodic stimulus (Fig. 2.54). Such characteristics can be found in auditory nerves, cochlear nuclei, superior olive complex, lateral lemniscus nuclei, inferior colliculus, and medial geniculate body; however, only for stimuli with lower frequencies, the closer the central part, the lower the upper frequency limit (for auditory nerves, their upper limit is about 4–5 KHz) and the fewer the number of

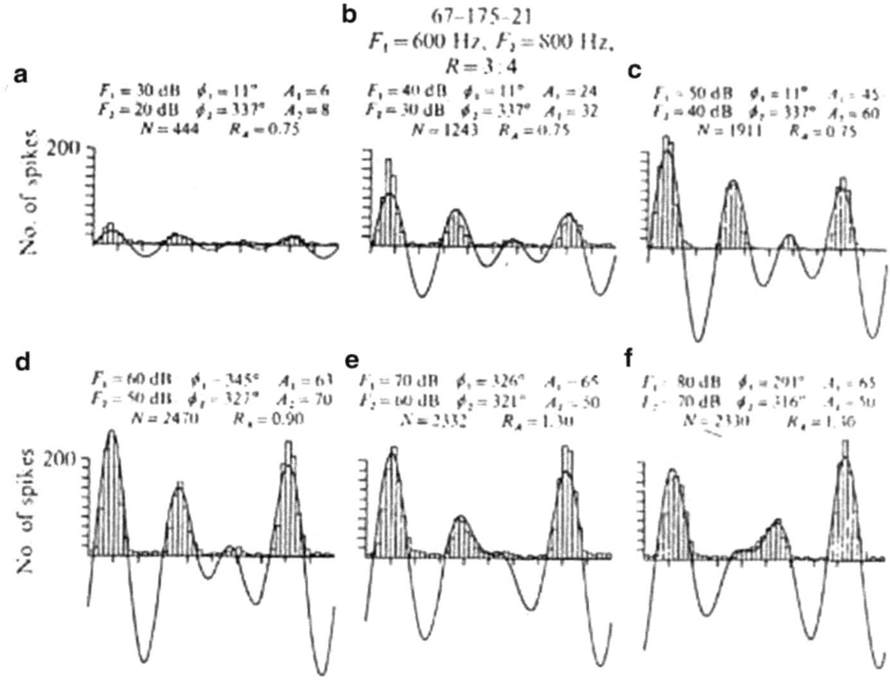

Fig. 2.54 Phase-locking phenomenon for periodic stimuli with complicated wave forms. Period histogram recoded from the auditory nerve fibers of a squirrel monkey stimulated by a complex periodic stimulus, which is composed of a summation of two pure tones with a fixed frequency ratio of 3:4. The amplitude ratio of these two pure tones of 10 dB keeps also unchanged within a range of 50 dB. All the amplitudes and phases of the two pure tones used are marked in the figure (Reproduced with permission from Aage R. Møller (1983) *Auditory Physiology*. Academic Press. Fig. 2.30)

neurons which are phase locking. There are very few phase-locking neurons in the medial geniculate body and no such neurons in the auditory cortex at all.

Phase-locking phenomenon is kept also for periodic stimuli with complicated wave forms; the shape of the period histogram is just the same as the stimulus waveform after half-wave rectification (Fig. 2.54). Besides, such waveforms keep unchanged over all the range of the stimulus intensity; no waveform top-cutting is observed owing to saturation of the intensity-rate curves.

Nevertheless, the frequency selectivity could not be elucidated completely by the temporal principle alone. Firstly, the temporal principle can only be used to the stimuli with frequencies lower than 4–5 KHz. There is no phase-locking phenomenon when the stimulus frequency is higher than 5 KHz. Secondly, errors may be made at every relays; thus synchronization becomes poorer and poorer, then how to keep temporal synchronization along the entire auditory ascending pathway may become a big problem.

In summary, it seems that the frequency selectivity can be thoroughly elucidated neither by place theory nor by temporal principle. The temporal principle cannot be used again when the stimulus frequency is higher than 4–5 KHz; maybe, the place theory is the only way to identify the frequency. However, for very low frequency, say, below 50 Hz, phase locking may be the only way to encode the frequency information. The place principle may divide the auditory pathway into several channels with different path bands when the stimulus frequency is lower than 5 KHz, thus decomposing sound signals with broad frequency bands into a series of narrower band limited signals with different central frequencies, to solve the first difficulty we have mentioned at the beginning of this section.

It should also be noted that the boundary value of 5 KHz has some special meaning in human hearing. Firstly, the main components in human speech are less than this value; secondly, the hearing incremental threshold increases abruptly for the stimuli with frequencies higher than 5 KHz. Thus, it may hint that temporal principle might play important role in fine discrimination of the frequency.

Intensity Coding Just as the other sensory modalities, sound intensity is coded by the firing rate of the auditory neurons. The stronger the sound, the higher the firing rate. However, for an individual neuron, its firing rate will saturate at a sound intensity 40 dB or so above its threshold, but the auditable range of the sound intensity is above 100 dB; this could not be reached by only one neuron. Thus, there must be other additional mechanism responsible for such wide intensity range; one possibility is that strong sound recruits more neurons with different thresholds to be activated.

Two-Tone Suppression In the above paragraph, the response of auditory neuron to single pure tone was discussed, and then what would happen if two tones with difference frequencies are given simultaneously? Using a pure tone with a fixed frequency as the stimulus, draw a threshold-frequency curve of an auditory neuron; then another pure tone with different frequency is used simultaneously. Vary the frequency and amplitude; two areas on both sides of the response area of the auditory nerve could be found; the response to the first stimulus would be inhibited when the second stimulus with its frequency and amplitude located in these two neighboring areas is given at the same time; this phenomenon is called two-tone suppression (Fig. 2.55). This means that there are interactions between different components of a given complex sound stimulus when the auditory system is stimulated by a complex sound; the frequency components with higher intensities can suppress the ones with lower intensities.

Coding of the Auditory System to Complex Sounds As what we have mentioned above, it is impossible to predict the response of an auditory neuron to complex sound based on its responses to "simple" sounds. We don't know what are the simplest elements of the stimulus sound; here we use the term "elements" to denote those sounds, if we knew the responses of the auditory system to these elements, then we could predict the response to any stimulus consisting of these elements. Owing to the fact that there are no such elements, we have to study the response of

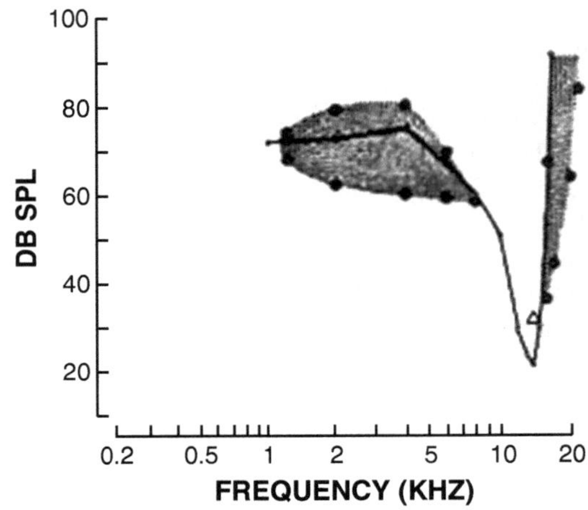

Fig. 2.55 Two-tone suppression. The curves connecting the *filled circles* are the second tone suppression area. The central curve is a frequency-threshold curve of an auditory nerve. The second pure tone will suppress the response of the tested neuron (shown as the *open triangle* within the response area) when its amplitude and frequency are within the two *shadowy regions* on both sides (Reproduced with permission from Borisyuk A et al. (2005) *Tutorials in Mathematical Biosciences I: Mathematical Neuroscience*. Springer. (Borisyuk A. Physiology and mathematical modeling of the auditory system. Fig. 10))

the auditory system to complex sound in each case. As the possibility of the variety of complex sound is infinite, we have to select those sounds with relative simplicity, such as frequency-modulated sounds or amplitude-modulated sounds (coincidently, some of such signals have functional meaning for some animals, e.g., the call of the bat is a kind of frequency-modulated sound) or those sounds with significant functional meanings, such as the calls of the experiment animals or the human speech. Although these studies are important for practical applications, it is difficult to extend them into general cases.

It is similar to visual systems; the more specific its function is, the closer to the cortex the neuron is. The auditory nerve is something like pass-band filters. Its firing pattern almost copies the stimulus sound wave after band-pass filtering and half-wave rectification.

The responses of neurons in the nuclei above the cochlear nuclei are varied. For a stimulus of sustained pure tone, some are similar to the response of auditory nerves, some only fires at the beginning of the stimulus, some has a violent firing at the very beginning then a pause and returns to a sustained firing with lower rate, and some has a decaying comb-like firing pattern (Fig. 2.56). As for the neurons in the nuclei closer to the auditory center, the closer the nuclei to the center, the weaker the response to sustained pure tones. Generally speaking, the neurons in the cochlear nuclei are sensitive to the frequency change of the stimulus. In the nuclei above the cochlear nuclei, some neurons in the inferior colliculus only respond to

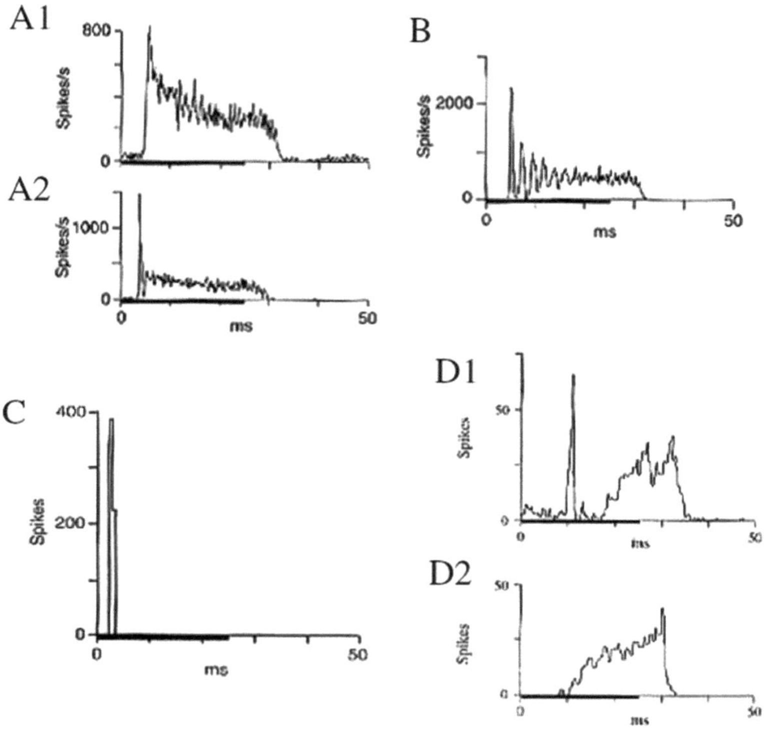

Fig. 2.56 Poststimulus time histogram (PSTH) of a variety of different kinds of neurons in the cochlear nuclei to tone burst stimulus. (**a**) Primary-like response. (**b**) Primary like with notch response. (**c**) Chopper response. (**d**) Single-spike onset response. (**e**) Pauser response. (**f**) Builder response. In all panels, *dark black line* under the axes is the tone presentation (Reproduced with permission from Borisyuk A et al. (2005) *Tutorials in Mathematical Biosciences I: Mathematical Neuroscience*. Springer. (Borisyuk A. Physiology and mathematical modeling of the auditory system. Fig. 13))

amplitude-modulating sound. It has been shown in some experiments that there may be feature detectors to their own call in these nuclei for some animals.

As for the problem if the auditory coding is place coding or temporal coding, Zhian Liang (1998) suggested some inspiring ideas. He suggested that the sound stimuli are infinite in variety; however, there are only ten thousands of auditory receptors. Thus, it is unimaginable to solve the problem of transmitting so rich information with such limited channels from the traditional point of view that considers the auditory system as some Fourier analyzer and one channel only for transmitting components in a narrow frequency band. He suggested that the same channel must be used in a variety of ways to transmit the rich information with finite channels. To verify his ideas, he designed some sophisticated experiments, in which he proved that a variety of frequency information could still be transmitted to the auditory center by some "nonspecific" pathways even if the specific pathways were totally blocked. His experiment can be briefly described as follows: As what we have mentioned above, the spectrum of the speech is below 5 KHz, and most of its components are below 3 KHz; he took pure tone with frequency higher than 5 KHz as carrier wave and used speech signal to modulate it either by frequency modulation or amplitude modulation. Obviously, there are no low-frequency components of the speech signal in the spectrum of such modulated sound; there are only the high-frequency components of the carrier and the difference between these components and the frequency components of the speech signal or their harmonics; thus, these components are still much higher than the highest frequency component in the speech signal. Therefore, if the transmission is totally based on place theory, these signals can only be transmitted by the high-frequency channels, and the listeners should not understand; however, to the contrary, the listeners can still understand the contents in the speech signal with such "modulated speech." To exclude the possibility that some frequency components in the speech spectrum may be produced owing to the nonlinearity of the auditory system so that these components might still be transmitted by the "normal" low-frequency channels, he used narrow band-passing noise with its central component focused on the speech spectrum to mask speech signal; the modulated speech could still be understood. However, if the central component of the noise is focused at the frequency of the carrier, the listener could not understand anything. These results suggest that the information in the speech with lower frequencies does transmit via the "high-frequency channel."

Thus, Liang proposed a concept of "multiple mechanism coordination" and noted that such principle might be used in practical communication. For example, when people communicate with each other in a noisy environment with noise spectrum close to the speech spectrum, then delivering speech signal superimposed by "modulated speech" may help the receiver to understand, as the environment noise at that time can only mask the speech signal, but not the information in the modulated speech.

2.7 Olfactory System

The olfactory system is one of the simplest sensory systems; thus, its study may provide some basic principles for other sensory systems.

Olfactory receptors, which are real bipolar neurons, are located in a thin layer of olfactory epithelium deep in the nasal cavity (left panel in Fig. 2.57). The axon of the olfactory receptor is the olfactory nerve (cranial nerve I), which composes the primary olfactory nerve (PON) and enters the olfactory bulb through a bony cribriform plate (right panel in Fig. 2.57). The olfactory receptor dendrite has only one ball-like end with many hairs, which are embedded in some flowing mucus, which is updated every 10 min. The inhaled air flows over the surface of this mucus layer; some odorant molecules are dissolved in the mucus and bound to

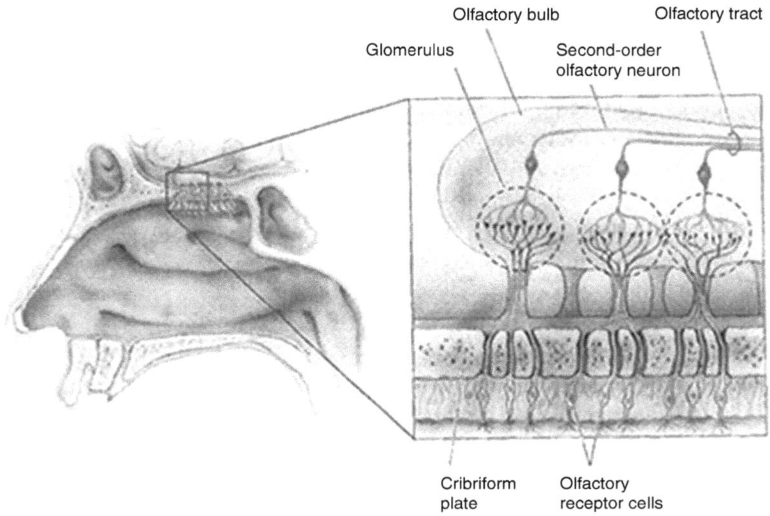

Fig. 2.57 Olfactory receptors and olfactory bulb. (*left*) The location of the olfactory epithelium; (*right*) the structure of the olfactory epithelium and olfactory bulb (Reproduced with permission from Bear MF, Connors BW, Paradiso MA (2007) *Neuroscience: Exploring the Brain* (3rd Edition). Lippincott Williams & Wilkins. Fig. 8.14)

the receptors at the surface of the hairs, which triggers a series of reactions and leads to the depolarization of the cell membrane to produce receptor potentials. The receptor potential will transmit to the trigger area of the receptor cell, a spike will be triggered if the potential exceeds a threshold, and the spike will propagate along the olfactory nerve into the olfactory bulb. This process will go to its end soon, owing to the diffusion of the odorant molecules in the mucus or resolving by some enzymes. Even if the odorant remains for a long time, the receptors will also adapt in about one minute.

Now it has been found that there are hundreds of olfactory receptor types for human beings. The olfactory epithelium could be separated into several regions, the olfactory receptors of different types are scattered randomly in every region. Every olfactory receptor can respond to a variety of different odors, although with different preference, while the same odor can stimulate a variety of different types of receptors. In addition, the higher the concentration of the odorant, the stronger the response. Thus, it seems that every single olfactory receptor can only give an ambiguous representation of the odor. Then the question is how can we use hundreds of different types of olfactory receptors to discriminate tens of thousands of different odors? It seems that a population coding is the only possible explanation. That means that every odor is represented by the tempo-spatial activities of a population of neurons. The molecular basis of the population coding of olfaction mentioned above was elucidated by 2004 Nobel Laurent Linda Buck and Richard Axel. It is shown that the activation maps on the olfactory bulb for different odors are different and reproducible; a technique of the so-called optical imaging was used to show this, in which some voltage-sensitive dyes were used on the surface of the olfactory bulb, so that a sensory map can be seen when the animal sniffs a special odor and activates different neurons in it. Interestingly, a hypothesis of similar population coding of olfaction was proposed much earlier by Walter J. Freeman using EEG technique.

Olfactory nerves end at ball-like structures, glomeruli in the olfactory bulb, while there are no interconnections between these nerves (right panel in Fig. 2.57). In every glomeruli, primary olfactory nerves converge to secondary olfactory neurons—mitral cells (M) to form excitatory axodendritic connections; the number of mitral cells is much less than olfactory nerves. Thus, every glomeruli receives inputs from a large piece of olfactory epithelium, and such projection is quite accurate; every glomeruli only receives the inputs from the same kind of sensory receptors. Periglomelular neurons (P) receive inputs from primary olfactory nerves on one hand and interact with each other either within the glomeruli or between them. Mitral cells receive inputs from both primary olfactory nerves and periglomelular neurons; there are also excitatory interactions, and their axons compose the lateral olfactory tract (LOT) entering the olfactory cortex (OC). There is another kind of interneurons in the olfactory bulb—granule cell (G); these cells inhibit each other. In addition, the effect of mitral cell to granule cell is excitatory, while the effect of the latter to the former is inhibitory. The lateral olfactory tracts end at the excitatory neurons within superficial pyramidal cells in anterior olfactory nucleus (AON) and prepyriform cortex (PC) (these excitatory neurons are briefly noted by E and A). There are also inhibitory neurons in anterior olfactory nucleus and prepyriform cortex (I and B). These excitatory and inhibitory neural populations compose negative feedback loops. The outputs of prepyriform cortex end at external capsule (EC) via deep pyramidal cells (C) on one hand and turn back to anterior olfactory nucleus and olfactory bulb on the other hand. Besides, there are also feedback pathways from the anterior olfactory nucleus to granule cells in the olfactory bulb. A diagram of the neural pathways in the olfactory system is shown in Fig. 2.58.

Based on the above considerations, Freeman proposed that the mesoscopic dynamics is a proper tool based on activities of neural populations to study the mechanism of olfactory information processing. He suggested that activities of many neural populations should be recorded simultaneously. He placed tens of electrodes (e.g., an array of 64 electrodes) on the surface of a specific brain region (say, olfactory bulb), and the distribution of these electrodes should cover a large part of the surface; then the local field potentials at the neighboring area to these electrodes could be recorded. These potentials represent the degree of excitation of the neural populations within these areas.

Freeman trained an animal (e.g., a rabbit) by conditional reflex to learn recognizing some specific odor. Before the experiment, the animal was deprived of drinking for a few days. In the experiment, he used two different odorants (e.g., banana oil and butyric acid) as stimuli; the banana oil is the conditional stimulus, while the butyric acid is the unrelated stimulus. Giving water is unconditional stimulus, which will induce a licking response. After giving water and the banana

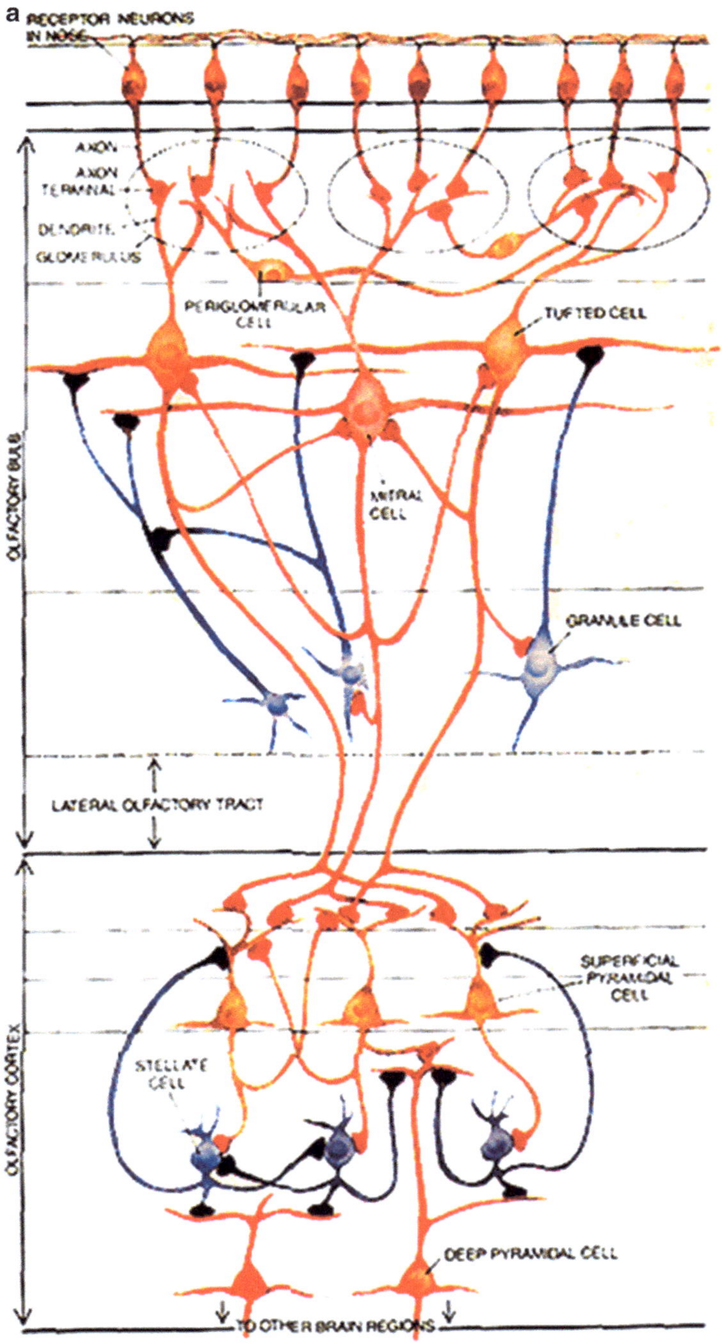

Fig. 2.58 Neural pathway of olfactory systems. (**a**) A diagram of the neural pathway of olfactory systems (Reproduced with permission from Freeman WJ (1991) The Physiology of Perception. *Scientific American*, 264 (2): 78–85; p. 81). (**b**) A schematic diagram of principal types of neurons, pathways, and synaptic connections in the olfactory mucosa, bulb, and cortex (Reproduced with permission from Freeman WJ (1975) *MASS ACTION IN THE NERVOUS SYSTEM – Examination of the Neurophysiologic Basis of Adaptive Behavior through the EEG*. Academic Press. Fig. 4.25)

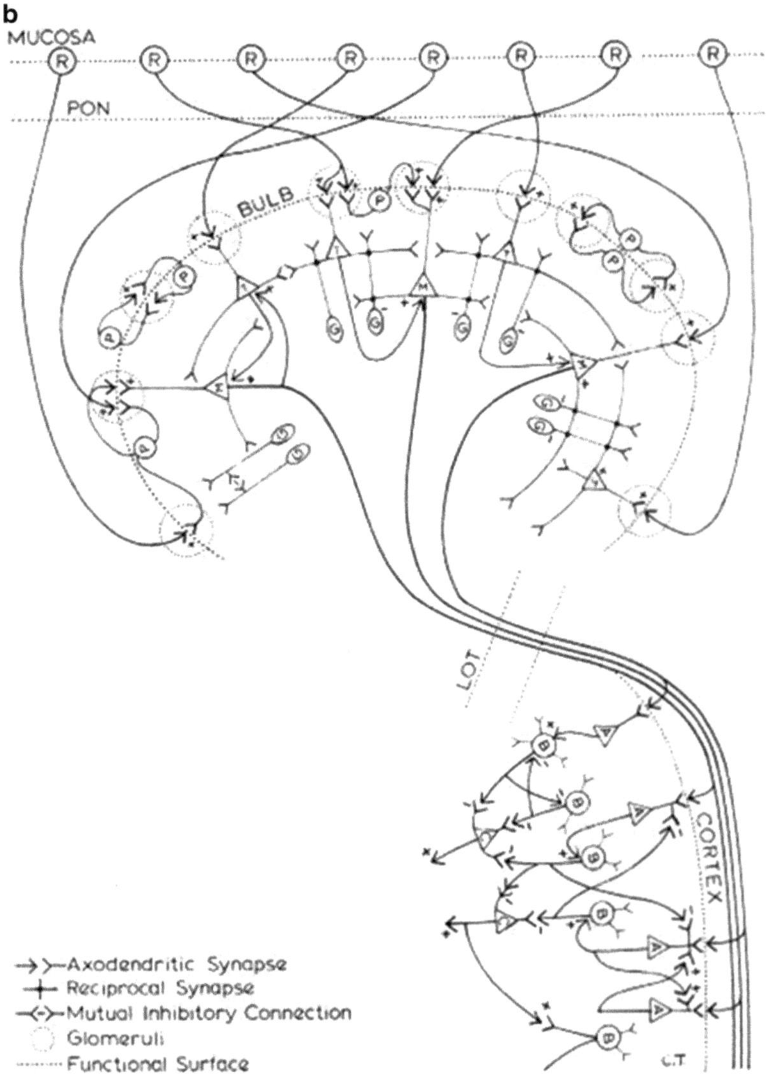

Fig. 2.58 (continued)

oil simultaneously for several times, even only giving banana oil without water, the animal would also lick, which means that the animal had learned to recognize the odor of banana oil. On the contrary, giving butyric acid had never been combined with giving water as a reward, the rabbit only sniffed rather than licked. In addition, only giving air was used as a control. In every trial, 64 channels of EEG were recorded for 6 s, within which two periods were divided into a control period and a testing period. In the two different periods, the waveforms recorded were quite

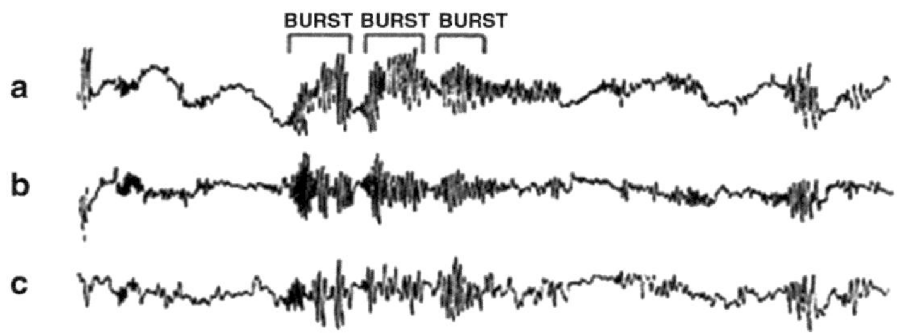

Fig. 2.59 EEGs recorded simultaneously at the olfactory bulb (**a**), anterior olfactory cortex (**b**), and posterior olfactory cortex (**c**). The bursts with high frequency and high amplitude presented when the animal recognized the odor it had learned (Reproduced with permission from Freeman WJ (1991) The Physiology of Perception. *Scientific American*, 264 (2): 78–85. p. 82 lower picture)

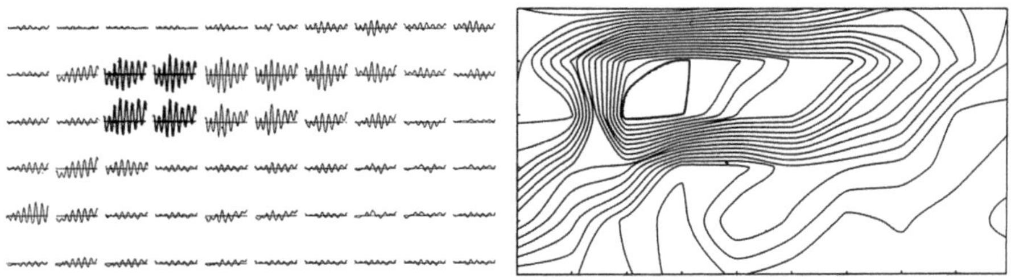

Fig. 2.60 (*Left*) The γ waves recorded from 60 channels simultaneously on the surface of olfactory cortex when the rabbit is trying to recognize some odor; (*right*) the contour map of average amplitudes of γ waves at the surface of the brain region based on the data from the *left graph* (Reproduced with permission from Freeman WJ (1991) The Physiology of Perception. *Scientific American,* 264 (2): 78–85. p. 84 top picture)

different. He found that when the air with acquainted odor was inhaled, EEGs suddenly became more regular, and their amplitudes and frequencies became higher; the range of frequencies was within 20–80 Hz, i.e., the EEGs were dominated by γ rhythms and formed the so-called burst (Fig. 2.59). It is also found that all the carrier waveforms of the γ waves recorded from the 64 channels were identical and only their amplitudes were different (the left graph in Fig. 2.60). If all the average amplitudes on the surface of the brain region are marked and all the points with same amplitudes are lined by isoclines, a contour map can be drawn (the right graph in Fig. 2.60). In this way, it was found that the spatial distribution pattern of γ wave amplitudes, i.e., the contour map, can be repeated again and again when the same odors are inhaled, in spite of the fact that their carrier waves vary every time. Therefore, it is hinted that the olfactory information about the odor is carried in the spatial pattern of the amplitudes of γ waves. Individual olfactory neuron could not identify a specific odor; only the activities of a large group of neurons together can recognize it. Thus, Freeman (2000) summarized his idea as

Fig. 2.61 (*Left*) EEG amplitude-modulated pattern to sawdust after the rabbit has learned to recognize its odor; (*middle*) EEG amplitude-modulated pattern to banana after its odor is used as the conditional stimulus and the rabbit has learned to recognize the new odor; (*right*) EEG amplitude-modulated pattern to sawdust when it is used as conditional stimulus again and after the rabbit has learned to recognize the odor again. Note, although in this case the sawdust is used as the conditional stimulus as in the left graph, their EEG amplitude-modulated spatial patterns are radically different (Reproduced with permission from Freeman WJ (1991) The Physiology of Perception. *Scientific American*, 264 (2): 78–85. p. 84 lower picture)

follows: "In brief, the reception of an odorant requires a sparse network of neurons, while the perception of an odor requires all the neurons in the bulb."

Now we have direct evidence for such spatial representation of olfactory information. Kauer (1991) presented three different odorants to a salamander, orange, banana, and pineapple, and added voltage-sensitive dye to its olfactory bulb; thus, the activities of the olfactory bulb neurons could be recorded optically. It was shown that complicated but repeatable spatial patterns could be found for the given odorants.

Especially interesting, if the conditional stimulus was switched between two different odorants, i.e., the original conditional stimulus became the unrelated stimulus, while the original unrelated stimulus became the conditional stimulus, although these odorants themselves did not change at all in their physical and chemical properties, the EEG amplitude-modulating spatial patterns changed completely (Fig. 2.61). Similarly, when some new odorant was used as stimulus during the experiment, the trained amplitude-modulated pattern changed. And even using several odorants as conditional stimulus one by one, after a run and the first one is used again, the existing pattern would also be changed to a new pattern. Thus, the olfactory amplitude-modulated pattern is not related to the stimulus directly, but related to its meaning, as what Freeman suggested (Freeman 1991).

From Freeman's work, a lesson we have learned is that cognition is not limited to extract features of the stimulus; it depends also on experience. Cognition could not be explained only by the individual neuronal properties, but synergetic activities of a large group of neurons are involved. This is not only true for olfaction but also for other sensations.

2.8 Somatosensory System

The somatic sensory system is responsible for detection of many kinds of stimulation, including touch, temperature, pain, and itch, as well as static position and movement of limbs and body. The somatic sensory system is widely distributed in almost all parts of the body, including skin, muscles, joints, and the viscera, and is involved in processing all sensory information in addition to those specific sensation such as visual, auditory, gustatory, olfactory, and vestibular. Receptor subtypes, neural transmission pathways, as well as activation of relevant cortical areas are responsible for the generation of various kinds of somatic sensory perception. In this session, we will discuss the mechanisms for tactile sensation and pain sensation and relevant information transmission pathways.

Touch The sensation of touch starts with the mechanoreceptors in the skin. These receptors are expressed with mechanically gated ion channels on their membrane, whose activity can be changed by membrane tension. The mechanoreceptors are therefore sensitive to distortions such as bending or stretching and can detect the stimulation from the skin. Because the mechanoreceptors on the skin are with different receptive field properties, as well as sensitive to different vibration frequency and with pressure selectivity, we can thus precisely identify the stimulations on the skin such as press, vibration, or insect sting.

Skin mechanoreceptors are illustrated in Fig. 2.62. The Pacinian corpuscles are located deep in the dermis, with their length being around 2 mm and diameter being around 1 mm. The Ruffini corpuscles are slightly smaller than the Pacinian corpuscles. These two types of receptors are both with relatively large receptive fields, which cover larger surface on the fingers or palm. The Meissner's corpuscles, with their size being only 1/10 of the Pacinian corpuscles, are located in the superficial

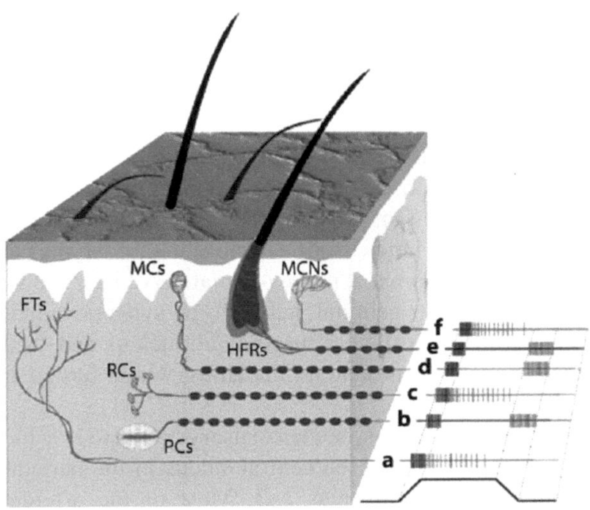

Fig. 2.62 Distribution of somatosensory receptors in the skin. *FTs* free terminals, *PCs* Pacinian corpuscles, *RCs* Ruffini corpuscles, *MCs* Meissner corpuscles, *HFRs* hair-follicle receptors, *MCNs* Merkel cell-neurite complexes (From Galizia and Lledo 2013, Springer, Neurosciences – from molecule to behavior, Fig. 16.6)

layer of the glabrous skin. The Merkel disks are located in the epidermis. These two types of receptors are with smaller receptive fields, which only cover a small area of several millimeters. Each type of the mechanical receptors is with different temporal properties in response to stimulation. The Meissner's corpuscles and the Pacinian corpuscles are with quick adaptation properties, so that they are sensitive to vibrations, while the Merkel disks and the Ruffini endings are with slow adaptation properties. In addition, hair-follicle receptors change their firing activities, while the hairs are bended; this type of mechanoreceptors can either be quickly adaptive or slowly adaptive. The receptor densities on the skin together with the receptive field architectures determine the ability of stimulation allocation on the body surface. In humans, the skin on the finger tips is with a high density of the receptors which have small receptive fields; therefore, finger tips are highly specified for spatial location identification.

Somatosensory receptors send sensory information through primary afferent axons to spinal cord via dorsal root (Fig. 2.63). Cell bodies of primary afferent axons are located in the dorsal root ganglion. Axons from skin receptors are termed as Aα, Aβ, Aδ, and C fibers according to their sizes (Fig. 2.64). Among these, C fibers are the finest and are unmyelinated, while all other axons are myelinated.

The neurons located in the dorsal horn which receive inputs from the primary afferent axons are called secondary sensory neurons. Aβ axons transferring skin touch sense form synapses with the secondary sensory neurons in the dorsal horn. Such connection allows for swift subconscious reflections. The ascending branches of Aβ axons join the dorsal column in the inner part of the ipsilateral dorsal horn, sending the information about touch and limb positions to the brain. The axons of the dorsal column terminate at the dorsal column nucleus located at the junction of the spinal cord and medulla. Cells in the dorsal column nucleus send their ascending axons projecting to the inner ventral part of the medulla and with the fibers from the left part of the body crossing to the right side and vice versa. The somatic sensory system above this level receives sensory projection from the contralateral part of the body. The ascending axons from the dorsal column nucleus travel in the medial lemniscus, passing through medulla, pons, and midbrain, and then have synaptic connections with neurons in the ventral posterior nucleus of thalamus. Axons from neurons in the ventral posterior nucleus further project to the primary somatic sensory cortex (SSC, S1) (Fig. 2.65).

In addition to tactile sensory inputs from the body, somatic sensation also includes sensory inputs from face and tongue via trigeminal nerve (cranial nerve V) and sensory input from ear, nose, and pharynx via facial nerve (cranial nerve VII), glossopharyngeal nerve (cranial nerve IX), and vagus nerve (cranial nerve X). Axons of the trigeminal nerve form synaptic connections with the secondary neurons in the ipsilateral trigeminal nucleus and then project to the ventral post nucleus of the contralateral thalamus, which further project to the somatosensory cortex (Fig. 2.66).

In humans, primary somatosensory cortex (S1) is located in the postcentral gyrus which is posterior to the central sulcus in the parietal lobe (Fig. 2.67). This area includes Brodmann areas 1–3. Most of the afferent fibers from the thalamus

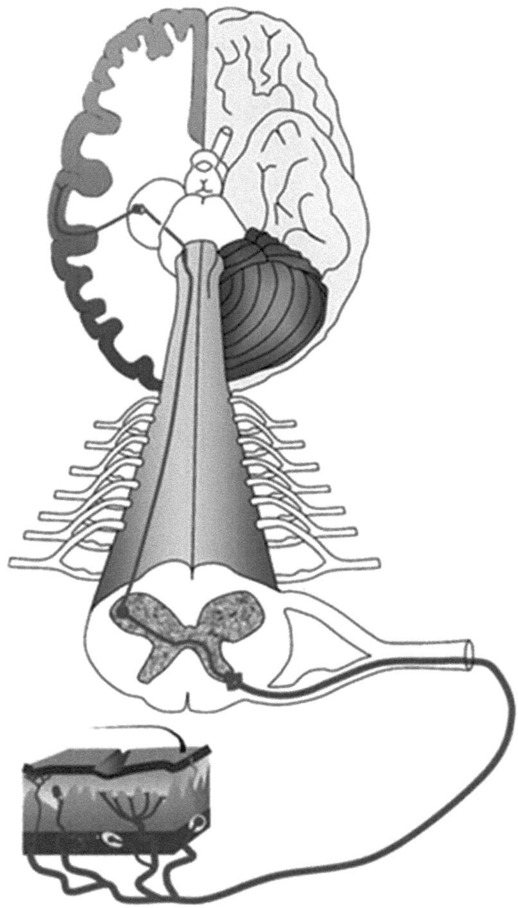

Fig. 2.63 Afferent fibers from the skin entering the spinal cord via dorsal root and being sent to the somatic sensory cortex (From Pfaff 2013, Springer, Neurosciences in the 21st century, Fig. 22.21)

terminate at area 3, the function of which is mainly related to texture, size, and shape of objects. Neurons in area 3 further project to areas 1 and 2, among which area 1 is to process texture information, while area 2 is related to size and shape of objects.

In the primary somatosensory cortex, each single neuron has a specific receptive field corresponding to a particular part of the body. These receptive fields form regular mapping in the cortex, which is called cortical somatotopy (Fig. 2.67). The arrangement of the body parts is not necessarily continuous in the somatotopy. Besides, each part of the body is not evenly represented in the somatotopy. In humans, hands and head/face occupy a large portion of the cortical areas, while trunk and limbs are only represented by a small portion of it. Such distribution maybe related to the amount of information sent from each part, but may also be related to the importance of the information. It is apparent that the importance of sensory information from each part of the body differs

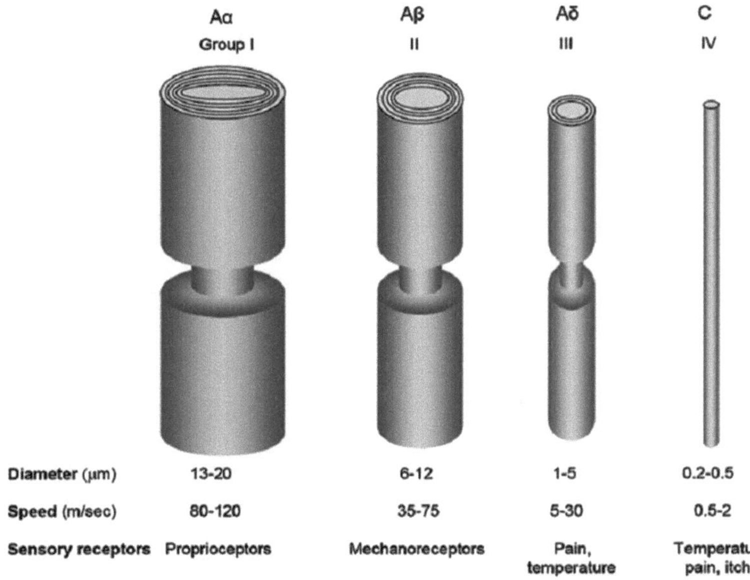

Fig. 2.64 Primary afferent axons and their classification (From Pfaff 2013, Springer, Neurosciences in the 21st century, Fig. 22.16)

tremendously among species. Whiskers on rodent face have large representative area in the primary somatosensory cortex (Fig. 2.68); sensory information from each whisker follicles is sent to a cluster of neurons, which form specified structure like "barrels" on stained slices. Another feature of somatotopy is plasticity. When sensory input from a certain part of the body is lost for some reasons (i.e., amputation), relevant somatic sensory cortical area will be altered to be responsive to stimulation coming from the adjacent body parts. This means that the cortical somatotopy is adjusted to some extent. Accordingly, when inputs from a certain part of the body are increased and sustained for a certain period of time, relevant cortical representative area will be expanded. These phenomena suggest that the cortical somatotopy is not stereotyped, but instead can be regulated according to the information influx.

Pain In addition to skin mechanoreceptors, there are also abundant nociceptors in the somatosensory system. Nociceptors are unmyolinated free nerve endings (see Fig. 2.62). The activation of nociceptors induces entirely different subjective perceptions as compared to the activation of mechanoreceptors, which is due to the tremendous difference in information transmission pathways. Pain is crucial to life, it reminds us to avoid possible hurt, and in the body, pain perception is independent to any other perceptions. Being different from other perception processes, the perception of pain can be interfered by brain activity and then can be controlled.

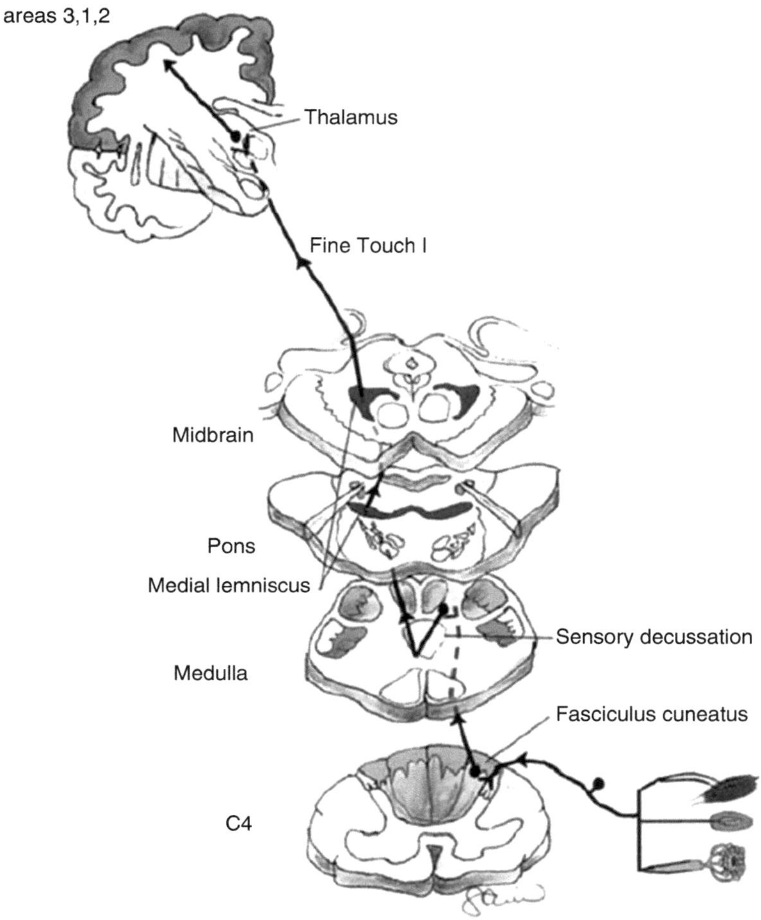

Fig. 2.65 Organization of the dorsal column-medial lemniscus pathway sending tactile and proprioperception information to the sensory cortex (From Jacobson and Marcus 2011, Springer, Neuroanatomy for the neuroscientist, Fig. 14.5)

When the body tissue is subject to severe mechanical stimulation, or in exposure to certain chemical environment, the body might be injured. The ion channels expressed on nociceptor membrane can be activated by such injury stimulations. The effect of mechanical stimulation can activate the mechanical gated ion channels on the membrane and cause membrane depolarization and initiate action potentials. At the meantime, the injured cells will release various chemicals, including potassium ions, adenosine triphosphate (ATP), and protein enzymes. It has been well acknowledged that the elevation of extracellular potassium concentration can directly induce the depolarization of neuron membrane and ATP can depolarize neuron via the modulation of ATP-gated ion channels, while protein

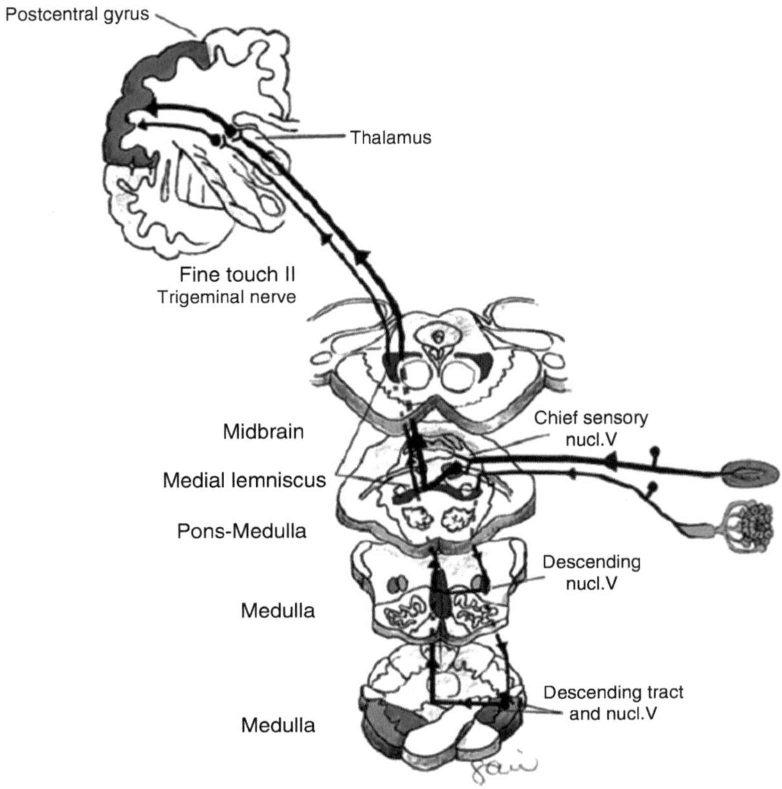

Fig. 2.66 Trigeminal pathway sending tactile sensation from the head (From Jacobson 2011, Springer, Neuroanatomy for the neuroscientist, Fig. 14.6)

enzymes can modulate the membrane ion channel activity via more complicated process to produce pain signal.

Extreme heat stimulation can also cause pain. When skin temperature is within the range 37–45 °C, warmth can be perceived via thermoreceptors in the human body. But when the skin temperature is over 45 °C, heat-sensitive ion channels on nociceptors can be activated to produce burning and pain perception. The perceptions of warm and hot are mediated by different neural mechanisms. In fact, temperature over 45 °C is harmful to the body. In addition, lactic acid generated during anaerobic metabolism can cause an enhancement in extracellular H^+ concentration, which will activate H^+-gated ion channels on nociceptors and produce pain perception. Insect sting also activates mast cells in the immuno-system, which leads to the release of histamine and then activates nociceptors to produce pain.

From the above introduction, we can see that pain can be induced by many factors. On the other hand, the nociceptors are selective to pain stimulations and can

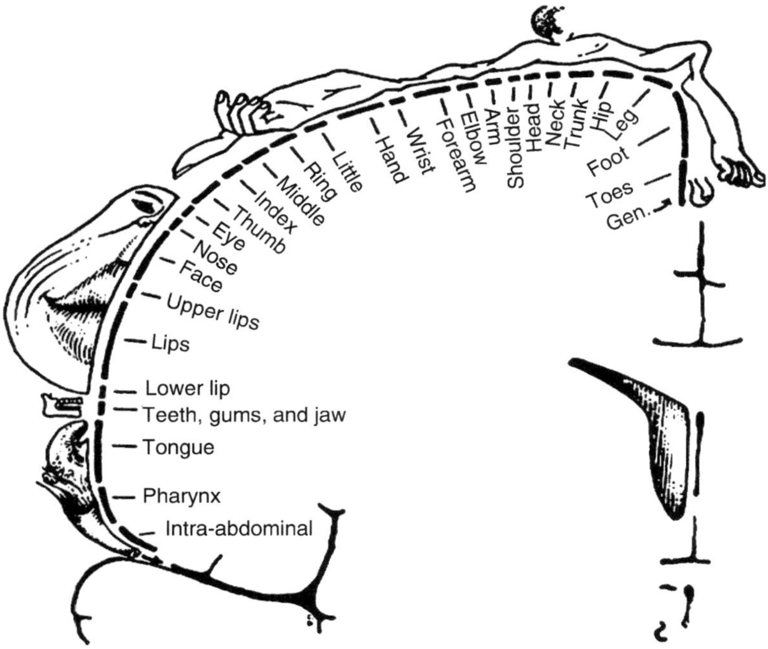

Fig. 2.67 The somatotopy in the primary somatosensory cortex (From Jacobson and Marcus 2011, Springer, Neuroanatomy for the neuroscientist, Fig. 14.1)

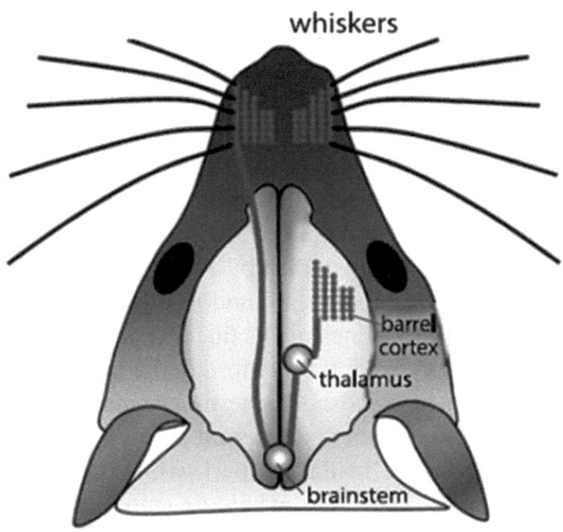

Fig. 2.68 Somatotopy of rodent facial whiskers in the somatic sensory cortex (From Galizia and Lledo 2013, Springer, Neurosciences – from molecule to behavior, Fig. 16.7)

be further classified as mechano-nociceptors, thermo-nociceptors, and chemo-nociceptors. But in the meantime, many nociceptors respond to more than one kind of stimulations and are called polymodal nociceptors. Nociceptors are distributed in many parts of the body such as skin, bones, muscles, and visceral organs.

Generally, nociceptors have activation threshold, i.e., they can only be activated when stimulation reaches a certain intensity, so as to produce pain perception. But when the body is injured, many chemical substances will be released, which includes bradykinin (which can enhance the activity of thermosensitive ion channels), prostaglandins (which can sensitize the nociceptors to other stimulations), and substance P (which can promote the release of histamine from mast cells). These substances can regulate the nociceptors' activities and raise their sensitivities to mechanical and chemical stimulations. Therefore, the threshold is often lowered in the injured tissue and its surrounding areas, which causes hyperalgesia. Furthermore, central neural mechanisms are also involved in hyperalgesia. In spinal cord, the interaction of touch and pain transduction pathways can also activate Aβ fibers, which will also elicit pain, while in normal condition, the activation of Aβ fibers is related to the transduction of mechanical sensory information.

Pain information is transmitted in the spinal cord via Aδ fibers and C fibers (see Fig. 2.64). Due to the difference in action potential propagating speeds in these two kinds of fibers, stimulation on skin can cause swift and sharp "first pain" mediated by Aδ fibers' activities and slow but sustained "second pain" mediated by C fibers' activities.

Same as Aβ fibers which transfer mechanical stimulation, the cell bodies of Aδ and C fibers mediating pain perception are located in the dorsal root, with their fibers projecting to dorsal lateral part of the dorsal horn of spinal cord, and then make synaptic connections with cells located in lateral part of dorsal horn in the substantia gelatinosa (Fig. 2.69).

Same as skin nociceptors, visceral nociceptors send their axons to the central part via the same pathway in the spinal cord. Pain information generated from these two sources is mixed in the spinal cord (Fig. 2.70). Therefore, sometimes the activity of visceral nociceptors can be perceived as pain stimulation from skin. Pain information from head and face are projected up to the thalamus via trigeminal lemniscus.

From the above introduction, we get to know that there are a number of differences between the tactile and pain information pathways. First of all, these two kinds of perception are mediated by different receptors; secondly, tactile stimulation is transferred by Aβ fibers, so it has a faster transduction speed, while pain is transferred by Aδ and C fibers and bears a slower transduction speed; and thirdly, the branches of Aβ fibers terminate deep in the dorsal horn, while branches of Aδ and C fibers travel through the lateral part of the dorsal horn and terminate at substantia gelatinosa. Furthermore, information about these two kinds of perception is sent to the brain via different pathways. Tactile information is sent via ascending pathway in the ipsilateral part, while pain information is sent through the contralateral part. Therefore, when the spinal cord suffers a unilateral injury, mechanical sensory deficit will occur in the same side of the injury, while pain deficiency will occur at the opposite side of the body.

Also being different from other sensory systems, the perception of pain can be modulated. We have mentioned that during hyperalgesia, a soft touch on the skin will cause pain perception. The activation of mechanical receptors (Aβ fibers) can

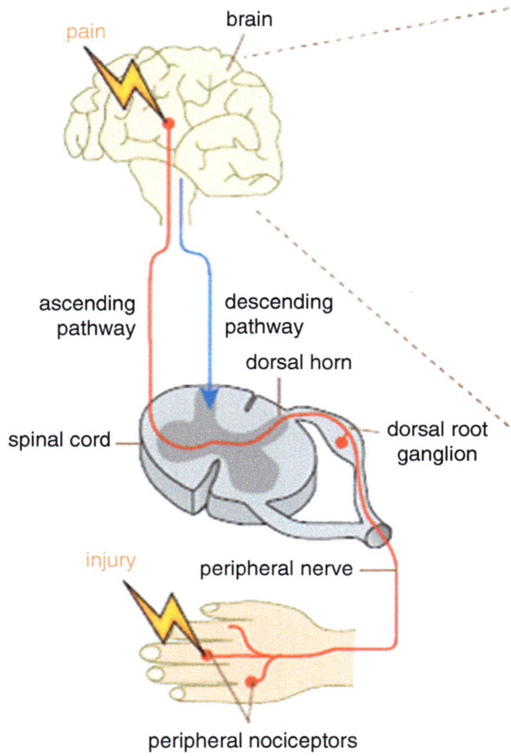

Fig. 2.69 Pain information transmission from nociceptor to the somatosensory cortex via the spinal cord (From Galizia and Lledo 2013, Springer, Neuroscience: from molecule to behavior, Fig. 21.2)

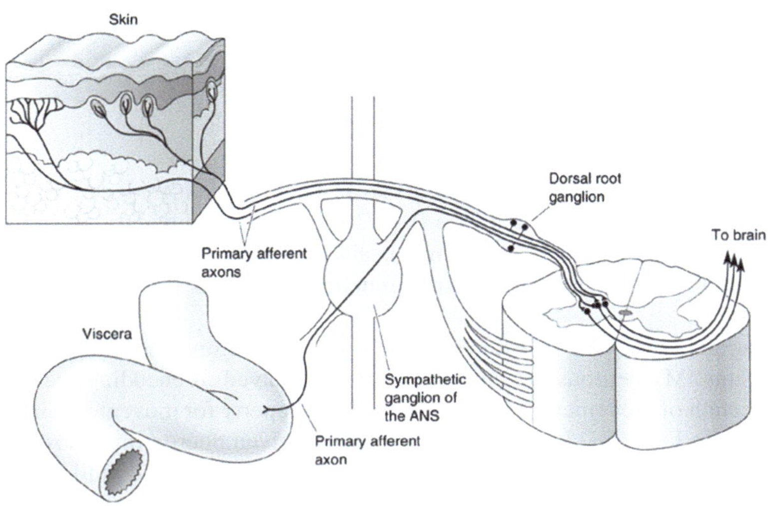

Fig. 2.70 Skin nociceptor and visceral nociceptor send their afferents to the brain via the same pathway in the spinal cord (From Galizia and Lledo 2013, Springer, Neurosciences – From molecule to behavior, Fig. 21.3)

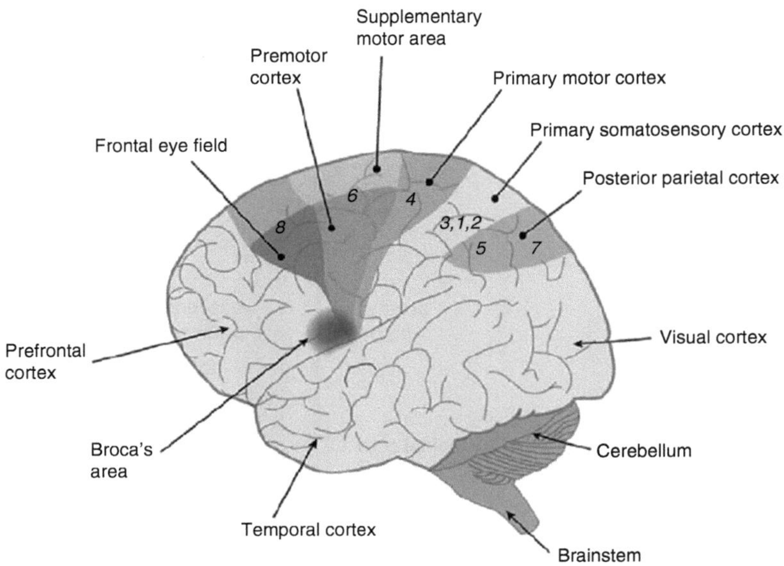

Fig. 2.71 The cortical areas involved in voluntary movements, among which area 4 and area 6 form the motor cortex (From Pfaff 2013, Springer, Neuroscience in the 21st century, Fig. 37.2)

effectively attenuate the pain initiated by nociceptors. Furthermore, the brain can produce endogenous morphine-like molecules which are called endorphine. Endorphine has its effects on various processes; it can reduce excitatory signal transmission via inhibiting presynaptic glutamate release, or generating inhibitory effect by hyperpolarizing postsynaptic membrane, etc.

2.9 Somatomotor System

In the brain, the motor cortex consists of areas 4 and 6 which are located in front of the central sulcus in the frontal lobe (Fig. 2.71). Motor commands are initiated in area 4 of the frontal cortex, so this area is also called the primary motor cortex, or M1. Area 6 is of higher-order motor function, which is responsible for motor planning.

Firing activities can be detected in M1 neurons before and during voluntary movements. M1 neurons are considered to be involved in encoding the intensity and direction of movement. M1 neurons' coding property for movement direction is illustrated in Fig. 2.72. In this example, a certain given neuron fires most strongly when the animal's forearm moves along a certain direction, and while the movement direction has a departure from the neuron's optimal direction, the cell firing activity is reduced to a certain lower rate, which means that although this neuron has direction selectivity, it is broadly tuned in general. This suggests that the

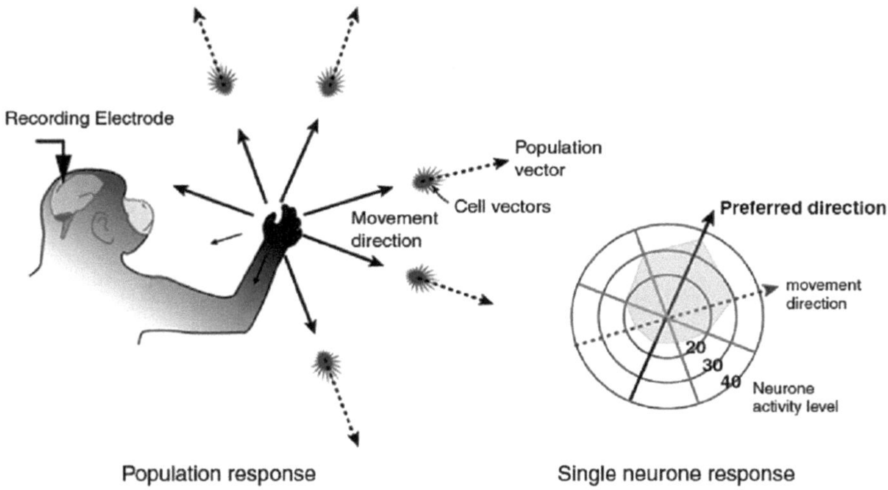

Fig. 2.72 Movement control and populations coding in M1 motor neurons. The movement of the animal arm is determined by the population activity of the M1 motor neurons (presented by the population vector which is a sum-up of the individual neuron's activity vector) (*left*), while each individual neurons are broadly tuned and fire with a certain intensity when the arm moves along a particular direction (which is represented by the single neuron's vector), and each M1 neuron has a "preferred direction" which is related to the strongest firing activity (*right*) (From Pfaff 2013, Springer, Neuroscience in the 21st century, Fig. 37.13)

direction of movement can hardly be encoded by any single Ml neuron's activity, but more likely be encoded by population activities.

To test such population coding hypothesis, Apostolos Georgopoulos in Johns Hopkins University and his colleagues recorded large population of neurons which contained several hundreds of M1 neurons and measured direction tuning curve for each recorded M1 cell. According to their work, each cell's firing intensity corresponding to a certain movement direction is represented by a certain vector, with its direction being the neuron's optimal movement direction and its length being related to the cell's firing rate. The population vector, which is the sum-up of all the individual vectors, represents the activity of the neuronal population and is along the same direction of the actual movement (Fig. 2.72). Therefore, it is accepted that during each movement command, many motor cortical neurons are activated, but the movement direction is not determined by any single neuron's activity but is instead determined by the overall spatial and temporal pattern of the neuron population.

Based on such population coding effect, it seems reasonable to infer that the cortical control to a particular movement is finer and more precise if the control is performed by a larger population of neurons. Actually in humans, movement control for face and hands is finer and more precise than for any other parts in the body, and from the cortical somatotopic map illustrated in Fig. 2.73, one can tell that the cortical areas related to these two parts are relatively large.

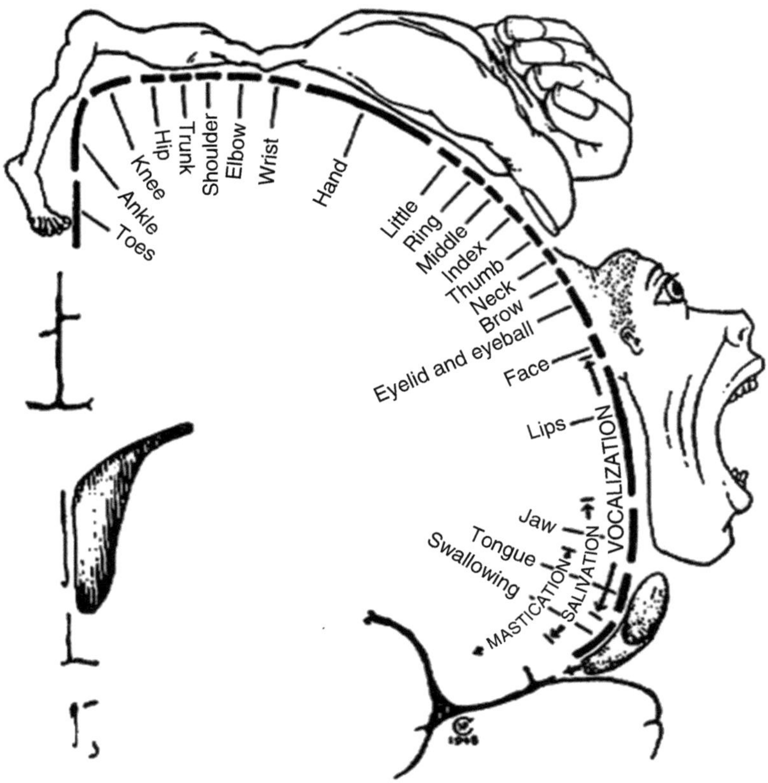

Fig. 2.73 Motor presentation in primary motor cortex in the human precentral gyrus (From Jacobson and Marcus 2011, Springer, Neuroanatomy for the Neuroscientist, Fig. 10.11)

On the other hand, motor planning involves neuronal activity in area 6. When the subject is asked to mentally imagine the movement process instead of performing an actual movement, neurons in area 4 will not be activated. Instead, neurons in area 6 (which is adjacent to area 4) will be activated in doing such a task. Anatomically, area 6 can be classified into two parts. The premotor area (PMA) is in the lateral part, and the supplementary motor area (SMA) is in the medial part (see Fig. 2.71). These two regions are with similar functions, but they are to control different muscle groups—the axons of SMA neurons innervate and control distal motor units directly, while PMA neurons are mainly connected to reticulospinal neurons that innervate proximal motor units. The activities of area 6 neurons are crucial for complex serial body movements.

Area 6 neurons mainly receive inputs from ventral lateral (VL) nucleus in the dorsal thalamus, while VL receives input from basal ganglia (including caudate nucleus, putamen, globus pallidus, subthalamic nucleus, and substantia nigra) sitting deep within the telencephalon (Fig. 2.74). On the other hand, the basal ganglia are the targets of the frontal, prefrontal, and parietal lobes. Therefore, the

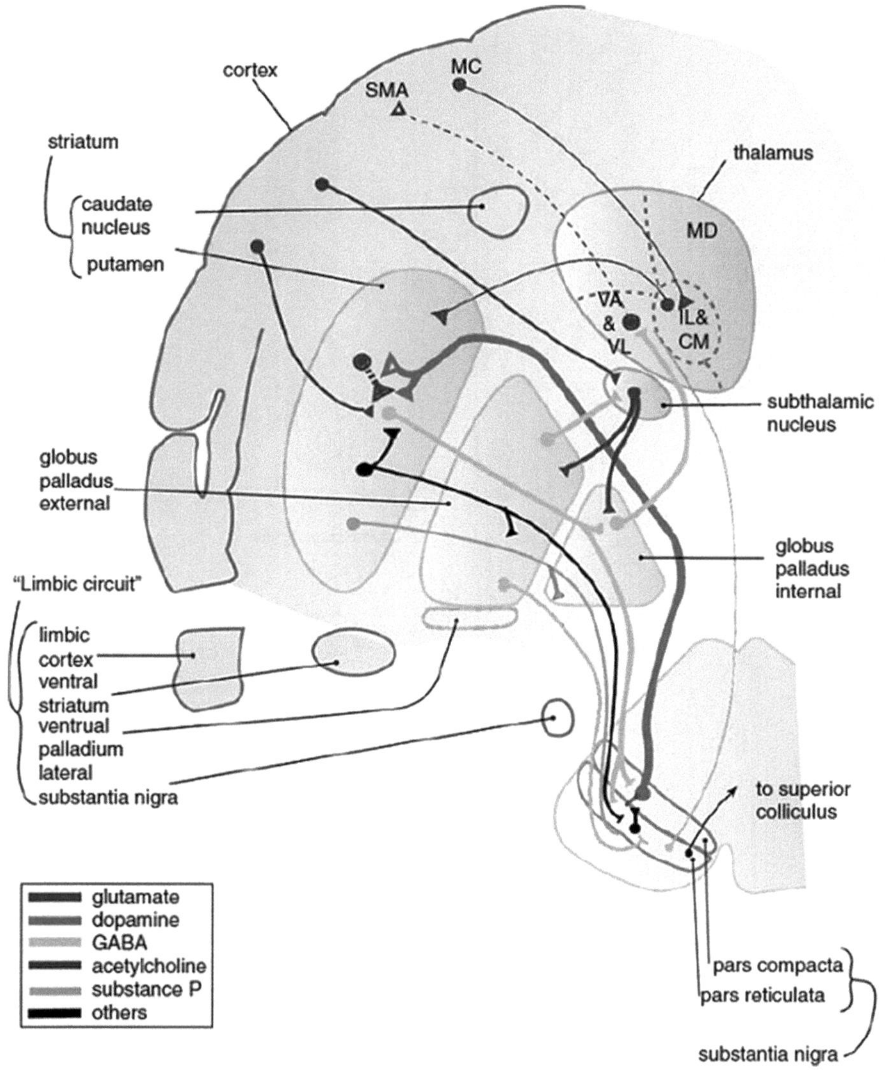

Fig. 2.74 The neural connections in the cortex, basal ganglia, and thalamus (From Pfaff 2013, Springer, Neuroscience in the 21st century, Fig. 33.2)

information pathway from the frontal cortex, prefrontal cortex, and parietal cortex to cortical SMA via basal ganglia and the thalamus is involved in initiating and modulating voluntary movement.

In the brain, the association areas of the neocortex together with the basal ganglia are responsible for identifying the goal of the movement and planning of the optimal strategy of the movement; the motor cortex and cerebellum are related

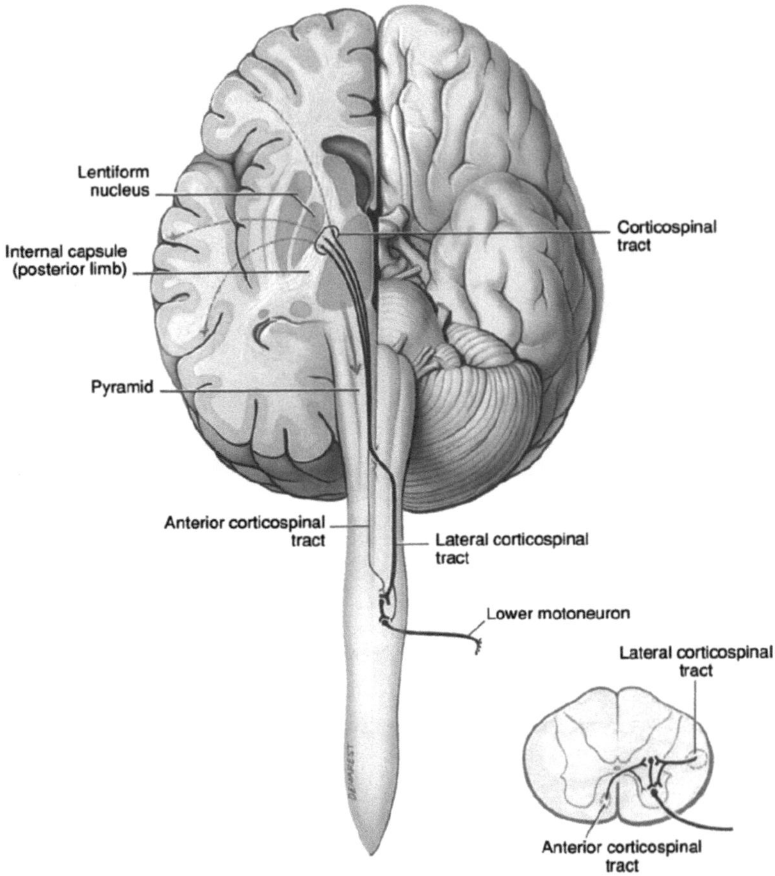

Fig. 2.75 The corticospinal tract located in the lateral pathways originated from the neocortex (From Noback et al. 2005, Springer, The human nervous system, Fig. 11.2)

to the spatial and temporal arrangement of sequential muscle contractions, while the brain stem and spinal cord are related to execution of the movement.

The axons from the cortical neurons travel through the spinal cord in two descending tracts, the lateral pathways and the ventromedial pathways. The lateral pathways are under direct cortical control and are involved in voluntary movement of the distal muscles. The corticospinal tract located in the lateral pathways is originated from the neocortex (Fig. 2.75), which is the longest and one of the biggest axon tracts (which contains 10^6 axon fibers). About 2/3 of the axons in the corticospinal tract are originated from areas 4 and 6 of the frontal cortex, while most of the remaining 1/3 axons originated from the somatosensory cortex in the parietal lobe. The corticospinal tracts from two hemispheres are crossed at the junction of the medulla and spinal cord, which means that movements of the right side of the body are governed by the nerve fibers from the left half of the brain and vice versa.

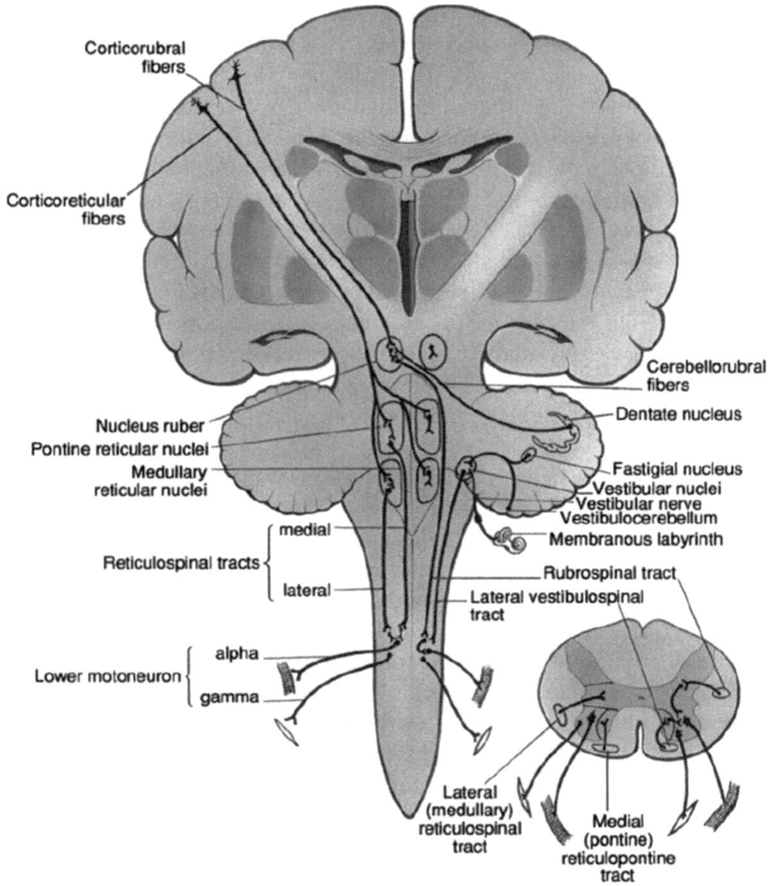

Fig. 2.76 The rubrospinal tract in the lateral pathways and the tracts in the ventromedial pathways (From Noback et al. 2005, Springer, The Human Nervous System, Fig. 11.3)

Another part of the lateral pathways is the rubrospinal tract, which originated from the red nucleus of the midbrain and receives neuronal inputs from the frontal cortex. Fibers from the rubrospinal tracts are first crossed in the pons and then join those in the corticospinal tract (Fig. 2.76).

The ventromedial pathways contain four descending fiber tracts which originated from the brain stem, including the vestibulospinal tract, the tectospinal tract, the pontine reticulospinal tract, and the medullary reticulospinal tract (Fig. 2.76). The functions of these tracts are to utilize information about balance, body position, and visual environment, to achieve the control for the trunk and proximal muscles, so as to adjust the body gesture and maintain the balance of the body.

Among these, the vestibulospinal tract originated from vestibular nucleus and transfers sensory information generated by the vestibular labyrinth located in the inner ear. The tectospinal tract originated from the superior colliculus in the midbrain, receiving direct retinal inputs. Head movements can be regulated according to the information from the visual and vestibular systems to maintain the balance of the head. The reticulospinal tract is originated from the reticular formation in the brain stem; it includes the pontine reticulospinal tract and the medullary reticulospinal tract, the functions of which are enhancing the antigravity reflexes of the spinal cord and the relieving the antigravity muscles from reflex control, respectively.

Motor neurons in the spinal cord receive inputs from the abovementioned descending pathways and sent output signals via nerve fibers generated from ventral root and merge with the sensory fibers generated from dorsal root to form spinal nerves. Spinal nerves are named according to the position they are originated, namely, cervical (C) 1–8, thoracic (T) 1–12, lumbar (L) 1–5, and sacral (S) 1–5 (Fig. 2.77). Each spinal nerve contains spinal motor fibers that belong to a spinal segment.

Skeletal muscles are innervated with motor neurons located in the ventral horn of the spinal cord (Fig. 2.78). These motor neurons are also called "lower motor neurons" (with those located in higher-order cortex being called "upper motor neurons"), the axons of which are clustered together and form the ventral root. The nerve fibers innervating arm musculature are originated from spinal segments C3-T1. Those fibers controlling leg musculature are clustering in spinal segments L1-S3. The distribution of lower motor neurons in the ventral horn is closely correlated to the spatial location of the muscles they innervate: the neurons innervating the axial muscles are located in the medial part and those innervating the distal muscles sitting in the lateral part, while the cells controlling flexors and extensors are distributed in the dorsal and ventral parts, respectively.

Lower motor neurons in the spinal cord can be further classified into α and γ subtypes. Among these, α motor neurons receive inputs from three different sources and provide direct control to the muscle contraction. The upper neurons in the motor cortex and brain stem provide input signals to α motor neurons to initiate and control voluntary movement. Another source of inputs is from dorsal root ganglion cells transferring information from muscle spindle (being innervated by γ motor neuron) which senses the muscle length; this provides a feedback signal from musculature to α motor neurons. The third input is from interneurons in the spinal cord, which can either be excitatory or inhibitory, and contributes to the spinal motor neuronal circuitry. An α neuron together with all the muscle fibers it innervates forms the elementary component of motor control, which is termed as motor unit. Muscle contraction normally involves activity of single or multiple motor units. Assembly of α neurons innervating a single muscle makes up a motor neuron pool. At neuromuscular junction, which is the specified synapse between α neuron and skeletal muscle fiber, the neuron releases neurotransmitter acetylcholine (ACh) and elicits excitatory postsynaptic potential (EPSP, sometimes also called end-plate potential) in the muscle fiber. When such EPSP exceeds the threshold, an action potential will be triggered and a muscle fiber contraction will be induced. Sustained muscle contraction needs sustained burst of action potentials.

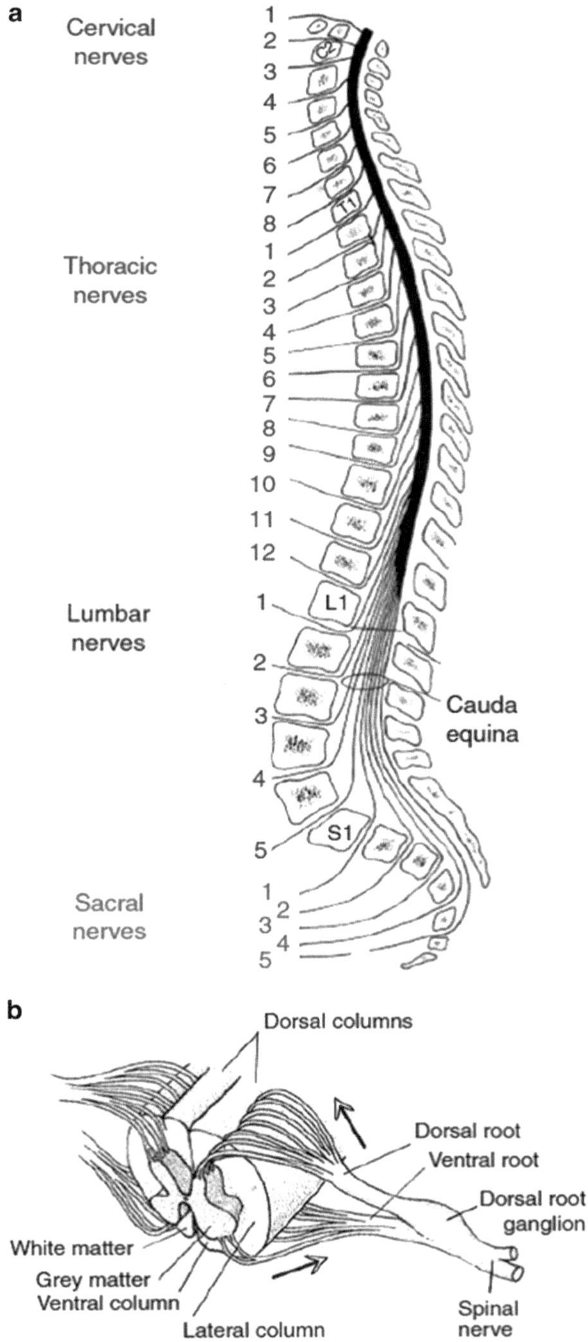

Fig. 2.77 Anatomy of spinal cord and spinal nerves. (**a**) Spinal nerves are named following the spinal segments they are originated. (**b**) The input–output circuit in the spinal cord and spinal verve fibers (From Pfaff 2013, Springer, Neuroscience in the 21st century, Fig. 46.1)

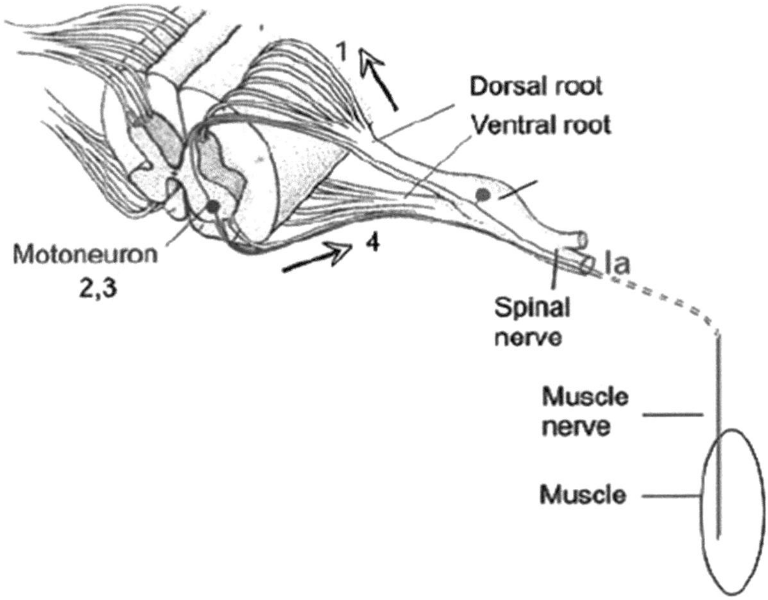

Fig. 2.78 Skeletal muscle is innervated by lower motor neurons located in the ventral horn of the spinal cord (From Pfaff 2013, Springer, Neuroscience in the 21st century, Fig. 46.3)

2.10 Cerebellum

The cerebellum is posterior to the cerebrum and is adjacent to the brain stem at the pons (Fig. 2.79). The cerebellum is widely connected with the cerebral cortex and the spinal cord, with its main function being motor control. While the activity of the cerebral motor cortex of each hemisphere is related to the motor control of the contralateral part of the body, the activity of cerebellum hemisphere is to control the ipsilateral part of the body. Although the cerebellum is of much smaller size as compared to the cerebrum and only occupies 10 % of the brain volume, it contains abundant neurons, the number of which is over 50 % of the whole central nervous system.

The cerebellum has a shape similar to a cauliflower. Its surface is covered by a thin layer of cortex which is folded repeatedly and forms many transverse shallow ridges called folia and is divided into 10 lobules by a set of deeper transverse fissures. Such structural properties greatly increase the surface area of the cerebellar cortex. In addition to the dense cortical neurons, there are several groups of neurons embedded deep in the white matter, forming the deep cerebellar nuclei, which relay the information transferred from the cerebellar cortex to a number of brain stem structures (Fig. 2.80).

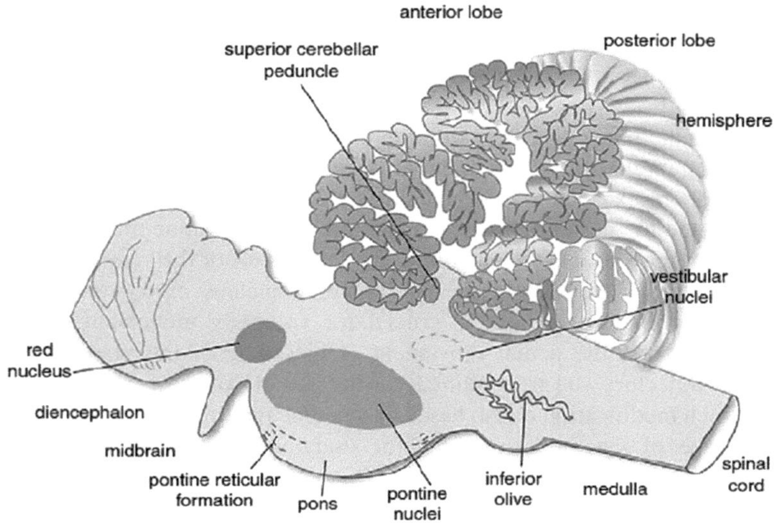

Fig. 2.79 Anatomy position of the cerebellum (From Pfaff 2013, Springer, Neuroscience in the 21st century, Fig. 35.1)

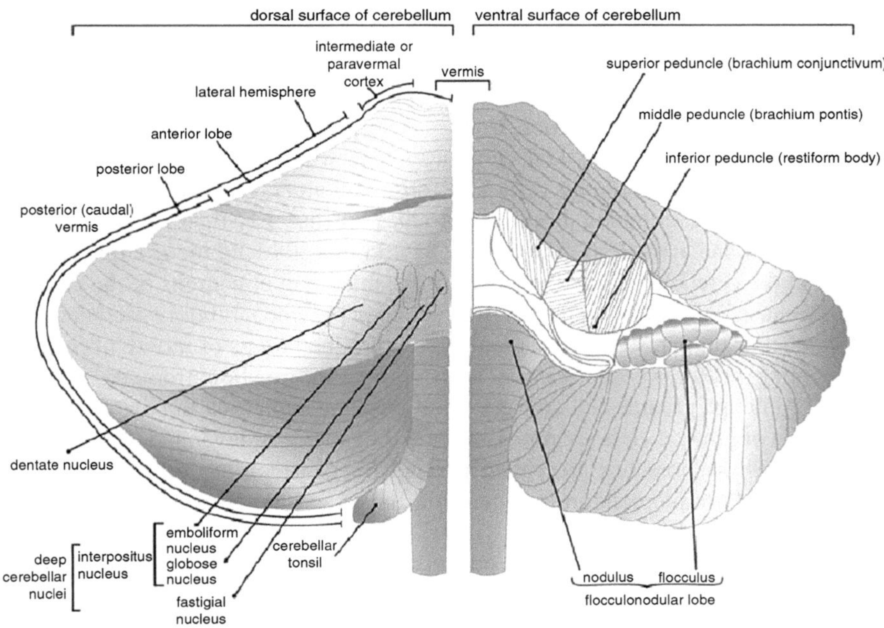

Fig. 2.80 Structure of the cerebellar (From Pfaff 2013, Springer, Neuroscience in the 21st century, Fig. 35.2)

The cerebellar is separated into two hemispheres by a ridge midline called vermis. The vermis and the hemispheres are endowed with different functions. While the vermis output is sent to the brain stem, running into the ventromedial descending pathways to control the axial musculature, the hemispheres are mainly connected to the brain structures forming the lateral descending pathways, particularly the cerebral cortex. The axons of layer V pyramidal cells of the cerebral sensorimotor cortex (including frontal areas 4 and 6, postcentral gyrus somatosensory areas, and the posterior parietal areas) are sent to pontine nuclei in the pons, which in turn project to the cerebellum. The lateral part of the cerebellum projects back to the motor cortex via the ventrolateral nucleus of the thalamus. This corticopontocerebellar pathway is crucial for voluntary movement control and modulation. When the actual movement is biased from the planned one, the movement trajectory can be modified via the activity of the cerebellar circuitry. Because such modification is also based on synaptic modification, the cerebellum is therefore one of the model systems for studying the synaptic connections in mammalian nervous system.

The cerebellar cortex contains two layers of cells: the Purkinje cells (for the memory of nineteenth century Czech physiologist Purkinje) and the granule cells. Between these cell layers and the pial surface, there is a molecular layer which contains a small amount of cell. The axons of the cerebellar granule cells form parallel fibers distributed in the surface of the molecular layer, and stellate cells are sitting at the bottom of the molecular layer. The Purkinje cells have their dendrites stretching into the molecular layer, with elongated dendritic fields (the area where the dendrites are stretched). The axons of Purkinje cells form the only pathway for the cerebellar cortex to be connected to other brain areas. Each Purkinje cell receives input from a single inferior olive neuron in the medulla, the function of which is to integrate information from muscle proprioceptors. The axons of the inferior olive neurons are called climbing fibers, which are widely connected to dendritic tree of the Purkinje cells and form excitatory synapses, so that action potentials from a climbing fiber can produce intense EPSP in its postsynaptic Purkinje cell and activate it. The axons of Purkinje cells send their output to deep nuclei in the cerebellum, and the latter are the main output neurons of the cerebellum. Because Purkinje cells are GABAergic, they provide inhibitory modulation to the cerebellar output. Cerebellar granule cells are the targets of mossy fibers from the pontine nuclei whose function is to convey information from the cerebral cortex. Although granule cell are very small, they are numerous and tightly packed. The axons of granule cells are processed into the molecular layer and form T-shape branches. These branches are called parallel fibers which are orthogonal to the dendrites of Purkinje cells in the molecular layer and are widely connected with these cells (Fig. 2.81).

The extensive connection between Purkinje cells and climbing fibers/parallel fibers and the structural properties lead to the inference that climbing fiber carries information about body position with its activity being related to the difference between the actual movement and the expected one, and such trajectory error can be corrected by adjusting the input from parallel fiber to Purkinje cell. Such motor

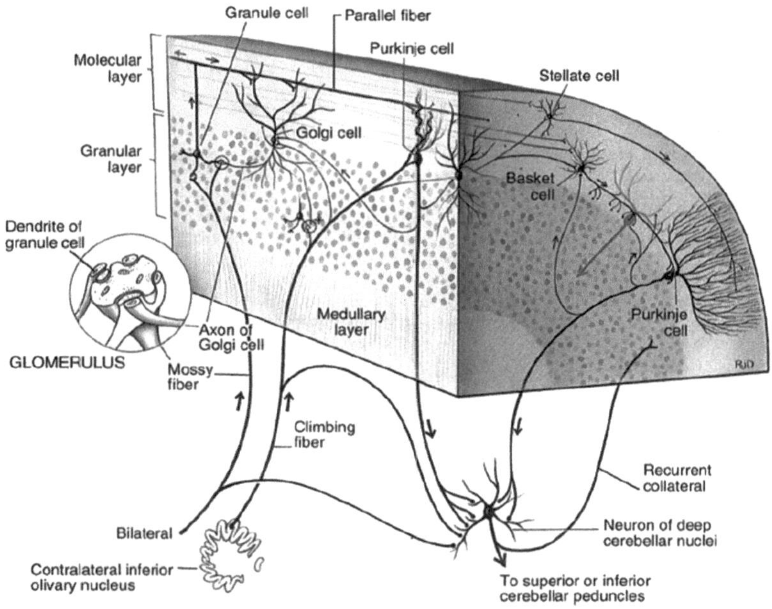

Fig. 2.81 Synaptic connections among cerebellar neurons (From Noback et al. 2005, Springer, The human nervous system, Fig. 18.3)

learning theory based on the synaptic plasticity between parallel fibers and Purkinje cells was independently forwarded by David Marr at Cambridge University and James Albus at Coddart Space Flight Center in Maryland in the 1970s and is called Marr–Albus theory of motor learning. This hypothesis was experimentally tested by Masao Ito and his colleagues at University of Tokyo, and it was verified that when parallel fiber-Purkinje cell synapse is activated in simultaneous with the depolarization of the Purkinje cell due to climbing fiber input, the synaptic connection between the parallel fiber and the Purkinje cell will be attenuated. The effect of such synaptic modification can last for up to one hour, which is termed as long-term depression (LTD) (Fig. 2.82). Such LTD only occurs at those parallel fiber synapses which are accompanied by climbing fiber activation.

Cerebellar Purkinje cell expresses two kinds of glutamate receptor subtypes: one is of AMPA (α-amino-3-hydroxy-5-methyl-4-isoxazole propionate) subtype and another is of metabotropic subtype which is coupled to the enzyme phospholipase C via G-protein. The activation of phospholipase C can produce a second messenger (diacylglycerol), which can activate protein kinase C. When presynaptic glutamate input is sent to the Purkinje cell, it induces Ca^{2+} influx and produces EPSP via the activation of AMPA receptors. In the meantime, the severe depolarization of the postsynaptic membrane activates the voltage-dependent sodium channels and voltage-dependent calcium channels on the postsynaptic membrane, which results in elevation of intracellular calcium concentration in the Purkinje cell. However,

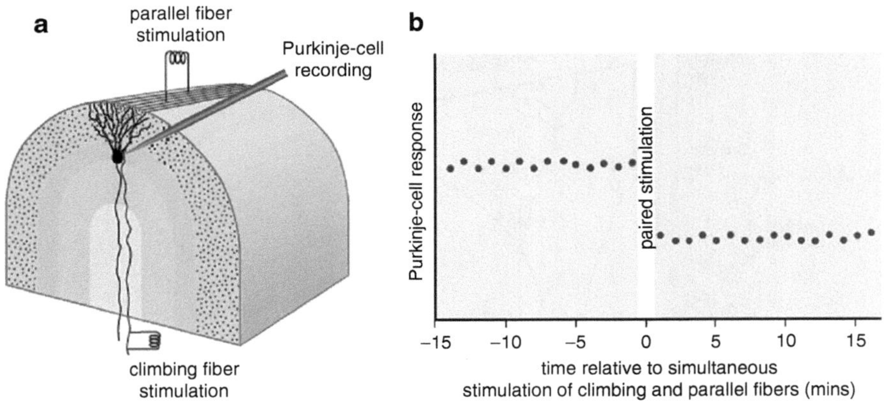

Fig. 2.82 LTD in synapse between cerebellar neurons (From Galizia and Lledo 2013, Springer, Neurosciences – from molecule to behavior, Fig. 26.15)

the activation of protein kinase C via the activation of metabotropic glutamate receptors and the result of protein phosphorylation cause a reduction in AMPA receptor channels on the postsynaptic membrane, which in turn reduces the synaptic sensitivity to glutamate input. In this process, the effect of AMPA receptor modulation is closely related to postsynaptic calcium concentration elevation. If the Purkinje cell is injected with Ca^{2+} chelator to prevent $[Ca^{2+}]$ increment, such LTD can be effectively blocked.

2.11 Hippocampus

The hippocampus is located in the medial temporal lobe and consists of two folded cell layers which are overlapped with each other (Fig. 2.83a). Among these, one layer is called dentate gyrus, and the other is Cornu Ammonis (CA). The dentate gyrus receives signal input from the entorhinal cortex via perforant path, while the axons of the dentate gyrus neurons are presynaptic to CA3 neurons. The axons of CA3 neurons are separated into two branches, one leaves the hippocampus via the fornix and the other which is called Schaffer collateral forms synapses on CA1 neurons. Due to its simple structure, also because that in vitro hippocampal slice can be maintained for several hours while being properly superfused, hippocampus is one of the ideal models for investigating synaptic transmission of the brain.

It is now well acknowledged that the long-term potentiation (LTP) of synaptic connectivity observed in the hippocampus is crucial for the formation of memory. In the hippocampus, CA1 neurons receive inputs from CA3 neurons; therefore, if we stimulate a presynaptic CA3 neuron with a brief current, an excitatory postsynaptic potential (EPSP) can be recorded from its postsynaptic CA1 neuron

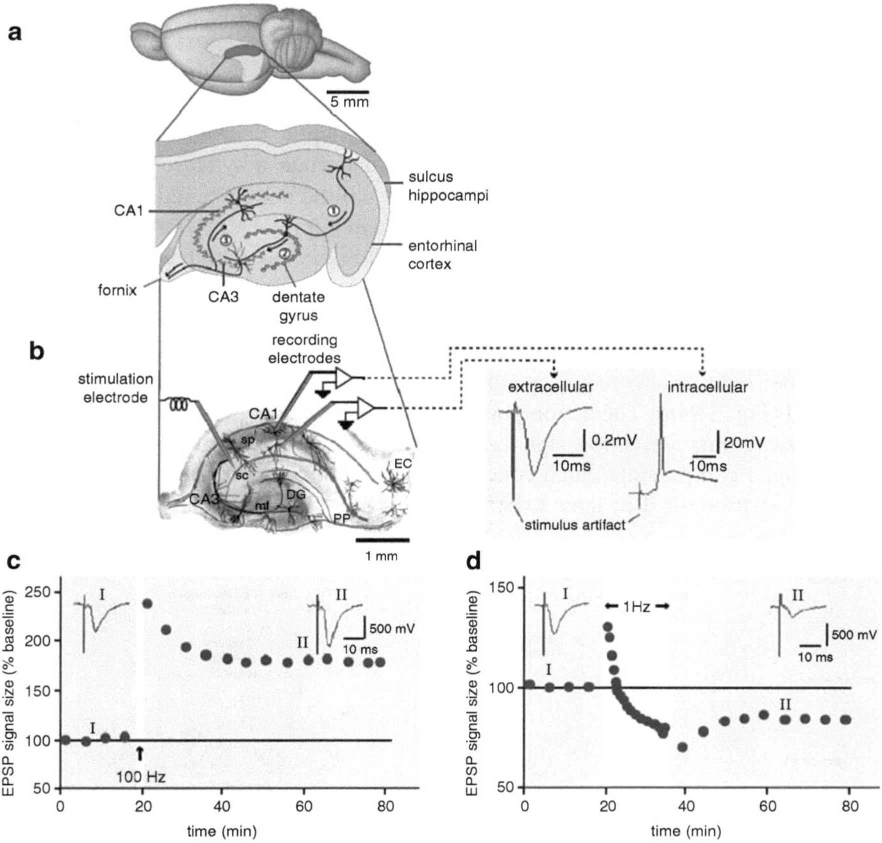

Fig. 2.83 The rodent hippocampus and synaptic plasticity. (**a**) The position of the hippocampus in the rodent brain and the cell layers. (**b**) Electrophysiology for measuring synaptic transmission between hippocampal neurons. (**c**) High-frequency tetanus stimulation (100 Hz) induces LTP. (**d**) Low-frequency stimulation (1 Hz) induces LTD (From Galizia and Lledo 2013, Springer, Neuroscience: from molecule to behavior, Fig. 26.11)

(Fig. 2.83b). If repeated electrical stimulation is given at a frequent of 1/min, steady recording can be obtained. But if a tetanus stimulation is given to the presynaptic neuron with high frequency (normally 50–100 stimulations with a frequency of 100 Hz), remarkably increased EPSP can be observed in the following recordings (Fig. 2.83c). Such brief tetanus stimulation-induced signal transmission enhancement effect can last for many hours, or even days.

This phenomenon is thus called long-term potentiation (LTP) of synaptic connectivity, and it was later demonstrated that high-frequency stimulation is not necessary for LTP. What is requested is the synaptic activation together with a remarkable depolarization of the postsynaptic CA1 neurons. Or in other words, LTP will only occur in the synapse(s) being stimulated and activated, but not those un-activated synapses.

In the hippocampus, the excitatory synapses are mediated by glutamate. When a presynaptic neuron releases glutamate and activates the postsynaptic AMPA subtype glutamate receptors, Na^+ influx is induced and results in EPSP at the CA3-CA1 synapse. When sufficient synapses are activated at the same time, an action potential can be generated in the postsynaptic neuron due to sufficient depolarization. Donald Hebb once proposed that when a synapse participates in eliciting the postsynaptic neuron, this synapse will be strengthened. In addition to AMPA receptors, hippocampal CA1 neuron also expresses NMDA (*N*-methyl-D-aspartate) receptors. Such glutamate receptor subtype has a high affinity to glutamate and is blocked by Mg^{2+} during rest.

When the postsynaptic membrane is depolarized, the Mg^{2+} blockage effect is relieved, which makes the membrane permeable to Ca^{2+} and mediates Ca^{2+} influx and further induces severe postsynaptic membrane depolarization and bursts action potential (Fig. 2.84a). The action potential burst on soma is in turn back propagated to dendrites, thus activation signals can be distributed synchronously at pre- and postsynaptic components. Such concept is the "Hebbian learning rule" of synaptic connection strength. But later experimental results revealed that the induction of

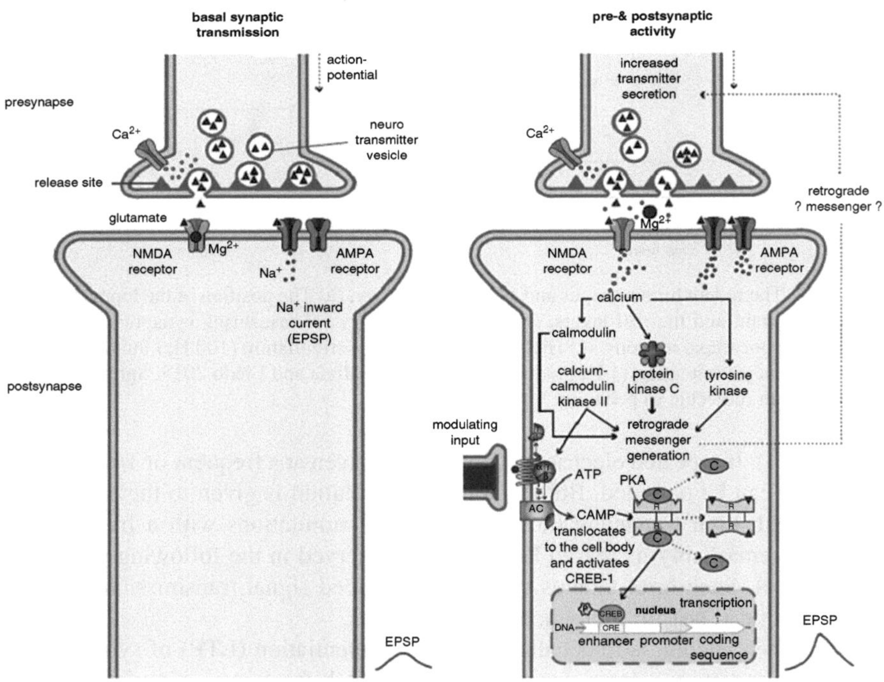

Fig. 2.84 synaptic mechanism for LTP. (**a**) Synaptic transmission during baseline stimulation. (**b**) Induction of LTP depends on the activation of pre- and postsynaptic neurons within the same time window and involves postsynaptic mechanisms (From Galizia and Lledo 2013, Springer, Neuroscience: from molecule to behavior, Fig. 26.12)

LTP is closely related to the time sequence of pre- and postsynaptic events. When a postsynaptic action potential is subsequent to an EPSP within a time window of 50 ms, the synapse will be significantly strengthened. But if the time order is shifted, the synaptic strength will be attenuated. This finding is called "spike-timing-dependent plasticity."

Many lines of evidence revealed that the generation of LTP is dependent on the significant enhancement of postsynaptic $[Ca^{2+}]_i$, while the elevated $[Ca^{2+}]_i$ causes the activation of two kinds of protein kinase: protein kinase C and calcium–calmodulin-dependent protein kinase II (CaMKII), which in turn phosphorelates the AMPA receptors and enhances the conductance of receptor channels. In addition to this calcium-induced AMPA receptor phosphorelation, another parallel process is that the activation of CaMKII will induce AMPA receptor insertion on the postsynaptic membrane (Fig. 2.84b). When NMDA receptors are chemically blocked, or Ca^{2+} chelator is injected to the postsynaptic neuron to prevent the enhancement of postsynaptic $[Ca^{2+}]_i$, no LTP will be induced. Protein kinase C inhibition or CaMKII inhibition by drug application will also inhibit LTP induction.

In addition to LTP, there is also long-term depression (LTD) existing in hippocampal neurons. Long-term tetanus stimulus with low frequency will cause input-specific LTD (Fig. 2.83d). What is interesting is that similar to LTP, the induction of LTD is also closely related to postsynaptic Ca^{2+} influx through NMDA receptors. Whether it induces LTP or LTD depends on the level of NMDA receptor activation. When the postsynaptic neuron is severely depolarized, the Mg^{2+} blockage effect on NMDA receptor is entirely eliminated; Ca^{2+} floods into the postsynaptic neuron through NMDA receptors. High $[Ca^{2+}]_i$ activates protein kinase in the postsynaptic neuron, which results in phosphorelation of AMPA receptors. But when the postsynaptic neuron is only slightly depolarized, the Mg^{2+} blockage effect on NMDA receptor is still partly remained; CA^{2+} influx is at a low level; only a small amount of Ca^{2+} flows into the postsynaptic neuron. The slowly mild $[Ca^{2+}]_i$ elevation activates the protein phosphatases, which induces AMPA receptor dephosphorelation. In addition, the induction of hippocampal LTD is also related to the internalization of AMPA receptors at the postsynaptic membrane.

The hippocampus is involved in various functions for learning and memory. Hippocampal injury will cause deficiency in learning and memory. Rat behavioral experiments showed that many neurons in the hippocampus have response to specific location in the environment. Such neurons are called "place cell," and the place in which most active response of a cell is induce is determined as the neuron's "place field" (which is similar with the "receptive field" of sensory neurons). In a certain environment, place cell's response is related to the position the animal is in. In maze experiment, it can be recorded that when a freely moving rat is moving to a particular area, relevant place cell will be firing with high frequency, but when the animal leaves this area, the firing frequency will be remarkably decreased, or even eliminated. Different place cells have different place fields. Therefore, the activity of place cells reflects the place coding of the nervous system. But on the other hand, another important feature of place cell is the dynamic modulation property of the place field. When the environment is changed, place cell will change its receptive field property and adapt to the environment. Therefore, damage of the hippocampus will influence the animal's memory to the environment.

Other experiments found that the response of rat hippocampal place cell is influenced by various cues including the additional visual and olfactory inputs, or even moving speed and direction, etc. Therefore, it is hypothesized that highly processed sensory information is fed to the hippocampus and other structures in the medial temporal lobe and is stored in parallel with other event.

Actually, in addition to simple space location, some particular hippocampal cells are responsive to more complex environmental stimulations. It was observed in experiments performed on freely moving mouse that some highly specified hippocampal neurons were activated or inactivated when the animal was put in a "nest" or a "bed," regardless of their locations or environments (Fig. 2.85). This suggests that the hippocampus is an important part of the hierarchical structure for knowledge generating in the brain.

2.12 Summary and Advanced Readings

Neurobiological background is necessary for understanding neural information processing. Traditional AI tried to mimic human intelligence just by sophisticated programming and symbol processing without any insight of the brain mechanism; this effort went to a dead end. Therefore, in a book with the main topic of elucidating neural information processing, an introduction to neuroanatomy and neurophysiology with enough details is necessary. This chapter is devoted to this topic.

In biology, the function of some organ is closely related to its structure; therefore, at the very beginning of this book, in Sect. 2.1, the gross anatomy of the nervous system is introduced.

As the structural and functional unit, the morphology of the neuron is introduced in Sect. 2.2, while its electrical property—the main characteristics related to neural information processing—is introduced in Sect. 2.3.

As the most advanced structure, cerebral cortex is described in Sect. 2.4. The argument that for some specific function if only a local area is related or the whole brain is needed is discussed. The outstanding features of its architecture—hierarchy and reentry, layer structure, and column structure—are also discussed.

In Sect. 2.5, visual system, which is supposed to provide most environmental information for primates and has been most intensively studied and understood during the past decades, is introduced, including both anatomical structure and physiological functions.

As an important sensation only next to vision, the auditory system is elucidated in Sect. 2.6. The argument between place theory and temporal theory is also given.

Olfaction, the oldest and simplest sensation, is introduced in Sect. 2.7. It is a good example for population coding.

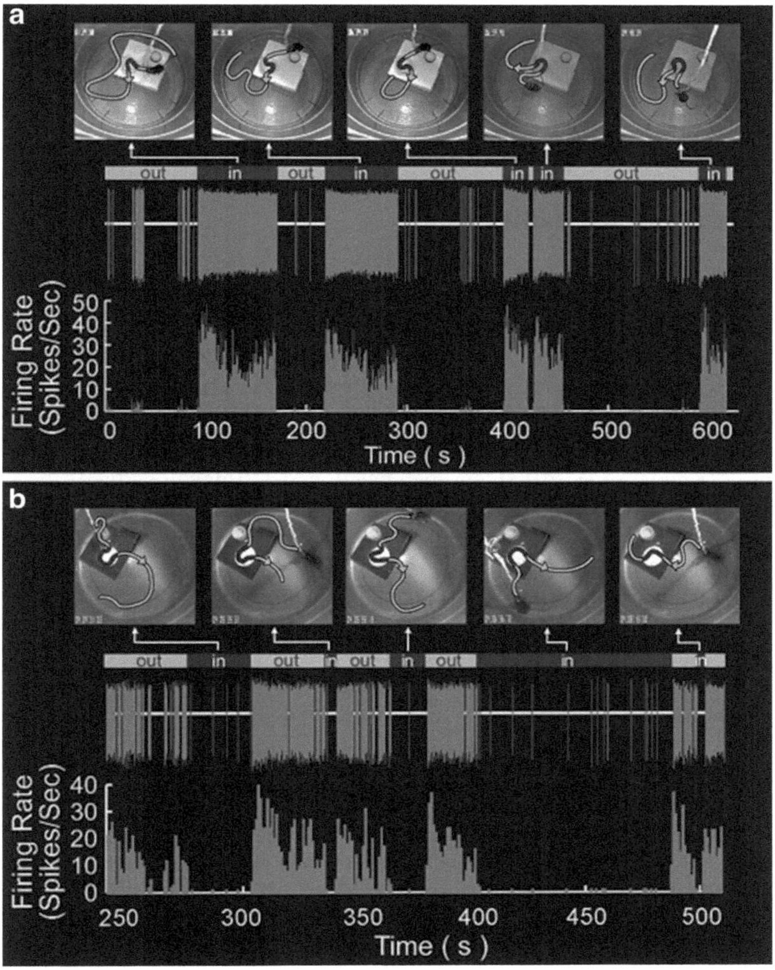

Fig. 2.85 Different types of nest-responsive cells. (**a**) A persistent on-cell fired robustly and persistently when the animal was inside the "nest." (**b**) A persistent off-cell ceased to fire when the animal entered the "nest" (From Lin et al. 2007 PNAS 104(14):6066–6071)

Sections 2.8 and 2.9 are, respectively, devoted to somatosensory and somatomotor systems, both of which have nicely organized cortical representation and organized communication with the peripheral parts via the spinal cord.

Section 2.10 is focused on the cerebellum, whose network connectivity has been well studied. Therefore, the significance for cerebellum study is twofold—one is for its function involved in motor control and the other is for studying the synaptic plasticity, particularly stimulation-dependent long-term depression of the synaptic connectivity.

The hippocampus, which is involved in function such as learning and memory, is introduced in Sect. 2.11. It is also a nice model for studying the synaptic plasticity, particularly stimulation-dependent long-term potentiation of the synaptic connectivity.

2.13 Advanced Reading

Bear et al.'s textbook "Neuroscience: Exploring the Brain (3rd Edition)" (Bear et al. 2007) and Dowling's monograph "Neurons and Networks: An Introduction to Neuroscience (2nd Edition)" (Dowling 2001) are strongly recommended to any beginner who wants to know more about neuroscience. These books are interesting and readable. For readers who want to consult some specific topic he or she is especially interested in, Kandel et al.'s textbook "Principles of Neural Science (5th Edition)" (Kandel et al. 2013) and the encyclopedia "Encyclopedia of the Human Brain" edited by Ramachandran (Ramachandran 2002) are strongly recommended. For those who wish to know more about the brain from a viewpoint of information processing, "The Handbook of Brain Theory and Neural Networks (2nd Edition)" edited by Arbib (Arbib 2002) could also be consulted.

References

Adams DL, Horton JC. Capricious expression of cortical columns in the primate brain. Nat Neurosci. 2003;6(2):113–4.

Abeles M. Corticonics. Cambridge: Cambridge University Press; 1991.

Arbib MA, editor. The handbook of brain theory and neural networks. 2nd ed. Cambridge: MIT Press; 2002.

Ascoli GA, et al. Petilla terminology: nomenclauture of features of GABAergic interneurons of the cerebral cortex. Nat Rev Neurosci. 2008;9(7):557–68.

Bear MF, Connors BW, Paradiso MA. Neuroscience: exploring the brain. 3rd ed. Philadelphia: Lippincott Williams & Wilkins; 2007.

Borisyuk A, et al. Tutorials in mathematical biosciences I: mathematical neuroscience. Berlin: Springer; 2005.

Bower JM, Parson LM. Rethinking the "lesser brain". Sci Am. 2003;289(2):48–57.

Braitenberg V. Brain. Scholarpedia. 2007;2(11):2918.

Buxhoeveden DP, Casanova MF. The minicolumn hypothesis in neuroscience. Brain. 2002;125:935–52.

Churchland PS, Sejnowski TJ. The computational brain. Cambridge: The MIT Press; 1992.

Conn PM, editor. Neuroscience in medicine. 3rd ed. Totowa: Humana Press; 2008.

Dowling JE (2001) Neurons and networks – an introduction to neuroscience, 2nd edn. The Belknap Press of Harvard University Press.

Druckmann S, Hill S, Schürmann F, Markram H, Segev I. A hierarchical structure of cortical interneuron electrical diversity revealed by automated statistical analysis. Cereb Cortex. 2012. doi:10.1093/cercor/bhs290.

Edelman GM, Tononi G. A universe of consciousness – how matter becomes imagination. New York: Basic Books; 2000.

Fields RD. The other half of the brain. Sci Am. 2004;2004(4):26–33.
Freeman WJ. Mass action in the nervous system – examination of the neurophysiological basis of adaptive behavior through the EEG. New York: Academic Press; 1975.
Freeman WJ. The physiology of perception. Scientific American. 1991; Feb:78–85.
Freeman WJ. Tutorial on neurobiology: from single neurons to brain chaos. Int J Bifurcation Chaos. 1992;2(3):451–82.
Freeman WJ. Neurodynamics: an exploration in mesoscopic brain dynamics. London: Springer-Verlag; 2000.
Galizia CG, Lledo PM, editors. Neuroscience – from molecule to behavior: a university textbook. Berlin: Springer Spectrum; 2013.
Hawkins J, Blakeslee S. On intelligence – how a new understanding of the brain will lead to the creation of truly intelligent machines. New York: Henry Holt; 2004.
Hawkins HL, McMullen TA. Auditory computation: an overview. In: Hawkins HL, McMullen TA, Popper AN, Fay RR, editors. Auditory computation. New York: Springer; 1996.
Hodgkin AL, Huxley AF. A quantitative description of membrane current and its application to conduction and excitation in nerve. J Physiol. 1952;117:500–44.
Horton JC, Adams DL. The cortical column: a structure without a function. Philos Trans R Soc Lond B Biol Sci. 2005;360(1456):837–62.
Hubel DH, Wiesel TN. Uniformity of monkey striate cortex: a parallel relationship between field size, scatter, and magnification factor. J Comp Neurol. 1974;158:295–305.
Jacobson S, Marcus EM. Neuroanatomy for the neuroscientist. 2nd ed. New York: Springer; 2011.
Kandel E, et al. Principle of neural science. 5th ed. New York: The McGraw-Hill Education; 2013.
Kauer JS. Contributions of topography and parallel processing to ordor coding in the vertebrate olfactory pathway. Trends Neurosci. 1991;14:79–85.
Koch C. The quest for consciousness – a neurobiological approach. Denver: Roberts and Company Publishers; 2004.
Kosslyn SM, Andersen R, editors. Frontiers in cognitive neuroscience. Cambridge: MIT Press; 1992.
Liang Z. Neural mechanisms of auditory sensation and discrimination. Shanghai Science& Technology Education Press; 1998 (In Chinese).
Lin L, et al. Neural encoding of the concept of nest in the mouse brain. PNAS. 2007;104(14):6066–71.
Maciokas JB, Goodman P, Kenyon J, Toledo-Rodriguez M, Markram H. Accurate dynamical models of interneuronal GABAergic channel physiologies. Neurocomputing. 2005;65–66:5–14. doi:10.1016/j.neucom.2004.10.083.
Markram H. The human brain project. Sci Am. 2012;2012(6):50–5.
Markram H, Toledo-Rodriguez M, Wang Y, Gupta A, Silberberg G, Wu C. Interneurons of the neocortical inhibitory system. Nat Rev Neurosci. 2004;5(10):793–807. doi:10.1038/nrn1519.
Marr D. Vision. A computational investigation into the human representation and processing of visual information. New York: W.H. Freeman and Company; 1982.
Miikkulainen R, et al., editors. Computational maps in the visual cortex. New York: Springer; 2005.
Møller AR. Auditory physiology. New York: Academic Press; 1983.
Mountcastle V. An organizing principle for cerebral function: the unit module and the distributed system. In: Edelman GM, Mountcastle VB, editors. The mindful brain. Cambridge: MIT Press; 1978. p. 7–50.
Mountcastle VB. The columnar organization of the neocortex. Brain. 1997;120:701–22.
Noback CR, et al. editors. The human nervous system. 6th ed. Totowa: Humana; 2005.
Pfaff DW, editor. Neuroscience in the 21st century: from basic to clinical. New York: Springer; 2013.
Ramachandran VS, editor. Encyclopedia of the human brain. San Diego: Academic Press; 2002.

Richardson MJE, Melamed O, Silberberg G, Gerstner W, Markram H. Short-term synaptic plasticity orchestrates the response of pyramidal cells and interneurons to population bursts. J Comput Neurosci. 2005;18(3):323–32.

Rose JE, Hind JE, Anderson DJ, Brugge JF. Some effects of stimulus intensity on response of auditory nerve fibers in the squirrel monkey. J Neurophysiol. 1971;34:685–99.

Shamma SA. Auditory cortex. In Arbib MA, editor. The handbook of brain theory and neural networks. 2nd ed. Cambridge: MIT Press; 2002.

Steward O. Functional neuroscience. New York: Springer; 2000.

Strominger NL, Demarest RJ, Laemle LB. Noback's human nervous system seventh edition: structure function. Totowa: Humana Press; 2012.

ten Donkelaar HJ, editor. Clinical neuroanatomy: brain circuitry and its disorders. New York: Springer; 2011.

Toledo-Rodriguez M, Gupta A, Wang Y, et al. Neocortex: basic neuron types. In: Arbib MA, editor. The handbook of brain theory and neural networks. 2nd ed. Cambridge: MIT Press; 2002.

Wagner H, Silber K. Physiological psychology. London and New York: BIOS Scientific Publisher; 2004.

Chapter 3
Single Neuron Models

Abstract Neurons are generally considered as both the structural and functional units of the nervous system and also as its basic units for information processing. A single neuron and its models have been well studied. As examples to show how computational neuroscientists use mathematical and informatic ideas and approaches to solve neurobiological problems, a variety of single neuron models are discussed in this chapter, including the Hodgkin–Huxley model, the multi-compartmental model, the simplified model, and some simplified or generalized models of the H–H model. Dynamic analysis of some models mentioned above is also given, to show how such analysis might give us a theoretical framework to elucidate the variability of neuron behaviors.

Keywords Single neuron models • Hodgkin–Huxley model • Multi-compartmental model • Simplified model • Neurodynamics

Although there are still some debates, neurons are generally considered as both the structural and functional units of the nervous system and also as its basic units for information processing. It is well acknowledged that the founder of the neuron doctrine Santiago Ramón y Cajal is the father of neuroscience. The key role of neurons playing in the neural information processing in the nervous system is undoubted. Single-unit and multiple-unit recording of neurons dominated neurophysiologic studies for most of the time in the last century. Neurons may be the objects which we know best in the nervous system and of which the first mathematical model was established in neuroscience history (Lapicque 1907; McCulloch and Pitts 1943). A lot of different neuron models have been developed further. Hodgkin and Huxley's model of neuronal membrane (Hodgkin and Huxley 1952) is regarded as the founding and a classical achievement of computational neuroscience and even remains the touchstone for neural modeling today. Thus, neurons may be the objects in nervous systems which have been most well studied and modeled. Therefore, models of single neurons have been picked up as an example to show how computational neuroscientists use mathematical and informatic ideas and approaches to solve biological problems. And, in a book about neural information processing, to introduce single neuron models at the very beginning might be proper, not only owing to the importance of neurons but also to that it might give readers classical examples to show how to use mathematical, physical, and

P. Liang et al., *An Introduction to Neural Information Processing*,
DOI 10.1007/978-94-017-7393-5_3

informatic ideas and tools to solve neurophysiologic problems. The mathematical modeling of neurons gives us insight to their underlying mechanism, suggesting new experiments to verify hypothesis assumed for modeling and predictions from simulation results. The mathematical analysis makes us realize the varieties of the neuron behavior in a systematic way, which may seem puzzling when they are only observed as experiment phenomena. Of course, on the contrary, it would be meaningless if such analysis could not predict any new phenomenon which can be verified by experiments or a neuronal model is built with some assumptions clearly violating basic biological facts or has nothing to do with the real neuron, if it is claimed to solve biological problems, such as neural coding. These are the topics which we will discuss in this chapter.

3.1 Hodgkin–Huxley Model

In the ancient time, people thought that there was some "spirit" or fluid flowing in the nerves to make muscle contract. Only after Luigi Galvani's classical experiments that it is known that it is some electrical signal propagating along the nerves, rather than the mysterious spirit (Finger 1994). However, how nerves produce electrical signals was still an open problem for a long time.

3.1.1 Bernstein's Membrane Theory

In 1840, Italian scientist Carlo Matteucci found that there were currents flowing from the section of a frog muscle to its undamaged surface. This current is called a demarcation current. A little later, German physiologist Emil du Bois-Reymond found that the potential difference between the injured surface and the intact surface decreased transiently when a stimulus was given and he called such phenomenon "negative variation." His student, German physiologist Julius Bernstein, continued his study and found that: (1) Such negative variation is transient and its duration is about 1 ms and independent of the stimulus intensity; (2) if the stimulus is strong enough, the variation amplitude can have an overshoot, i.e., surpass the amplitude of the resting potential. However, he observed such phenomenon only in nerves and neglected this important fact when he tried to explain the underlying mechanism of action potentials; later, this phenomenon contradicts his theory. (3) The propagating speed of spikes along the nerve is about 28.7 m/s, which coincides with the value estimated by the German scientist Hermann von Helmholtz from a nerve-muscle preparation in 1850 (Seyfarth 2006).

To elucidate the mechanism of a resting potential, Bernstein considered the cellular membrane as a semipermeable membrane. As it was known at that time, the potassium ion concentration is higher in the intracellular fluid, while the sodium ion concentration and chlorine ion concentration are higher in the extracellular fluid.

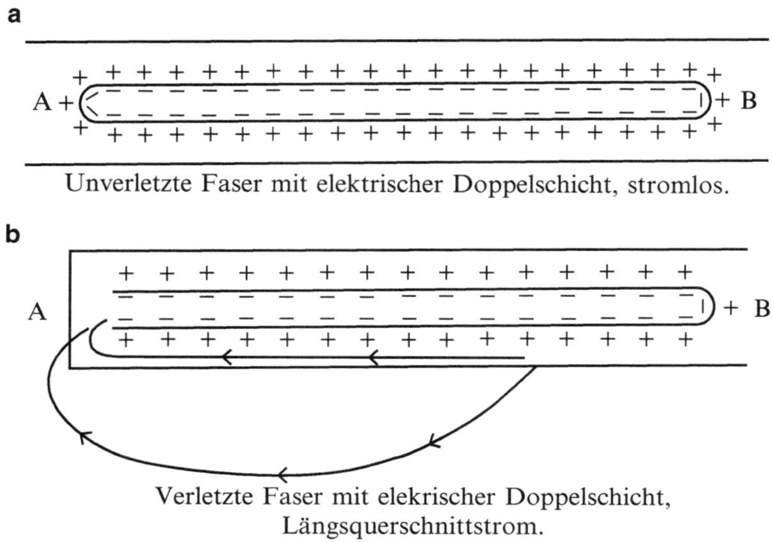

Fig. 3.1 *A diagram elucidating the mechanism of the demarcation current by Bernstein.* The *upper* diagram shows an intact muscle, and there are negative charges and positive charges accumulated on the inside and outside of its membrane; the *lower* diagram shows a muscle with a damaged section at its left end so that its inside space is exposed to the extracellular fluid and which makes a shortcut to let the outside positive charges flow into the inside; this is just the demarcation current (Reproduced with permission from Seyfarth (2006). Fig. 6)

He calculated the equilibrium potential with these data using the Nernst equation;[1] the result was approximately the same as the injure potential recorded experimentally. As the injured section is exposed to the intracellular fluid directly and the intact membrane surface is outside of the neuron, so the injure potential can be considered the same as the resting potential (Fig. 3.1).

Thus, Bernstein proposed his famous membrane theory in 1902: (1) the cellular membrane separates the intracellular space from its outside environment, and the membrane has selective permeability to different ions in intracellular and extracellular fluids; (2) there is a constant potential difference across the membrane without any stimulus (at the resting state), and this difference increases linearly with absolute temperature; (3) when the cell is activated, the membrane permeability increases dramatically for all ions, especially for potassium ions, and thus the

[1] The Nernst equation is deduced by the following ideas: Owing to the difference of the concentrations of a kind of ions inside and outside a semipermeable membrane, ions will flow from the higher concentration side to the lower one. However, as the ions are charged with electricity, the accumulated charges at the lower concentration side will resist such flow until some equilibrium is reached, when the concentration gradient force is balanced by the electricity gradient force. The potential difference across the membrane then is called the Nernst potential of this ion and the net flow is equal to zero at that time.

potential difference decreases and this is the cause of the negative variation (Bernstein 1902).

Surprisingly, although Bernstein did observe the phenomenon that the amplitude of an action potential is much bigger than the resting potential, which contradicts his theory, he just neglected this fact. In addition, his theory can neither explain why the action potential can propagate along the axon. These are the biological problems which British physiologists Alan Lloyd Hodgkin and Andrew Fielding Huxley tried to solve about half a century later.

3.1.2 Hodgkin and Huxley's Ionic Theory

Hodgkin entered the Trinity College in Cambridge in 1932; his major was physiology, but he also took mathematics and physics as elective courses. On the contrary, his student and colleague Huxley's major was physics but took physiology as an elective course in 1935. In 1938, Hodgkin did an experiment: he sank axons of crabs and lobsters into oil and put one electrode at the section of the axon and another electrode on the uninjured surface of the axon, and he found that the amplitude of the action potential was almost twice as the resting potential. This result definitely indicated that the action potential is not just abolishing the selectivity of permeability to different ions to reduce the resting potential. There is a reverse of the polarity of the membrane potential, which could not be explained by Bernstein's theory. At the same year, he was invited by K. S. Cole to work in the latter's laboratory in Woods Hole, USA. He learnt how to measure the membrane conductance change during spiking there. Their experiments showed that the membrane permeability to ions was greatly enhanced, but it was still not clear at that time which ions were responsible for such conductance increase. Huxley joined his group in 1939 and an idea struck Hodgkin that the giant axon of Atlantic squid was an ideal preparation for their study, as the diameter of such axon is almost 1 mm, while for majority of the animals, their axon diameters are only several μm. Thus, they could insert a thin electrode into the axon for 10–30 mm without touching the membrane wall; the nerve fiber was intact except for only a little damage at the end where the electrode was inserted into. The other electrode was placed in the bathing solution, so that the electrical activities could be recorded accurately (Fig. 3.2). Unfortunately, the Second World War broke out and they had to suspend their studies until the end of the war (Hodgkin 1963).

To explain the overshoot during the spiking, considering the fact that the concentration of potassium ions in the intracellular fluid is much higher than the one in extracellular fluid and, on the contrary, the sodium ionic concentration is much higher in the extracellular fluid, Hodgkin and Huxley proposed a hypothesis that the membrane permeability to sodium ions must be much bigger than that of potassium ions during spiking, so that there are more positive sodium ions pouring in than the potassium ions leaving out and the electrical polarity inside the nerve can even be reverted to a positive value near the sodium Nernst potential. The key

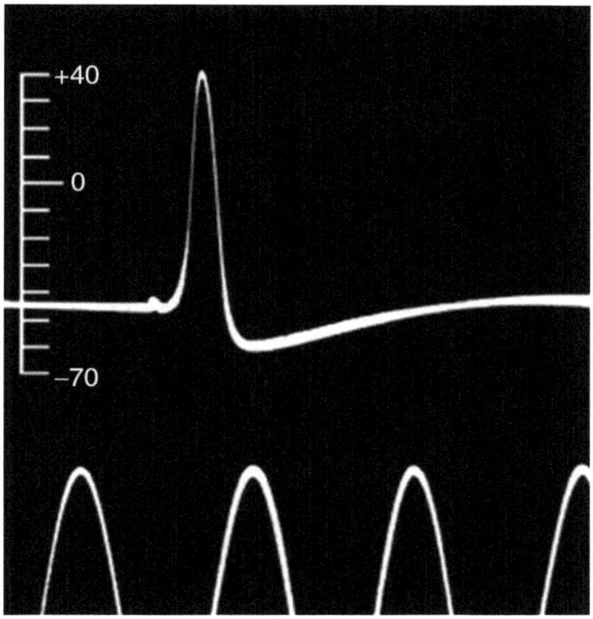

Fig. 3.2 *The waveform of action potentials recorded by Hodgkin and Huxley in the giant axon of a squid in* 1939. The *bottom* semi-sinusoid wave is a timing mark (Reproduced with permission from Kandel et al. (2000). Fig. 2–3)

point here is that the permeability of the nerve cell membrane to sodium and potassium ions is not constant but depends on the voltage difference across the membrane and the time.

They summarized their ideas with an electrical circuit model (Fig. 3.3). As what we know now, the cell membrane is composed of an impermeable lipid bilayer with embedded proteins in it. The lipid bilayer can be equivalent to a capacitor and the embedded proteins can be equivalent to several changeable resistors depending on the voltage difference across the membrane and also on the time. In addition, as there are ionic concentration differences between the inside and outside of the membrane and thus there are the Nernst potentials of such specific ions across the membrane, the difference between the membrane potential and the Nernst potential will push the ion to penetrate the membrane in the opposite direction to the one by the concentration gradient force; the corresponding Nernst potential is equivalent to a battery. Hodgkin and Huxley assumed that there were potassium channel, sodium channel, and some leak channel at the cell membrane, although these channels hadn't been identified at that time and they were thoroughly verified by patch-clamping technique only about a half of a century later. As far as the chlorine channel is concerned, its Nernst potential is very close to the membrane resting potential, so its effect could be neglected at the first approximation.

The Nernst potential could be calculated by the following Nernst equation:

$$E = \frac{RT}{ZF}\log_{10}\left(\frac{[C]_o}{[C]_i}\right)\ \text{mV} \tag{3.1}$$

where $[C]_o$ is the ionic concentration outside the cell, $[C]_i$ is the one inside the cell, Z is the valence of that ion, T is the absolute temperature, F is the Faraday constant,

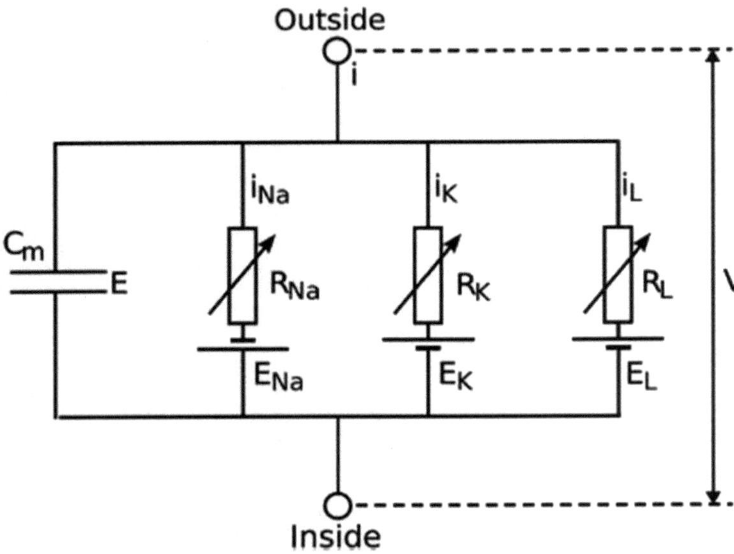

Fig. 3.3 *The circuit model of neuronal membrane*. Where C_m is the membrane capacitance, R_{Na} is the sodium resistance, R_K is the potassium resistance, R_L is the leak resistance, i_{Na} is the sodium current, i_K is the potassium current, i_L is the leak current, E_{Na} is the sodium equilibrium potential, E_K is the potassium equilibrium potential, E_L is the equilibrium potential for leak current, V_m is the voltage across the membrane, *inside* is the intracellular space, and *outside* is the extracellular space (Reproduced with permission from Keating and Novère (2012). Fig. 15.3)

and R is the gas constant. When the temperature is about the room temperature, the RT/F is near 58. However, the Nernst equation is only for the case if there is only one type of ion with different concentrations on both sides of the membrane. In the case of the neuronal membrane where there are several different ions with different concentrations, an extended version of the Nernst equation—the Goldman equation—is needed to be used. The formula of the Goldman equation is as follows:

$$V_m = \left(\frac{RT}{F}\right) \ln\left(\frac{P_K[K]_o + P_{Na}[Na]_o + P_{Cl}[Cl]_i}{P_K[K]_i + P_{Na}[Na]_i + P_{Cl}[Cl]_o}\right) \tag{3.2}$$

where P_i is the permeability of the membrane to the ion i. As these two equations are not related to the topic of information processing directly, to save the space, the deduction of these equations is neglected here. The deduction of these equations can be found in many books on electrophysiology (Bahill 1981; Plonsey and Barr 2007).

Based on the above considerations, the neuronal membrane can be modeled by the circuit shown in Fig. 3.3.

According to the above circuit model and Kirchhoff's law, the equation of the model can be written as follows:

$$C_m \frac{dVm}{dt} = g_L(E_L - V_m) + g_{Na}(E_{Na} - V_m) + g_K(E_K - V_m) + I \tag{3.3}$$

where g_K, g_{Na}, and g_L are the conductance of the potassium channel, sodium channel, and leak channel, respectively; C_m is the membrane capacitance; V_m is the voltage across the membrane; E_{Na} and E_K are the Nernst potential of the sodium ion and potassium ion, respectively; E_L is the reversal potential[2] of the leak current; and I is the injected current from outside.

There is still one problem left; as there are several different ions inside and outside the cell, the membrane voltage calculated by the Goldman equation, which is also the voltage recorded experimentally, is different from the Nernst potentials of these ions, i.e., for every ion, the force which drives the ion in one direction owing to the concentration gradient cannot be balanced by the electric field force, which drives the ion in the opposite direction; thus, the net flow of ions cannot be zero, and the ions must flow across the membrane in some direction. In case the situation is just what we described above, then how can the resting potential be maintained for a long time? In other words, if the membrane is not absolutely impermeable, then how can the concentration difference of an ion across the membrane be kept? The key point here is that besides the ion channels, there is another specific protein—the so-called ionic pump—which can move the ion from one side of the membrane to the other side against its concentration gradient. The cost is to consume some energy from biochemical reactions. However, this pump is neglected in the above model with an implied hypothesis that the ionic concentration difference can be maintained unchanged.

The key point in Hodgkin and Huxley's theory is that they assumed there are potassium channels and sodium channels in the neuronal membrane; both of the channels have changeable conductance, depending on the voltage across the membrane and time, although these channels hadn't be found experimentally at that time. Now the further problem for modeling is to make sure how the potassium conductance and sodium conductance change with the voltage across the membrane and time quantitatively. To measure conductance, the easiest way is to measure their corresponding currents, as the conductance can be calculated by Ohm's law, if the corresponding current and voltage are given.

To abolish the capacitance component in the current, the membrane potential should be clamped at a fixed value, thus owing to $dV_m/dt = 0$, $C_m\, dV_m/dt = 0$, i.e., there will be no current to charge or discharge the capacitor. It is easy to clamp the voltage with modern technology, i.e., to measure the difference between the practical membrane potential and the one demanded and then control the injecting current according to the difference to make the difference approach to zero. To guarantee that there is only a radial current across the membrane and no current flowing along the axis of the axon, it is demanded that all the potentials along the axon are the same and spatial clamping is also needed. As the diameter of the giant axon can be bigger than 300 μm, it is convenient to insert a total silver electrode into the axon and a silver cylinder electrode surrounding the axon; in this way, the membrane potential does

[2] The reversal potential is the potential value at which the net ionic current through the membrane is zero, if there is only one kind of ion that exists inside and outside the membrane.

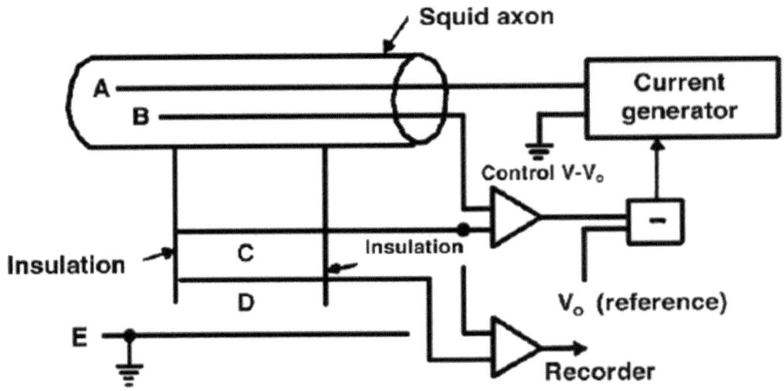

Fig. 3.4 *A schematic diagram of an experimental set for voltage clamping* (Reproduced with permission from Plonsey and Barr (2007). Fig. 5.7)

not depend on its location. Thus, the current is only a temporal function independent of the space. The whole experiment set is shown in Fig. 3.4.

After the above consideration, there are only the ionic currents across the membrane recorded. Owing to the fact that organic anions cannot pass the membrane, they have no contributions to the recorded ionic currents. As for the three main ions left, the Nernst potential of chlorine ions is close to the resting potential; thus, chlorine ions are close to its equilibrium state around the resting potential and there is little chlorine current flowing, while at the peak of the action potential, the permeability of the membrane to chlorine ions is very low, so that there is also little chlorine ionic current and it can be neglected in the first approximation. Therefore, only two ions should be considered further: potassium ions and sodium ions. Now the problem is how to separate these two ionic currents.

For a voltage-clamped axon, a three-phase membrane current curve I_M is observed, when a step depolarization stimulus is applied, as shown in Fig. 3.5. At the very beginning, there is a current impulse sustaining a few μs to discharge the membrane capacitor and then there is current entering the cell. If there is no voltage clamping, this current would depolarize the membrane further and change the potential difference across the membrane. The magnitude and duration of this inward current depend on both the initial membrane potential and the amplitude of the step depolarization. After 1–2 ms, the direction of this current becomes outward and it would sustain, if the clamping voltage keeps constant. The magnitude and duration of the outward current also depend on the initial membrane potential and the amplitude of the step depolarization. However, the outward current would decay quickly when the membrane repolarizes.

To be sure if the inward flowing ions are sodium ions only, 90 % of the sodium ions in the bathing seawater are replaced by large impermeant choline ions; in such a case, the bathing solution keeps its isotonicity unchanged. However, the sodium ion concentration inside and outside the membrane is almost the same; therefore,

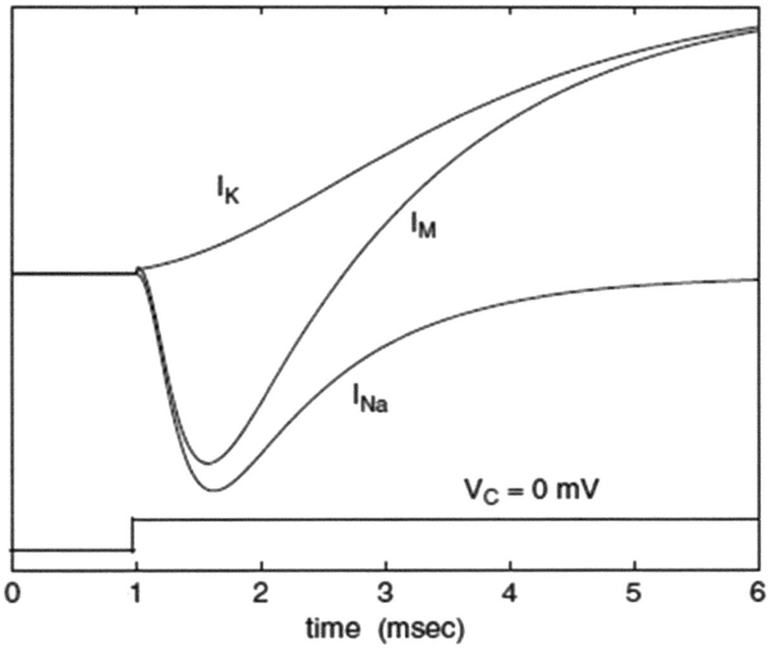

Fig. 3.5 *Analysis of ionic membrane current by the voltage-clamping technique*. The recorded ionic membrane current when the bathing solution is the seawater (curve I_M). The recorded ionic membrane current when 90 % of the sodium ions in bathing seawater were replaced by choline ions (curve I_K). The curve I_{Na} is the difference subtracted curve I_K from the curve I_M; it represents the net sodium ionic current, when the membrane potential rises from −60 to 0 mV before and after voltage clamping (the bottom trace) (Reproduced with permission from Ermentrout and Terman (2010). Fig. 1.8)

there will be no net sodium ionic flow, no matter if the sodium channels are open or closed, and the current recorded under such condition should almost be pure potassium ionic flow. Repeat the experiment under such condition; the inward component of the current flow is abolished, and even at the very beginning of the step depolarization, only an outward current appears, as shown as the curve I_K in Fig. 3.5, which is the pure potassium ionic flow. This hints that the initial inward current is due to sodium ions, while the outward current is due to potassium ions. If it is assumed that the sodium channel is independent of the potassium channel and vice versa, i.e., if there is no interaction between potassium ionic flow and sodium ionic flow, then a net sodium ionic flow could be got from subtracting the curve I_K from the curve I_M, as shown as the curve I_{Na} in Fig. 3.5. This independency hypothesis was verified by a series of experiments later, especially those experiments with ion channel blockers. For example, the chemical agent tetrodotoxin (TTX) can almost totally block the sodium channel but has no influence on the potassium channel; on the contrary, tetraethylammonium (TEA) could block the potassium channel but has no effect on sodium channels.

These experiments showed that the potassium conductance increases gradually and reaches an inflection point and then approaches the steady-state value when the membrane is voltage-clamped with depolarization. However, if the membrane is clamped with repolarization or hyperpolarization, the potassium conductance decreases to its steady state without any inflection point, as shown as the lower trace in Fig. 3.6. The rising rate and the steady-state value will be higher with the amplitude of the step depolarization. The above curves can be fitted with the following function:

$$g(t) = (a + be^{-ct})^d \tag{3.4}$$

The function $g(t)$ increases monotonically when b is a negative constant, while it decreases when b is positive.

With curve fitting, the parameters a, b, c, and d are estimated as follows: $d=4$, $a=1$, $b=-1$, and $c=-1$ to fit curve I_K in Fig. 3.6. However, the form of the function $g(t)$ depends on the initial state. Thus a more general form would be better. As it is easy to know that function $a + be^{-ct}$ can be considered as a solution of a first-order differential equation, the dynamics of the potassium channel can be described by the following equations:

$$g_K = \overline{g}_K n^4 \tag{3.5}$$

$$\frac{dn}{dt} = \frac{1}{\tau_n}(n_{ss} - n) \tag{3.6}$$

where $\overline{g}_k$ is a constant representing the maximal value of potassium conductance; the values n_{ss} andτ_n depend on the membrane potential, and they represent the normalized values of steady-state value and time constant of the potassium conductance, respectively. Therefore, the potassium conductance is a variable conductance, depending on both the membrane potential and time. The differential equation itself means that the potassium conductance is time dependent, while its membrane potential dependence is implied in the parameters n_{ss} andτ_n, as shown in Fig. 3.7b. The quantitative expression of such relationship can be got with curve fitting of these experiment data.

Hodgkin and Huxley gave the variable n the following physical explanation: They assumed that the potassium channel could be either closed to the passage of all ions or open to the passage of potassium ions. The channel is controlled by four independent gating particles, each of which can be in either an open or closed position, and only when all four gating particles are at the open position that potassium ions can pass through the channel. The gating variable n is the probability of a single potassium gating particle being in the open state and n^4 is thus the probability of the potassium channel being open.

Now let us turn to the relationship between sodium conductance and membrane potential and time. As what is shown in Fig. 3.6, when a depolarization voltage clamping is given, the sodium conductance rises quickly to a peak through an inflection point and then decreases to a low value, even while the clamped voltage

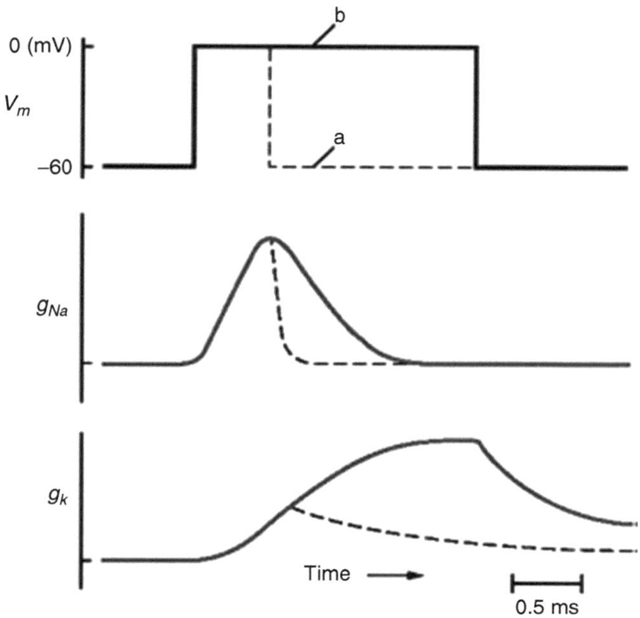

Fig. 3.6 *Time course of sodium conductance g_{Na} (middle curve) and potassium conductance g_K (bottom curve) when a depolarization of 60-mV square wave stimulus is given (top curve). Solid lines* for the stimulus *b* and *dashed lines* for the stimulus *a*. Note that for stimulus *b*, both sodium conductance and potassium conductance increases when the stimulus is on; both channels are activated. After this stimulus is off, potassium conductance decreases to its resting state, i.e., the potassium channel is deactivated. However, for sodium conductance, it decreases to its resting state even before the stimulus is off, i.e., there is an inactivating factor operating for sodium channel (Reproduced with permission from Kandel et al. (2000). Fig. 9.7)

remains in a sustained depolarizing step; this is the main difference between sodium conductance and potassium conductance and is termed inactivation instead of deactivation; while a hyperpolarization voltage clamping is given, the sodium conductance decreases quickly without any inflection point. Similar to the considerations in the case of potassium conductance, the behavior of sodium conductance can be described as follows:

$$g_{Na} = \overline{g}_{Na} m^3 h \tag{3.7}$$

$$\frac{dm}{dt} = \frac{1}{\tau_m}(m_{ss} - m) \tag{3.8}$$

$$\frac{dh}{dt} = \frac{1}{\tau_h}(h_{ss} - h) \tag{3.9}$$

where the parameter $\overline{g}_{Na}$ is a constant, representing the maximal value of sodium conductance, while the dependence of the parameters m_{ss}, τ_m, h_{ss}, andτ_h on

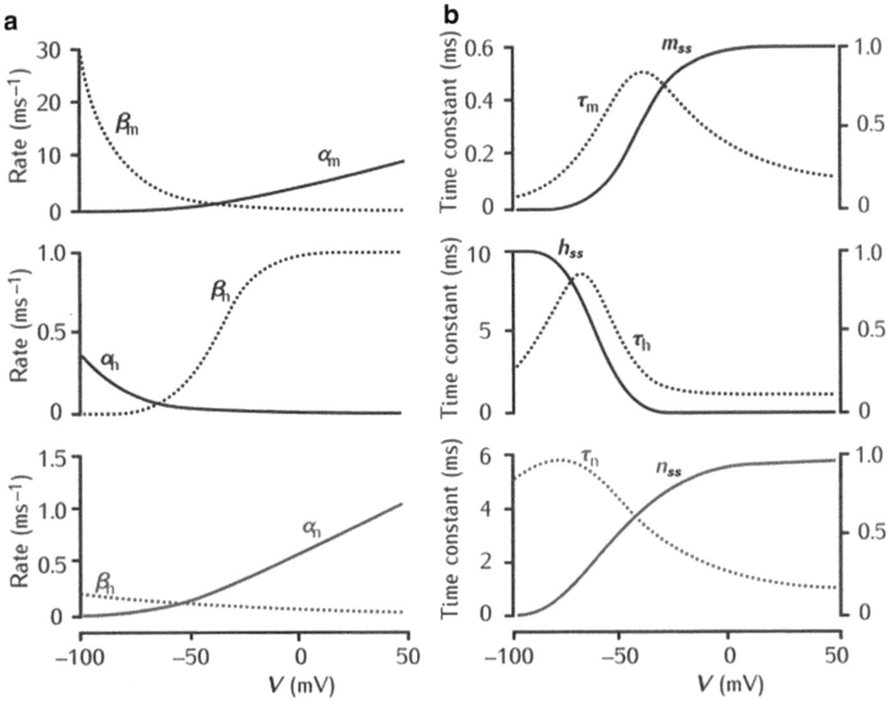

Fig. 3.7 *The relationship between the parameters in Hodgkin–Huxley model and the membrane potentials.* (**a**) Graphs of α_m, α_h, α_n (*solid lines*) and β_m, β_h, β_n (*dashed lines*). (**b**) The equivalent graphs for m_{ss}, h_{ss}, n_{ss} (*solid lines*) and τ_m, τ_h, τ_n (*dashed lines*) (Reproduced with permission from Sterratt et al. (2011). Fig. 3.10)

membrane potential is shown in Fig. 3.7b. The physical meaning of the variables m and h can also be given similar to the gating variable n for the potassium channel. However, for the sodium channel, m is for activation, while h is for inactivation, as implied in Eqs. (3.8) and (3.9).

The genesis of the action potential can be realized from Fig. 3.7 and Eqs. (3.3), (3.5), (3.6), (3.7), (3.8), and (3.9). Let us assume that a current is injected to raise the membrane potential V, and then m_{ss} will rise with V, and m will chase m_{ss} with a slight delay depending on τ_m. A rise of m leads to a rise of g_{Na} and a rise of I_{Na} too. The additional inward positive current will raise the voltage further, so this is a positive feedback. However, h_{ss} will drop with V at the same time with an even longer delay depending on τ_h, a negative feedback. And at the same time, n_{ss} will rise with V with a much larger time constant τ_n, so that the rise of n turns on the potassium channel more wildly too and pulls the potential down toward the negative value of the potassium Nernst potential, another negative feedback. It is the positive feedback which decides the rising phase of the action potential and it is the negative feedbacks which decide the restoring phase of the spike.

Modifying the forms of the first-order differential equations a little bit and determining the parameter values by curve fitting, the whole Hodgkin–Huxley equations could be described as follows:

$$C_m \frac{dV_m}{dt} + \overline{g_k} n^4 (V_m - E_k) + \overline{g_{Na}} m^3 h (V_m - E_{Na}) + g_L (V_m - E_L) = I \tag{3.10}$$

or

$$C_m \frac{dv_m}{dt} + \overline{g}_k n^4 (v_m - v_k) + \overline{g}_{Na} m^3 h (v_m - v_{Na}) + g_L (v_m - v_L) = I \tag{3.10'}$$

$$\frac{dn}{dt} = \alpha_n (1 - n) - \beta_n \tag{3.11}$$

$$\frac{dm}{dt} = \alpha_m (1 - m) - \beta_m m \tag{3.12}$$

$$\frac{dh}{dt} = \alpha_h (1 - h) - \beta_h h \tag{3.13}$$

$$\alpha_n = \frac{0.01(10 - v_m)}{e^{\frac{10 - v_m}{10}} - 1} \tag{3.14}$$

$$\beta_n = 0.125 e^{\frac{-v_m}{80}} \tag{3.15}$$

$$\alpha_m = \frac{0.1(25 - v_m)}{e^{\frac{25 - v_m}{10}} - 1} \tag{3.16}$$

$$\beta_m = 4 e^{\frac{-v_m}{18}} \tag{3.17}$$

$$\alpha_h = 0.07 e^{-\frac{v_m}{20}} \tag{3.18}$$

$$\beta_h = \frac{1}{e^{\frac{30 - v_m}{10}} + 1} \tag{3.19}$$

where $v_m = V_m - V_r, v_K = E_K - V_r, v_{Na} = E_{Na} - V_r, v_L = E_L - V_r, V_r$ is the resting potential; g_L is the leak conductance, which is a constant; and v_L is the potential when the leak current is zero. The unit of the potential is *mV*, current density $\mu A/cm^2$, specific conductance *m•mho*/cm^2, capacitance $\mu F/cm^2$, and time *ms*.

Equation (3.11) is a rewritten form from Eq. (3.6), in which the coefficients $\alpha_n = -(1/\tau_n n_{ss})$ and $\beta_n = 1/\tau_n + 1/\tau_n n_{ss}$ have clearer meaning. They are rate coefficients of the processes when the gating particles transfer their states from closed to open and from open to closed, respectively. The similar formulae and words can be said to the parameters α_m, β_m, α_h, and β_h.

Although Eqs. (3.11), (3.12), (3.13), (3.14), (3.15), (3.16), (3.17), (3.18), and (3.19) are deduced from curve fitting with the data obtained from voltage-clamping experiments, adopting the form of differential equations (3.11), (3.12), and (3.13) makes the model not depend on the initial states and the amplitude of the step stimulus again. Now the value of V_m may not be a constant again as in the voltage-clamping experiments; it is a function of both time and space. This is a key step to extend the model to a more general form. Now the problem is if Eqs. (3.10), (3.11), (3.12), (3.13), (3.14), (3.15), (3.16), (3.17), (3.18), and (3.19) can predict the generation of action potentials and its propagation, which have not been assumed in the hypothesis of the model. In other words, the problem is whether its solution can be similar to the real waveform and will it be able to propagate with a reasonable speed. In one word, the problem is that whether the model can not only describe the biological phenomena quantitatively from which the model is built but also predict new phenomena which have not been considered in the assumptions for building the model. If the answer is "yes," then we have enough reason to convince that the model has grasped the essence of the law governing the electrical behavior of the membrane with voltage-sensitive channels. This is similar to the Maxwell equation in electrodynamics in some way, which combines Coulomb's law, Gauss's law, Faraday's law, and Ampère's law together to predict the electromagnetic wave and its propagating speed, which are not implied in any of the four laws individually.

Now let us consider the problem how to solve the Hodgkin–Huxley equation and study the properties of its solution. Assume that the axon is a cylinder and the diameter of its section is far smaller than its length; thus it can be assumed that the potential and current are independent of radial directions, so that they are functions of time t and only one spatial variable x (the location along the axon axis). With these assumptions, the following cable equation can be used to describe the axon:

$$i = \frac{1}{r_1 + r_2} \frac{\partial^2 v_m}{\partial x^2} \tag{3.20}$$

where i is the membrane current per unit length of the axon, r_1 and r_2 are the resistances per unit of the length outside and inside the membrane, and x is the distance along the axis of the axon. If the axon is assumed to be bathed in a large quantity of seawater, then r_1 is much less than r_2 and can be neglected. Hence,

$$i = \frac{1}{r_2} \frac{\partial^2 v_m}{\partial x^2} \tag{3.21}$$

and this can be rewritten as

$$I = \frac{a}{2R_2} \frac{\partial^2 v_m}{\partial x^2} \tag{3.22}$$

where I is the current density, a is the radius of the axon, and R_2 is the specific resistance[3] of the axon plasma. Substitute Eq. (3.22) into Eq. (3.10′):

[3] A specific resistance is the resistance of a piece of material with a constant cross section divided by the volume of this resistor.

$$\frac{a}{2R_2}\frac{\partial^2 v_m}{\partial x^2} = C_m\frac{\partial v_m}{\partial t} + \overline{g}_K n^4(v_m - v_K) + \overline{g}_{Na} m^3 h(v_m - v_{Na}) + \overline{g}_L(v_m - v_L) \tag{3.23}$$

The above equation is a partial differential equation; however, as we know, if the model is correct, then its solution should have the property of an action potential, i.e., it should have a traveling wave solution taking the form as follows:

$$v_m(x,t) = v_m(x - \theta t) \tag{3.24}$$

where θ is the propagating speed; hence,

$$\frac{\partial^2 v_m}{\partial x^2} = \frac{1}{\theta^2}\frac{\partial^2 v_m}{\partial t^2} \tag{3.25}$$

Thus,

$$\frac{a}{2R_2\theta^2}\frac{d^2 v_m}{dt^2} = C_m\frac{dv_m}{dt} + \overline{g}_K n^4(v_m - v_K) + \overline{g}_{Na} m^3 h(v_m - v_{Na}) + \overline{g}_L(v_m - v_L) \tag{3.26}$$

Now the above equation is an ordinary differential equation, although the parameter θ is unknown; an estimation can be assumed and the above equation can be solved with the Euler method, Runge–Kutta method, or other algorithms to get a numerical solution. Digital computers were rare at that time; Hodgkin and Huxley used a hand-operated calculator to solve the equations for 3 weeks, which would take only a few milliseconds to do the same job with a microcomputer now! When the estimation value θ is too big or too small, the corresponding solutions will diverge to a positive or negative infinity, only when the value of θ is within a proper range; the function value v_m has the shape of an action potential and then restores to its resting value. The upper trace in Fig. 3.8 is a solution of the Hodgkin–Huxley equation, when the temperature parameter value is 6 °C and the initial depolarization is 15 mV, while the lower trace is an action potential recorded from a squid giant axon when the temperature is 9.1 °C. The coincidence between the curves suggests that the model is reasonable and the estimation value of θ is near the practical spike propagation speed. The simulation also shows the phenomena such as refractory period, threshold, etc., which are not assumed in the assumptions to deduce the model, and their underlying mechanism can be explained by the model (Sterratt et al. 2011). Therefore, people call the Hodgkin–Huxley equation the Maxwell equation in neurophysiology!

To elucidate the contribution of sodium ions in the extracellular fluid to the action potential, Huxley calculated the solutions of the Hodgkin–Huxley equation with different extracellular sodium ionic concentration and found that the amplitude of the action potential decreases with the decrease of extracellular sodium

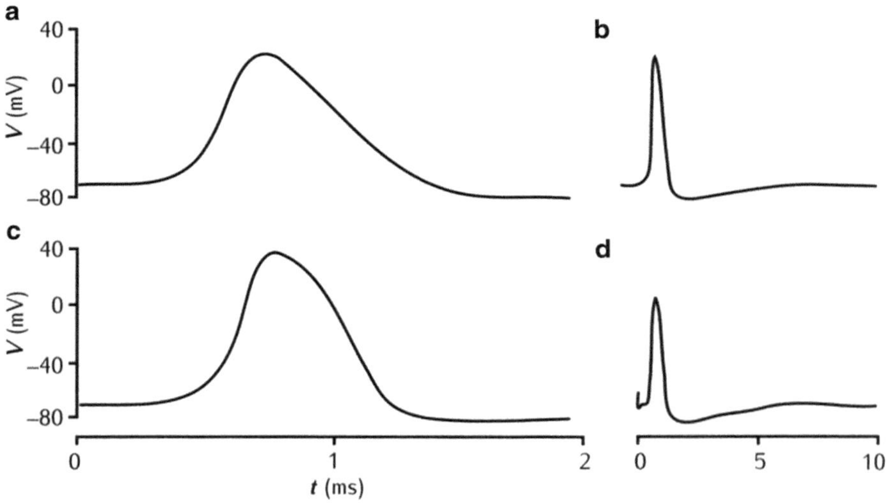

Fig. 3.8 *Comparison of a solution of H–H equation with the experiment result.* The *upper trace* is a solution of Hodgkin–Huxley equation; while the *lower trace* is an action potential recorded from a squid giant axon. The *right* graphs are the same as the corresponding *left* graphs with a slower time scale (Reproduced with permission from Sterratt et al. (2011). Fig. 3.14)

ionic concentration, which was confirmed by Hodgkin with biological experiments and against Curtis and Cole's report that the amplitude of the action potential was independent of the extracellular sodium ionic concentration.

In summary, the Hodgkin–Huxley model can not only fit those experiment data well, with which the model is established, but also predict phenomena which are not involved in their assumptions for developing their model. Thus the model has perfectly solved the problem of the mechanism of how action potentials generate and propagate; it becomes the founding and classical work to use mathematical modeling and simulation to solve neurophysiologic problems and promotes the birth of computational neuroscience—one of the frontiers in the modern sciences. Owing to their great contributions to understanding the neuronal mechanism, they won the 1963 Nobel Prize in Physiology or Medicine.

In addition, the success of their model hinted that their hypothesis about the potassium channel and sodium channel is reasonable, long before these channels were verified by experiments directly, which has stimulated many scientists to devote themselves to discovering these channels, and these channels were really found using patch-clamping techniques and in addition even more ion channels were found on the neuronal membrane. In the classical H–H model, only persistent K^+ current and transient Na^+ current are considered, which are responsible for the generation and propagation of a single action potential. Other slowly changing ionic currents such as Ca^{2+} current are neglected, which will modulate prolonged spiking processes and generate a variety of different firing rhythm patterns. We will discuss this problem in the following sections. Thus, the

Hodgkin–Huxley model can be generalized; in the generalized models, other ion channels are considered, and all of such kind of models are called conductance-based models (Skinner 2006).

Although the H–H model has made great success, it should be kept in mind that some assumptions which the model bases on are only approximately correct, such as the following assumptions: Each channel type is permeable to only one type of ion; the independence principle for different ionic currents, the linear instantaneous *I-V* characteristic, the independence of gating particles, and the gating current are not considered in the model (Sterratt et al. 2011). Therefore, the H–H model is an approximate model of squid giant axon, even if the model has told us so much about the general principle of the generation and propagation of action potentials.

Finally, a debating problem should be mentioned here. For the author, the Hodgkin–Huxley model is a model of the neuronal membrane with voltage-sensitive potassium channel and sodium channel on it, especially in squid giant axon preparation, and nothing more. It is not a whole neuron model but a part of it, although a typical example of such parts. Some modification to it has to be made to model different parts of the neuron. The H–H model considers such membrane as a system composed of a membrane capacitor, potassium channel, sodium channel, leak channel, etc., as its elements. It elucidates the emergent properties of such membrane—the generation and propagation of action potentials based on these elements. This is the job it does. However, if the phenomena above the neuron level are considered, then under normal conditions (say, the firing rate is not too high and there are not a lot neurons firing synchronously, etc.), if the all-or-none law of the action potential can be kept, the amplitude and shape of the action potential play a little role in further information processing, and timing is the only important factor. The details of the spike waveform are not important for the neural information processing at the higher level at all. Some scientists even argue that the activity of an individual neuron is not important for higher-level neural information processing owing to its stochastic essence; the key factor is the activities of a population of neurons! Therefore, for the authors, the Hodgkin–Huxley model is important for the problems at the neuronal level and subneuronal level, considering that the H–H model might not be necessary to solve problems above the neuronal level or at least above the neural circuit level, as big efforts would be spent to calculate the waveform of the action potential repeatedly which plays little essential role in the higher function, as this will be shown in the next section, where a model of a central pattern generator in lamprey is considered. The results show that the simulation of the rhythmic activity of the central pattern generator with neuron models based on different ion channels as elements is no better than much simplified neuron models.

As what we mentioned above, the Hodgkin–Huxley model is essentially a model of neuronal membrane with voltage-sensitive potassium and sodium channels on it. In addition, the membrane is considered as a single iso-potential electrical compartment. However, a neuron is far more complicated than such a compartment; it has a soma, an axon, and a complex dendrite tree and the voltages at these different parts are different. Only the parts with approximately identical voltage could be considered as a compartment; therefore, a neuron is a multi-compartmental system. In the next section, a multi-compartmental neuron model with more ion channels will be considered as an example.

3.2 Multi-compartmental Model

As what we mentioned above, the Hodgkin–Huxley model is only a model of neuronal membrane with voltage-sensitive potassium channel and sodium channel. Actually, it cannot be called as a neuron model. A neuron is composed of many different parts; the properties of these parts may be different from each other. For example, for some membrane at the soma or dendrites, there are transmitter-sensitive channels. And the membrane at different locations of a neuron may have a different voltage. Thus only a small patch of membrane which has the same ion channels and has the identical voltage could be considered as a compartment or as a basic unit of the neuron. A neuron could thus be considered as a multi-compartmental system and such model is often called as the biophysical model of the neuron. In this section, we will introduce a model developed by Ekeberg et al. (1991) as an example of such type of neuron models. Here, the important matter is not their concrete results but the ideas and approaches they used. It exemplified how to use Hodgkin–Huxley-like models as compartments to construct a biophysical neuron model. Now there are a variety of such models; however, their approaches are similar.

To establish a central pattern generator model of lamprey, Ekeberg et al. (1991) developed a neuronal biophysical model first. In the model, they simplified the morphology of the neuron, and they assumed that there was only one dendrite dividing into three series-coupled different segments; thus, their model is a model with four compartments in series, as shown in Fig. 3.9. Of course, for another neuron, according to its morphology, there may be a more complicated dendrite tree, say, there may be multiple branches; thus, one compartment may be connected to many other compartments with different resistors. Suppose that every compartment is iso-potential. The main difference between the soma compartment and dendritic compartments is the difference between the channels at their membranes. All the channels at the axon hillock and axon are voltage-sensitive channels; in a first approximation, it could also be assumed for soma, while the ion channels at the dendritic compartments are transmitter-sensitive channels. For simplicity, the effect of the propagation of the spike along the axon from the initial segment to the terminals is equivalent to an adjustable delay parameter. The neighboring compartments are coupled by a resistor representing the cytoplasmic axial resistance between the centers of the two connected compartments. The membrane potential in every compartment is denoted by E without obvious subscript indicating which compartment is concerned.

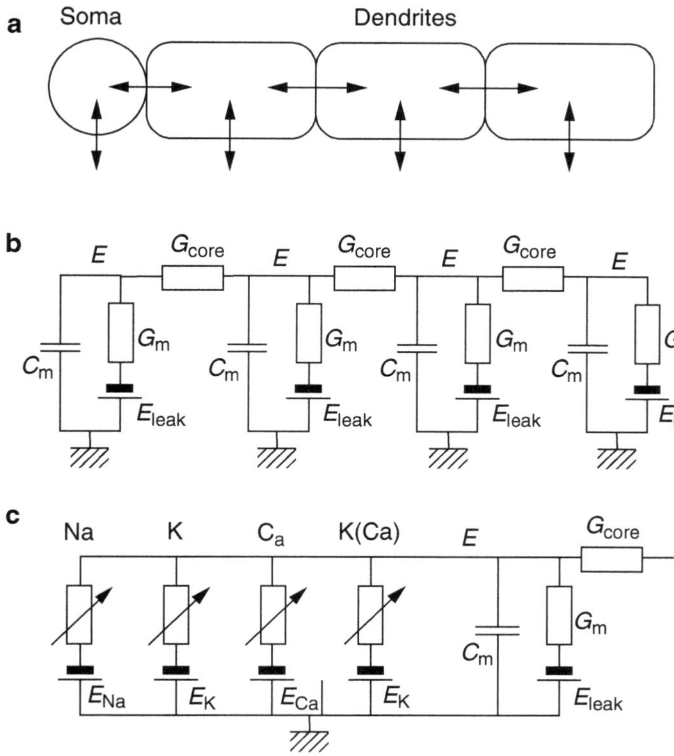

Fig. 3.9 *A biophysical model of neuron developed by Ekeberg et al.* (**a**) There are only one soma compartment and three dendritic compartments in series in this model. (**b**) The electrical circuit model of the neuron when only passive membrane properties are considered. (**c**) The electrical circuit model of the soma compartment when the ion channels in it are considered (Reproduced with permission from Ekeberg et al. (1991). Fig. 1)

Based on the electrical circuit model shown in Fig. 3.9 and Kirchhoff's law, a differential equation can be written for every compartment as follows:

$$\frac{dE}{dt} = \frac{(E_{leak} - E)G_m + \sum \left(E_{comp} - E\right)G_{core} + I_{channels}}{C_m} \tag{3.27}$$

where E_{leak} is the reversal potential of the leak current (mainly, the chlorine current), which is close to the resting potential, G_m is the leak conductance, E_{comp} is the membrane potential of its neighboring compartments, G_{core} is the coupling conductance between neighboring compartments, and C_m is the membrane capacitance. The summation symbol Σ means to sum for all the neighboring compartments. $I_{channels}$ is the summation of all the ionic currents through the corresponding ion channels. There may be different ion channels for different compartments; this will be discussed in detail later when the soma compartment and dendritic compartment are discussed. In case there is some injected current from the outside, then a term of injected current should be added at the numerator on the right side.

For the soma compartment, as what we mentioned above, the ion channels in it are sensitive to voltage, mainly, the potassium channel and sodium channel. Ekeberg et al. used a Hodgkin–Huxley-like model to model this compartment with only a little modification of the parameters to fit the experimental data they got for their preparations. Thus, the equations for the sodium current and potassium current can be listed as follows:

$$I_{Na} = (E_{Na} - E_{soma})G_{Na}m^3h \tag{3.28}$$

$$\frac{dm}{dt} = \alpha_m(1-m) - \beta_m m \tag{3.29}$$

$$\alpha_m = \frac{A(E_{soma} - B)}{1 - \exp((B - E_{soma})/C)} \tag{3.30}$$

$$\beta_m = \frac{A(B - E_{soma})}{1 - \exp((E_{soma} - B)/C)} \tag{3.31}$$

$$\frac{dh}{dt} = \alpha_h(1-h) - \beta_h h \tag{3.32}$$

$$\alpha_h = \frac{A(B - E_{soma})}{1 - \exp((E_{soma} - B)/C)} \tag{3.33}$$

$$\beta_h = \frac{A}{1 + \exp((B - E_{soma})/C)} \tag{3.34}$$

$$I_K = (E_K - E_{soma})G_K n^4 \tag{3.35}$$

$$\frac{dn}{dt} = \alpha_n(1-n) - \beta_n n \tag{3.36}$$

$$\alpha_n = \frac{A(E_{soma} - B)}{1 - \exp((B - E_{soma})/C)} \tag{3.37}$$

$$\beta_n = \frac{A(B - E_{soma})}{1 - \exp((E_{soma} - B)/C)} \tag{3.38}$$

where E_{soma} is the membrane potential of the soma. A, B, and C are constants, which may take different values in different equations. No subscripts denoted here is only for convenience; however, we should realize that they are different parameters with omitted subscripts.

As what we mentioned in the last section, the above equations can simulate the generation and propagation of the action potential and its realistic waveform. However, such model can only generate a spike train with equal intervals. To make the firing rate coincide with the one practically observed, a much slower calcium ion channel and calcium-dependent potassium channel, which can modulate the rhythm of the spike train, must be considered.

Similar to the way to establish a sodium ionic current and potassium ionic current equation, an equation for a calcium ionic current is listed as follows:

$$I_{Ca} = (E_{Ca} - E_{soma})G_{Ca}q^5 \tag{3.39}$$

$$\frac{dq}{dt} = \alpha_q(1-q) - \beta_q q \tag{3.40}$$

$$\alpha_q = \frac{A(E_{soma} - B)}{1 - \exp((B - E_{soma})/C)} \tag{3.41}$$

$$\beta_q = \frac{A(B - E_{soma})}{1 - \exp((E_{soma} - B)/C)} \tag{3.42}$$

The intracellular calcium ions can influence the membrane potential by activation of calcium-dependent ion channels. Here a key factor is the intracellular calcium ionic concentration $[Ca_{AP}]$; an equation for calcium ionic concentration can be written as follows based on the law of matter conservation:

$$\frac{d[Ca_{AP}]}{dt} = (E_{Ca} - E_{soma})\rho_{AP}q^5 - \delta_{AP}[Ca_{AP}] \tag{3.43}$$

where ρ_{AP} and δ_{AP} is the rate of the calcium ions entering the cell and flowing out the cell or uptaking by the organelle, respectively.

In Ekeberg's model, another calcium-dependent potassium channel is considered; it is assumed that its conductance is proportional to the calcium ionic concentration and thus

$$I_{K[Ca]} = (E_K - E_{soma})G_{K[Ca]}[Ca_{AP}] \tag{3.44}$$

This current mainly affects the after-hyperpolarization phase and thus modulates the firing rate. After considering all the above factors, the electrical circuit model of the soma compartment can be shown in Fig. 3.9c. The simulated action potential waveform and the spike train are shown in Fig. 3.10.

Now let us turn to the dendritic compartment; as what we mentioned above, there is a large quantity of transmitter-sensitive ion channels at the postsynaptic dendritic membrane, and these channels open only for a certain period when there is a spike reaching the presynaptic membrane and there are specific neurotransmitters releasing into the synaptic cleft. Ekeberg et al. combine the synaptic delay and the spike conducting time along the presynaptic neuronal axon as one delay parameter. Thus the synaptic current could be written as follows:

$$I_{syn} = \left(E_{syn} - E\right)G_{syn}s \tag{3.45}$$

where I_{syn} is the synaptic current, E_{syn} is the synaptic reversal potential, G_{syn} is the maximal conductance of the channel, and s is a square wave function, the value of which can only be 0 or 1; only a spike is fired at the presynaptic neuron and, after a

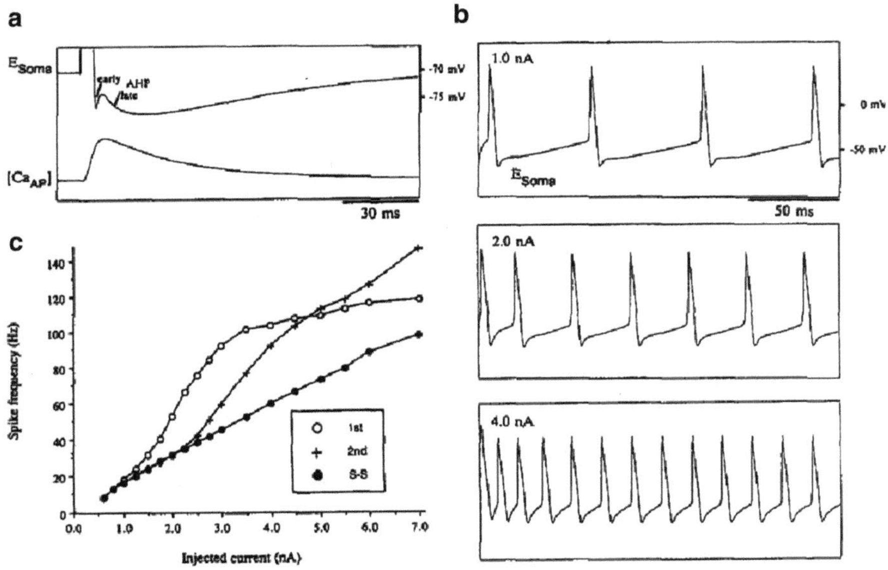

Fig. 3.10 *Simulation results of Ekeberg's neuron model*. (**a**) The simulated waveform of action potential when calcium channel and calcium-dependent potassium channel are considered (*upper trace*); please note the after-hyperpolarization phase in the waveform and the time course of calcium ionic concentration (*lower trace*). (**b**) The simulated spike train, when tonic currents with different intensities are injected; note that there is also adaptation in the train. (**c**) Simulated rate–current injected curves. The symbol *1st* means that the rate is calculated as the inverse of the interval between the first two impulses; *2nd* means the inverse of the interval between the second spike and the third spike; *s-s* means the inverse of the interval between two neighboring spikes when the firing rate reaches its steady state (Reproduced with permission from Ekeberg et al. (1991). Fig. 3)

delay mentioned above, the value can be 1, which can only be kept for a fixed period (the open time of the channel), and then returns to 0 again.

Finally, a specific kind of channel at the postsynaptic membrane is considered, which can be opened only when there is a specific neurotransmitter released in the synaptic cleft and at the same time its conductance depends on the membrane potential either. The NMDA (*N*-methyl-D-aspartate) channel is such a channel, the equations of which are as follows:

$$I_{NMDA} = (E_{NMDA} - E)G_{NMDA}ps \tag{3.46}$$

$$\frac{dp}{dt} = \alpha_p(1-p) - \beta_p p \tag{3.47}$$

$$\alpha_p = Ae^{E/C} \tag{3.48}$$

$$\beta_p = Ae^{E/C} \tag{3.49}$$

Besides, the NMDA channel is also permeable to calcium ions and activates calcium-dependent potassium channels. Thus, two additional equations should also be considered as follows:

$$\frac{d[Ca_{NMDA}]}{dt} = (E_{NMDA} - E)\rho_{NMDA}ps - \delta_{NMDA}[Ca_{NMDA}] \tag{3.50}$$

$$I_{K(Ca)} = (E_K - E)G_{K(Ca)}([Ca_{AP}] + [Ca_{NMDA}]) \tag{3.51}$$

Even after significant simplification of the morphology of the neuron mentioned above has been made, the nonlinear differential equations describing the neuron are very complicated. As what we described above, the model can accurately simulate the behavior of the neuron; thus, it is important for understanding the underlying mechanism of such neuron itself. Ekeberg et al. used the model as a unit to construct a neural network model of the central pattern generator of the lamprey based on the real connections in this biological neural circuit. This model can simulate the interrupted oscillation patterns of the neurons in this biological neural circuit accurately. Thus, they established a model across three levels—from the subcellular level to the neural circuit level. However, later they used a simplified neuron model without considering any ion channel mechanism as a unit to construct a network with the same connections; the results were similar, except for the waveform of the spike. Nevertheless, owing to the all-or-none law of the action potential, the waveform of the spike plays little role in information processing above the cellular level, say, at the circuit level. Thus, for simulation of the biophysical model of the central pattern generator, most of the computing time is spent on calculating the waveform of action potentials repeatedly without giving insights to the mechanism of the neural circuits. This is why the majority of mathematical models about nervous systems are limited to two levels; even some factors in the lower level should be considered, and authors often imply them in some parameters in the higher levels. However, this is still an open problem. Some author emphasized that the network model with biophysical neuronal models as its units may realize a more complicated function (Rall 2002). They pointed out that the simplified neuronal model often considers the neuron as only a node and the complicated morphology and function of dendritic tree are neglected; thus, some complex function of the network model with biophysical neuronal models as its units may be lost when they are replaced by simplified neuronal models. Only time can give a final conclusion about the problem and how detailed the neuron model should be, when higher brain function is considered.

Ekeberg's model is only an example of a biophysical model of neurons. Now, based on different morphologies and different kinds of ion channels at different compartments, a variety of biophysical neuronal models have been developed. Even so, most of such models are still simplified in some way. In the above model, only transmitter-sensitive channels are considered for dendritic compartments; no pure voltage-sensitive channels are considered in these compartments, and all the voltage-sensitive channels are concentrated in the soma compartment, which may be true for some neurons, e.g., the motor neuron in the spinal cord. However, now it has been discovered that some dendritic membrane is not purely passive; it cannot be treated as some pure passive cable only transmitting synaptic currents to the axon hillock to be summarized linearly. Recent studies suggested that some of the dendritic membrane is active; there are quite a lot of voltage-sensitive calcium channels, sodium channels, and potassium channels, which may magnify the remote postsynaptic potential, owing to the fact that the decayed EPSP may trigger the voltage-sensitive sodium channel opening to allow more sodium ions to pour in to enhance the synaptic signals. In addition, these channels may carry out a multiplying operation. There is also some evidence from stimulation of a brain slice that the action potentials initiated from initial segment propagate not only along the axon forward but also to the dendrites backward. Therefore, dendrites are much more complex than what people thought before (Koch 1997). In addition, in the similar models like the above one, no dendrodendritic synapse is considered, at which the dendrite may not only receive input from other dendrites but also send output to other dendrites. Such reciprocal graded interaction at the dendrodendritic synapse has not been considered in the above kind of models (Rall 2002).

3.3 FitzHugh–Nagumo Model and Hindmarsh–Rose Model

Although the Hodgkin–Huxley model successfully describes the behavior of squid giant axon membrane with voltage-sensitive potassium channel and sodium channel and elucidates the underlying mechanism of the generation and propagation of the action potential, however, the model needs a fourth-order system of nonlinear ordinary differential equations to describe. Although such a system can be solved numerically with computers, only a solution under specific conditions (such as some fixed parameter values, specific initial condition, and certain specific input current) can be obtained. Using such an approach, it is difficult to have a general view of its dynamics. In addition, at the early days when the model had just been proposed, digital computers were not as powerful and popular as today so to solve the above complicated equation system is time consuming and difficult, especially if the parameters and the initial conditions need to be changed systematically to have a global view of the system under various conditions. Therefore, a demand to reduce the dimension of the model by reasonable simplification but where its essential dynamics could be still kept was raised. In this way, the simplified model can be studied in the phase plane intuitively, and the dependence of its dynamic behavior on its parameters can be shown clearly, and a deeper insight of the underlying mechanism of such membrane can be got. This is the topic of Sect. 3.3.1. However, as what the proverb says, "There's no such thing as a free lunch." The cost for such simplification is that it could not predict the behavior as accurate as what the Hodgkin–Huxley model does when the specific parameters and the initial condition are given. There is always some "uncertainty principle" between accuracy and universality.

The above is the one side on modification of the H–H model, simplification; on the other side, as it was mentioned in Sect. 3.1, there are only two kinds of voltage-sensitive channels, the potassium channel and sodium channel, which have been considered in the H–H model, and the dynamics of these channels are quite fast, though the dynamics of potassium is a little slower than the one of sodium, and they are responsible for the generation and propagation of the action potential. Other ion channels with much slower dynamics, such as the calcium channel and so on, are neglected. However, these slower but long-lasting channels could modulate the timing of the spike, thus deciding the rhythmic pattern of the spike train as shown in Sect. 3.2. And as what we mentioned above, an individual spike might not be so important for neural information processing; the most important thing is the timing pattern of the spike train and therefore, except for the channels with fast dynamics, some slower channels must also be considered. Although the last section has given us such an example, however, it has the same problem as what the H–H equations have, and a new way to show its dynamic behavior on phase plane needs to be developed. Such an example generalized from the model described in Sect. 3.3.1 will be the topic of Sect. 3.3.3.

3.3.1 FitzHugh–Nagumo Model

The FitzHugh–Nagumo model, named after Richard FitzHugh (1922–2007) who proposed the model in FitzHugh (1961) and J. Nagumo et al. (1962) who created an equivalent circuit in the following year, is a simplified version of the H–H model with two dimensions (Izhikevich and FitzHugh 2006).

The basis for such simplification is the fact that the membrane potential $V(t)$ and sodium activation variable $m(t)$ change almost synchronously during the spiking period, which can be shown clearly if both variables are normalized and plotted at the same diagram (Fig. 3.11a); thus the two variables can be lumped into a single "activation" variable V (Fig. 3.11c); while the responses of sodium inactivation $1-h(t)$ and potassium activation variable $n(t)$ to a step current stimulus also change synchronously with a slower time scale (Fig. 3.11b), thus these two variables could also be lumped into a single "accommodation" or "refractoriness" variable W (Fig. 3.11d).

While the Hodgkin–Huxley Model is more realistic and biophysically sound, only projections on a 2D plane of its four-dimensional phase trajectories can be observed. The simplicity of the FitzHugh–Nagumo model permits the entire solution to be viewed at once. This allows an intuitive and global geometrical explanation of important biological phenomena related to neuronal excitability and spike-generating mechanism.

Considering the periodic oscillation pattern of the responses of these variables to a step stimulus, the behavior is similar to the one of the Van der Pol oscillator qualitatively, which was proposed by Dutch electrical engineer and physicist

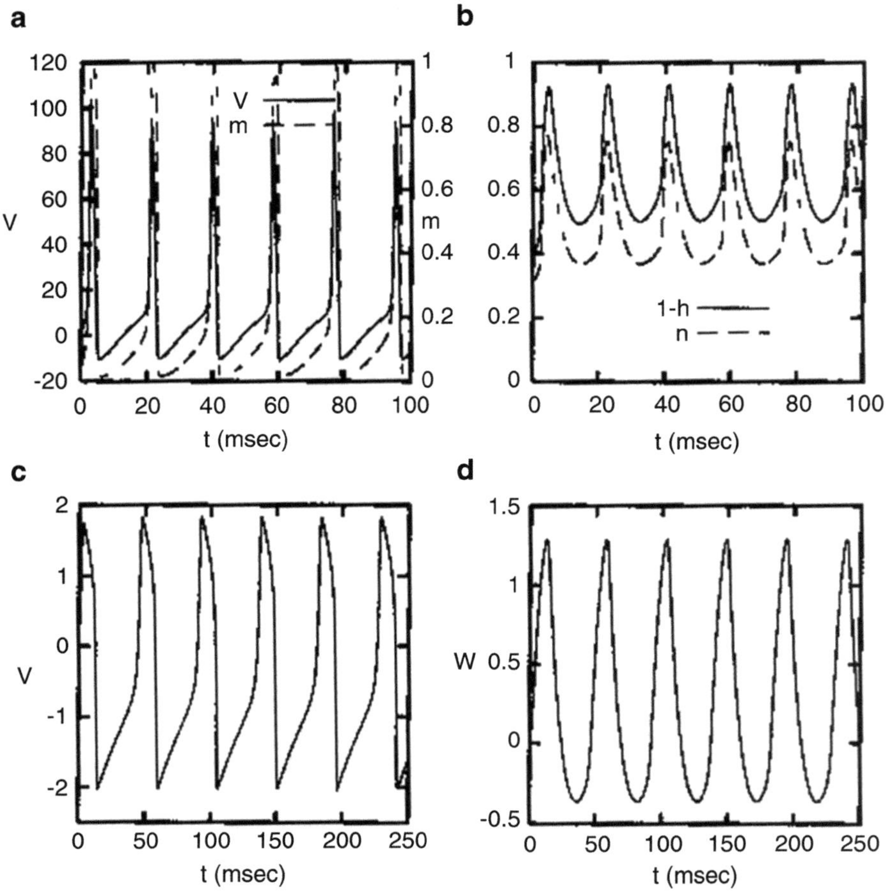

Fig. 3.11 *Reducing the H–H model to the F–N model*. In (**a**) and (**b**), the input stimulus current is equal to 0.18 *nA* for the H–H equation, while in (**c**) and (**d**), $I = 0.35$ for the F–N equation. (**a**) The waveforms of $V(t)$ and $m(t)$. (**b**) The waveforms of $1-h(t)$ and $n(t)$. (**c**) The solution $V(t)$ of the F–N equation. (**d**) The solution $W(t)$ of the F–N equation (Reproduced with permission from Koch (1999). Fig. 7.1)

Balthasar Van der Pol (1920) (Van der Pol and Van der Mark 1927) when he studied the electrical circuits applying vacuum tubes; his model equation is as follows:

$$\begin{aligned} \dot{x} &= \mu\left(x - \frac{1}{3}x^3 - y\right) \\ \dot{y} &= \frac{1}{\mu}x. \end{aligned} \tag{3.52}$$

As the nullclines of the Van der Pol equation are a vertical line and a cubic, so they intersect in a single rest point which is always unstable. Therefore, the above model should be modified in some way to make the unique rest point basically stable. FitzHugh realized that he could get such result by rotating the linear nullcline of the Van der Pol equation; for this, he added a linear term 0.7–0.8 W to the second equation of the Van der Pol model. FitzHugh restricted the slope of the linear nullcline to the set of cases so that there is only one intersection with the cubic.

Adding a constant term to the second equation allowed him to shift the resting point along the cubic. That resting point is now always stable on the ascending parts. However, to make the model a bit more palatable for physiologists, both equations got a constant term: I in the first equation, mimicking the experimental injection of external current into the membrane; the second equation got another, mathematically redundant constant 0.7. It ensures that the resting point for $I = 0$ (no stimulation) lies on the right ascending branch and is stable.

Thus, the FitzHugh–Nagumo model (F–N model) of the neuronal membrane with voltage-sensitive channels can be described as follows:

$$\begin{aligned} \dot{V} &= V - V^3/3 - W + I \\ \dot{W} &= 0.08(V + 0.7 - 0.8W) \end{aligned} \tag{3.53}$$

where V is a membrane potential-like variable, I is the intensity of the stimulus current, and W is a refractory variable. From Eq. (3.53), it is easy to see that the change of W is slower than the one of V, approximately slower by about $1/0.08 = 13.5$ times.

Although the F–N model is a simplified version of the H–H model with only two dimensions, the equation is still heavily nonlinear; it is still difficult to find an analytic solution. However, according to the qualitative theory of differential equations, the property of the solutions could be studied in a phase plane qualitatively, i.e., to plot the two nullclines, which are the curves with the equations of (3.53) when the left sides are set to zero, in the V–W phase plane (Fig. 3.12). Thus, the whole plane is separated into several regions; in every region, it can be known if the V and W will increase or decrease according to Eq. (3.53) and the trajectory of the system with time could be estimated approximately. The intersection of the two curves (if there is such an intersection) will be the equilibrium point of the system and its stability could be estimated by the evolutionary trajectories around this point. If set $I = 0$ in Eq. (3.53), its equilibrium is at -1.20 and -0.625. In case that the system is at that state without any outside disturbance, it will stay at that state forever; this is the resting state for a neuron. However, in the realistic world, it is rare that the system happens to be at that state, or even if it does, it will deviate from this state owing to outside disturbances. In case that after the system deviates from its equilibrium state and there is no further disturbance, if the system can return to its equilibrium at last, then the equilibrium state is stable; in such a case, the trajectory will move toward the equilibrium point; otherwise, the equilibrium point is unstable. In the case of the model (3.53), the system is stable. To prove this theoretically, a coordinate transformation should be made to move the original point to the equilibrium point and then all the nonlinear terms in the equation could be expanded by a Taylor series, and all the higher-order terms could be neglected to linearize the equation and the real parts of all the eigenvalues should be calculated

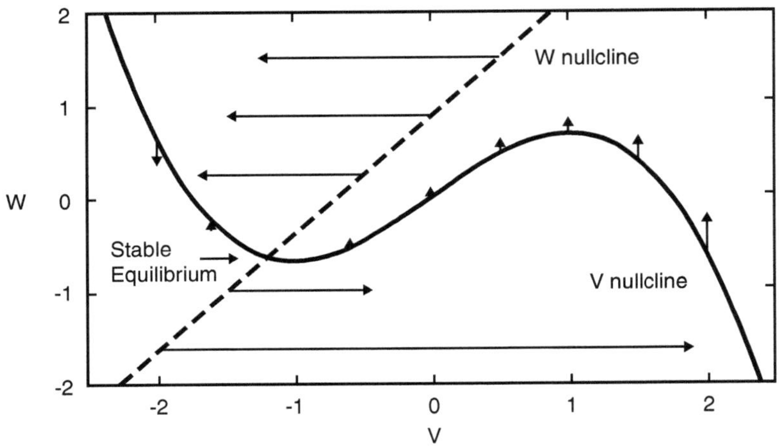

Fig. 3.12 *Phase plane portrait of the F–N model.* The nullcline for the V variable is a cubic polynomial and the W nullcline is a straight line. In this figure, I is taken to be zero. The *arrows* are proportional to dV/dt and dW/dt and indicate the direction and rate of change of the system (Reproduced with permission from Koch (1999). Fig. 7.2)

to see if all of them are negative and then a judgment can be made if the system is steady near that point. We will not discuss this matter in detail; the readers who are interested in such topics can find the materials in Koch's book (Koch 1999).

In the following, we will exemplify how to use the F–N model to explain spike generation when a stimulus is strong enough. Let us suppose that an instantaneous current $I(\mathrm{t}) = Q\delta(t)$ (with $Q > 0$) is applied; the solution is equivalent to the same equation with $I(t) = 0$, but the initial condition becomes $V = \mathrm{Q}$ and $W = 0$. Under such conditions, the trajectories in the phase plane portrait were shown in Fig. 3.13(a). Changing V from its initial value of -1.2 to -0.8 or -0.65 only causes quick excursions of the voltage around the equilibrium point with the system rapidly returning to rest, which corresponds to subthreshold responses of the neuron. If the current is large enough so that V exceeds -0.64, a stereotyped "all-or-none" sequence is triggered: V rapidly increases to positive values then dives below its resting value before finally coming back to that value. However, the amplitudes of the spikes are different a little bit, when the amplitudes of stimuli are only a little bit greater than the one to cause V exceeding -0.64. That means that the all-or-none law is not rigorous. If horizontal line is drawn through the intersection (shown as the thin dashed line in Fig. 3.13a), in which all the W values keep unchanged, the line could be considered as a phase line of a reduced system with only one variable V. The V phase line through the resting point has three intersections with the three branches of V nullcline, and the middle intersection is unstable and represents a threshold, as injecting current $I(\mathrm{t}) = Q\delta(t)$ is equivalent to shifting the initial condition to the value Q without any input, and if Q is big enough to reach the unstable middle intersection, a spike will be generated.

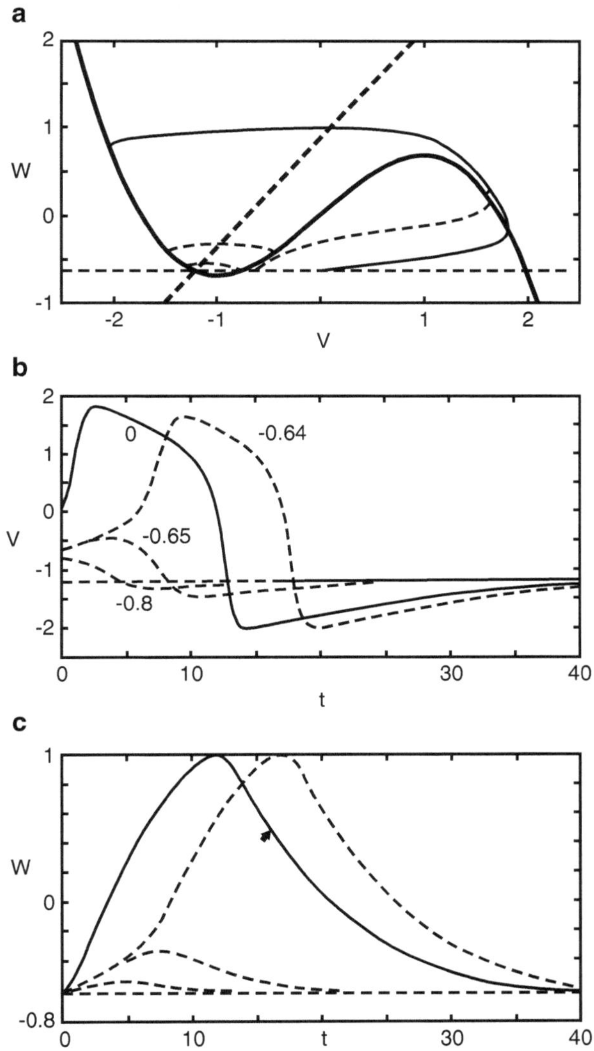

Fig. 3.13 *Response of the F–N model to current pulses*. (**a**) The trajectories in the phase plane portrait. (**b**) Waveforms of V. (**c**) Waveforms of W (Reproduced with permission from Koch (1999). Fig. 7.4)

Although the F–N model has made much simplification to the H–H model, it can still show many characteristics of the real neuron, such as excitation block, anodal break excitation, spike accommodation to slowly increasing stimulus, and so on and so forth. The approaches used to analyze these phenomena are similar to the one mentioned above for spike generation, so we will not discuss them in detail. If the reader is interested in these questions, he or she may find explanations in Izhikevich and FitzHugh's (2006) excellent review.

When FitzHugh and Nagumo developed their model, digital computers were still not so popular. To solve their equation, FitzHugh had to use an analog computer with vacuum tubes; the machine was so huge and radiated so much heat that FitzHugh had to take off his shirt and wear shorts to be comfortable in the hot summer days, as the heat from all these tubes sometimes overloaded the air conditioning. As for Nagumo and his colleagues, they simply designed an equivalent circuit including a tunnel diode and an inductor to mimic the behavior of their model. It should be noted that the circuit, as a matter of fact, is only a black box model of the membrane, a tool to solve their equation which was based on biological facts, just like FitzHugh using an analog computer to calculate his equations. Their equations express a model of the neuronal membrane, but the analog computer is not; the same words could also be said to Nagumo's circuit. Therefore, the F–N model is reasonable; however, this is not a reason for people to propose a circuit with some elements, no significant counterparts of which could be found in living neurons, as a key element of their neuron models and then deduce something from such an unrealistic model to elucidate the mechanism of some function of the neuron. Such work might be sophisticated in mathematics, but just as what Francis Crick said: "I don't see what it has to do with the brain" (Ramachandran 2004).

3.3.2 Hindmarsh–Rose Model

Although the F–N model gives a two-dimensional simplification of the H–H model, so that it can be analyzed in a phase plane intuitively and globally, it has the same deficit of the H–H model itself, which cannot simulate the pattern of spike train in a variety of conditions. For example, the interval between neighboring spikes is comparable with the spike duration itself as shown in Fig. 3.11c. However, for the real neuron, the former is much longer than the latter. This is not surprising, as the F–N model is deduced from the H–H model and the latter itself has the similar problem as shown in Fig. 3.11a. For the H–H model, the main task is to elucidate the mechanism of generation and propagation of the action potential but not the pattern of the spike train! In science history, all the complicated phenomena could only be studied step by step; no one can solve the problem once at all. The F–N model is a qualitative model based on the facts shown in Fig. 3.11a and b, a modified version of the Van der Pol oscillator. No further experiment data were used to verify the model quantitatively. British mathematician J. L. Hindmarsh and physiologist R. M. Rose used their physiological data and mathematical analysis to generalize the F–N model, not only overcome the deficit of short interval between neighboring spikes (Hindmarsh and Rose 1982) but also consider a new slow ion channel to elucidate the mechanism of isolated burst and periodic bursting (Hindmarsh and Rose 1984), just as what Ekeberg et al. did for a more complicated model, which can only be studied with numerical solution and lack of an intuitive phase plane analysis. This is the topic of this subsection.

Hindmarsh and Rose generalize the F–N model to the following form:

$$\frac{dx}{dt} = -a(f(x) - y - z) \tag{3.54}$$

$$\frac{dy}{dt} = b(g(x) - y) \tag{3.55}$$

where a and b are constants, x is the membrane potential, and y is ionic current. In the F–N model, x and y are symbolized as V and W and z as I. $f(V)$ is a cubic function, and $g(V)$ is a linear function. Now Hindmarsh and Rose identified the functions $f(x)$ and $g(x)$ by data fitting with the data they recorded from a large cell from the visceral ganglion of the pond snail, *Lymnaea stagnalis*, which was clamped to a range of different voltages (x_p). When they clamped the neuron voltage, the initial ($z_{xp}(0)$) and steady-state ($z_{xp}(\infty)$) values of the clamping current were measured (Fig. 3.14a), and they varied the clamping voltage (x_p) and thus their current–voltage plots can be plotted in Fig. 3.14b.

As the cell has spontaneous firing, the membrane potential was biased initially by adjusting z so that the cell just stopped firing. This corresponds to an artificial stable equilibrium point and will be used as the origin ($x=0$, $y=0$) in the phase plane. Therefore, at the equilibrium position, $x=y=dx/dt=dy/dt=z=0$. The time course of the current, after the onset of a step in voltage to x_p, is given (from Eq. (3.54) with $dx/dt=0$) by

$$z_{xp}(t) = f(x_p) - y(t) \tag{3.56}$$

where y satisfies Eq. (3.55) with $x=x_p$.

Therefore,

$$y(t) = g(x_p)\left(1 - e^{-bt}\right), \text{ and } z_{xp}(t) = f(x_p) - g(x_p)\left(1 - e^{-bt}\right) \tag{3.57}$$

and hence

$$z_{xp}(0) = f(x_p), \text{ and } z_{xp}(\infty) = f(x_p) - g(x_p) \tag{3.58}$$

or

$$f(x_p) = z_{xp}(0), \text{ and } g(x_p) = f(x_p) - z_{xp}(\infty) = z_{xp}(0) - z_{xp}(\infty) \tag{3.59}$$

However, $z_{xp}(0)$ and $z_{xp}(\infty)$ can be estimated by curve fitting with the data shown in Fig. 3.14, in which $z_{xp}(0)$ can be fitted by a cubic function and $z_{xp}(\infty)$ by an exponential function.

The time constant b was estimated from voltage-clamp current records (Fig. 3.14a) and a can be obtained from the slope of the voltage change immediately after the onset of the current-clamp step. This is given by Eq. (3.54) with $f(0)=0$ and $y(0)=0$:

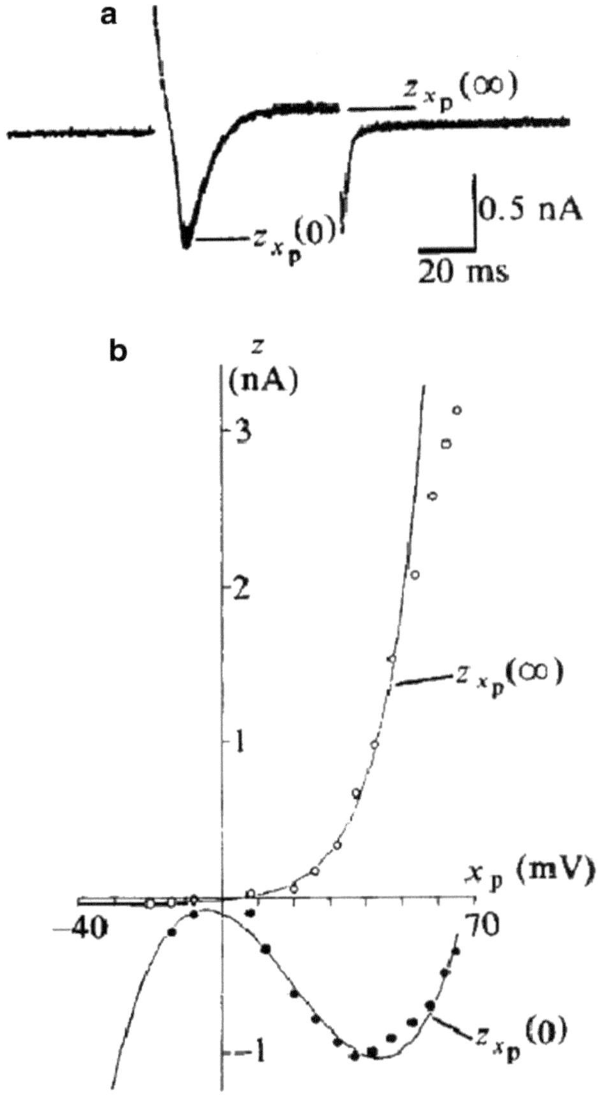

Fig. 3.14 *Voltage-clamp data on which H–R model is based*. (**a**) Current change recorded for a voltage step (x_p) of 25 mV from the equilibrium point. The initial ($z_{xp}(0)$ or *peak inward*) and final ($z_{xp}(\infty)$ or steady state *outward*) currents are indicated. (**b**) Initial and final currents values of clamp potential (x_p) measured from the equilibrium point. Negative values of z are inward and positive values are outward currents (Reproduced with permission from Hindmarsh and Rose (1982). Fig. 1)

$$\frac{dx}{dt}(0) = az_p \tag{3.60}$$

In summary, Hindmarsh and Rose summarized their equations as follows:

$$\frac{dx}{dt} = -a(f(x) - y - z) \tag{3.61}$$

$$\frac{dy}{dt} = b(f(x) - qe^{rx} + s - y) \tag{3.62}$$

where $f(x) = cx^3 + dx^2 + ex + h$ and a, b, c, d, e, h, q, r, and s are all constants, which can be estimated by their data. Thus the above two equations can be solved numerically and their solution $x(t)$ with $z = 0.008$ is shown in Fig. 3.15b and a comparable experimental data is shown in Fig. 3.15a.

Compared with the F–N model, the main advantage of the above model is that it has a reasonable inter-spike interval and this can also be seen and explained in its phase plane plot as shown in Fig. 3.16, in which the $dx/dt = 0$ curve is quite similar to the one of the F–N model, while the $dy/dt = 0$ curve is significantly different; it is not a linear function again but similar to a quadratic function, which is very close to the dx/dt curve in the lower left-hand quadrant. The limit cycle is divided into an action potential phase ($A \rightarrow B \rightarrow C$) and a recovery phase ($C \rightarrow A$), during which the phase point moves slowly in a narrow channel between the x and y nullclines. On the contrary, in the F–N model, the x and y nullclines diverge markedly on both sides of the diagram, so that the phase point moves almost with the same speed during both phases.

For convenience, let us rewrite Eqs. (3.61) and (3.62) as follows:

$$\frac{dx}{dt} = y - p(x) + I \tag{3.63}$$

$$\frac{dy}{dt} = q(x) - y \tag{3.64}$$

From Fig. 3.16, it is easy to see that $p(x)$ is a cubic function and $q(x)$ is a quadratic function. Thus, Eqs. (3.63) and (3.64) could be rewritten as follows:

$$\frac{dx}{dt} = y - ax^3 + bx^2 + I \tag{3.65}$$

$$\frac{dy}{dt} = c - dx^2 - y \tag{3.66}$$

As what is shown in Fig. 3.16, although the above model can simulate spike trains with reasonable inter-spike interval, however, this model will give a sustained spike train even if only a short current pulse is given.

To find the equilibrium points of the system, the intersection of the x and y nullclines should be decided. They can be solved by the equation

$$ax^3 - bx^2 = dx^2 - c \text{ or } x^3 + px^2 = q \tag{3.67}$$

where $p = \frac{d-b}{a}$ and $q = \frac{c}{a}$ and three intersections can be solved by (3.67). The properties of these intersections can be analyzed by standard stability analysis, which will be omitted here for conciseness. For those readers who are interested in it, the reference Hindmarsh and Rose (1984) should be consulted.

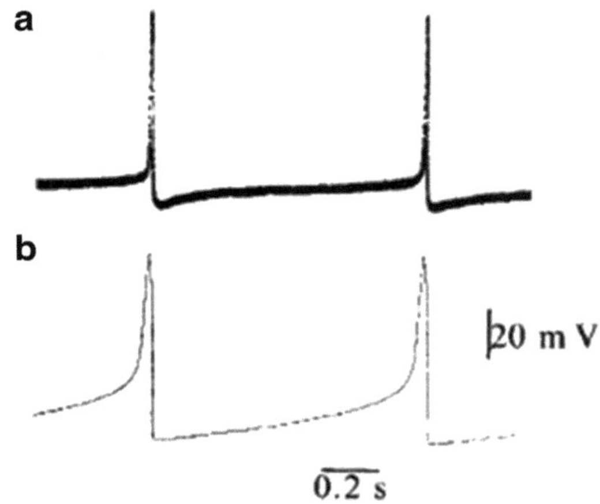

Fig. 3.15 *A comparison of real spike train with simulated one by Eqs.* (3.61) *and* (3.62). (**a**) Physiological experiment record. (**b**) Simulation result (Reproduced with permission from Hindmarsh and Rose (1982). Fig. 2)

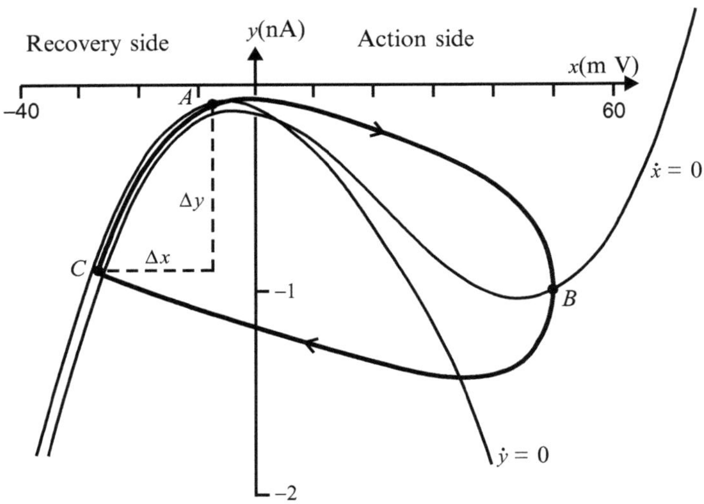

Fig. 3.16 *Phase plane plot of the limit cycle solution to Eqs.* (3.61) *and* (3.62). Compared with F–N model, the $dx/dt = 0$ curve is similar; however, the $dy/dt = 0$ curve is quite different (Reproduced with permission from Hindmarsh and Rose (1982). Fig. 3)

A phase plane plot of the system (3.65) and (3.66) is shown in Fig. (3.17) with $a = 1$, $b = 3$, $c = 1$, and $d = 5$. The equilibrium points A, B, and C is a stable node point, an unstable saddle point, and an unstable spiral point, respectively. A corresponding current–voltage curve is shown above the phase plane (Hindmarsh and Rose 1984).

If the initial conditions for x and y are chosen to be close to the unstable equilibrium point C on the right, the phase point spirals out until it enters a limit cycle, which means that the neuron fires a sustained spike train. On the other hand, if the point is started close to the saddle point B, the dashed line divides the phase plane into two regions. On one side of this line, the phase point moves toward the limit cycle, whereas on the other side it is deflected downwards the stable node, which corresponds to the resting state of the neuron. Thus, the model can behave as silence or sustained oscillation depending on its initial condition.

However, real neurons such as the cells in *Lymnaea* rarely behave in the above way; some neuron gives a variety of burst patterns, such as an isolated burst, periodic burst, or even adapted sustained firing. A simple way to terminate the oscillation may be introduction of a slow hyperpolarizing current. Thus a third variable z and its equation should be added into the above system as follows:

$$\frac{dx}{dt} = y - ax^3 + bx^2 + I - z \tag{3.68}$$

$$\frac{dy}{dt} = c - dx^2 - y \tag{3.69}$$

$$\frac{dz}{dt} = r(s(x - x_1) - z) \tag{3.70}$$

where x_I and y_I are the coordinates of the leftmost equilibrium point of the model without adaptation. This has the result that $(x_I, y_I, 0)$ is a stable equilibrium point of the model with adaptation. Thus, in the above model, x is a variable related to membrane potentials. Sodium and potassium ions are transported by fast ion channels and their transport is measured by $y(t)$, which is often called the spiking variable. Other ions are transported by slow channels and are taken into account with $z(t)$, which is called the bursting variable.

Although the new system is three dimensional, as the r value usually is very small (say, 0.001), thus z changes very slowly compared with x and y and could be considered as a slowly changing parameter of the system (3.68) and (3.69). The phase plane plot shown in Fig. 3.17 still could be used to analyze the behavior of the new model.

Assume that the initial x value is x_I and an impulse I is given, so that the phase point moves to the limit cycle basin and quickly approaches to an oscillation. However, now each time when an action potential fires, the adaptation current (z) will increase and the x nullcline will thus be upward on successive cycles. The firing rate will decrease because the narrow channel becomes even narrower. At the same time, the equilibrium points will move with the upward move of x nullcline. Eventually the saddle point displaced upward so far that the limit cycle trajectory will cross the saddle point separatrix and the phase point enters the basin of the stable equilibrium node and the firing ceases. And a burst is observed.

In Fig. 3.18, the numerical solutions of Eqs. (3.68), (3.69), and (3.70) with $a = 1$, $b = 3$, $c = 1$, $d = 5$, $r = 0.001$, and $s = 4$ for (a) $I = 0.4$, (b) $I = 2$, and (c) $I = 4$ are shown. At low levels of steady depolarizing current ($I = 0.4$), the model generated an isolated burst followed by an after-hyperpolarizing wave which slowly recovered to the starting value (x_I). At a higher level of current ($I = 2$), the model generated a long burst initially in response to the current step and this adapted and terminated to give the periodic burst pattern. At a still higher level of current

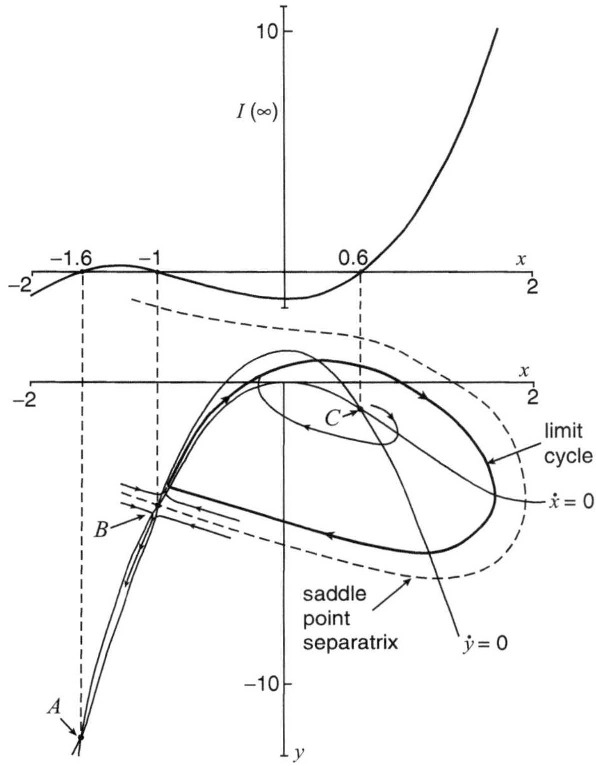

Fig. 3.17 *A phase plane plot of the system* (3.65) *and* (3.66) with $a = 1$, $b = 3$, $c = 1$, and $d = 5$. The phase plane plot is shown at the *bottom* diagram. A corresponding current–voltage curve is shown *above* the phase plane (Reproduced with permission from Hindmarsh and Rose (1984). Fig. 4)

($I = 4$), there was a continuous high-frequency discharge, with the frequency declining from the onset of the step to the steady repetitive firing. All the similar phenomena are observed in living bursting neurons; thus the H–R model has been a classical model for studying bursting behavior; the key point in this model is that there are two subsystems in this model, one fast subsystem to generate action potentials and a slow subsystem to modulate the spiking pattern.

3.4 Simplified Neuron Models

In Sect. 3.2, a biophysical neuron model, based on the knowledge about neuronal membrane, a variety of ion channels, etc., at the level below the cellular level, was given; it could simulate neuronal behavior quite accurately and thus it can give us insight to the underlying mechanism of neuronal behavior. However, if such model is used as a unit to construct a neural network, then a huge number of nonlinear differential equations are needed and the simulation is time consuming; much time would be spent on calculating the waveform of action potentials repeatedly while

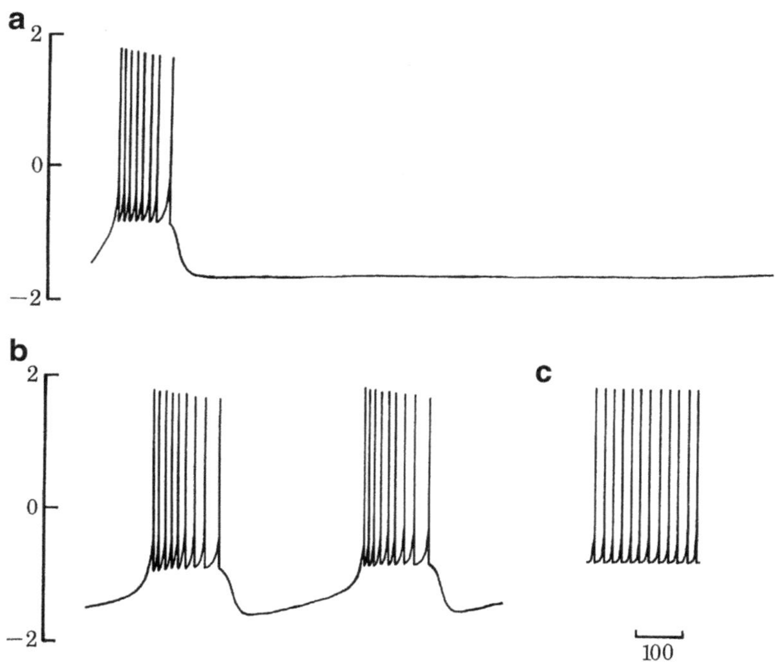

Fig. 3.18 *Burst generation*. Numerical solution of Eqs. (3.68), (3.69), and (3.70) with $a = 1$, $b = 3$, $c = 1$, $d = 5$, $r = 0.001$, and $s = 4$, for (**a**) $I = 0.4$, (**b**) $I = 2$, and (**c**) $I = 4$; (**b**) starts at 700 time units after the onset of I step and (**c**) after 1000 time units of continuous firing (Reproduced with permission from Hindmarsh and Rose (1984). Fig. 6)

such waveform tells little information about the behavior of the neural circuit. To construct a model of neural circuits or the one at even higher levels, the underlying mechanism of the neuron itself may not be necessary; only the input–output relationship of the neuron is important. This is similar to analyzing an electronic circuit; to know the input–output characteristics of its element (say, a transistor) is important, as for the mechanism of the movement of the charge carriers in that element can be neglected. Thus, at the very beginning of the history in neuron modeling study, such a black box model was proposed first, for which the neuron model is only an element, the complicated morphology of dendrites was neglected, all the synapses were simplified as inputs with different weights, the ionic mechanism inner the neuron was also neglected, and only the input–output relationship of a neuron was left. Such simplification may be helpful to study the problem above the cellular level on the one hand, just as what Christof Koch said: "Understanding any complex system always entails choosing a level of description that retains key properties of the system while removing those nonessential for the purpose at hand. The study of brains is no exception to this" (Koch 1999). However, on the other hand, as what Thomas P. Trappenberg pointed out: "It is, of course, an important part of every serious study to verify that the simplifying assumptions are appropriate in the context of the specific question to be analyzed via a model" (Trappenberg 2002). We must be very careful to say anything about modeling the human brain using such simplified neuron models as elements.

Some simplified neuronal models could be very simple, even with some simple properties which the living neuron does not have; the aim of such model is to be used as an element to construct artificial neural networks with some specific function to solve a practical engineering problem and not to be related to elucidating the mechanism of living neural circuits. Such model could be called as an artificial neuron, which does not relate to the main topic of this book; thus, it is omitted here. On the contrary, some simplified neuronal models although do not base on the underlying mechanisms of the subcellular level, and the main assumptions are realistic based on the knowledge we know about the behavior of living neurons; maybe it could be called as realistic simplified neuron model, which will be discussed in this section. In spite of the fact that a lot of such models have been developed, they could be categorized into two classes according to the way how they code: The first class is based on temporal coding, i.e., the information carried by their output depends on the timing of its spike train, that means that their output is a series of discrete spikes or$\Sigma_i \delta(t-t_i)$, if the waveform of the action potential is considered meaningless for information processing which we are interested in; the most important factor is every moment $\{t_i\}$ when a spike occurs. The earliest neuron models, such as Louis Lapicque's integrate-and-fire model (Lapicque 1907) and the formal neuron model proposed by the founders of the neural network theory, McCulloch and Pitts (1943), are all the examples of such class. However, the mathematical analysis of the discrete spike train is not so easy. The second class is based on the idea that the neural coding is rate coding, i.e., the output of the neuron is a continuous variable—firing rate or frequency. The Hopfield model (1994) is an example of such category. Owing to the good mathematical properties of the sigmoid function, such model is easier for mathematical analysis, and so such models dominated neural network studies before the late last century; however, owing to the fact that rate coding does not work in many cases for a living neuron and it is energy consuming, the spike neuron model has been revived in recent years. We will discuss a few typical models for both categories in this section.

3.4.1 Integrate-and-Fire Model

The earliest neuronal model in the neuroscience history may be the integrate-and-fire model proposed by Louis Lapicque (1907) Although it was not clear how the spike generates at his time, he grasped the following key factors about the excitability of the neuron: subthreshold passive potential will be integrated gradually until reaching a certain threshold, and then a stereotypical impulse will be fired and its potential restores to its resting state again.

According to the law about capacitance in physics, $Q = CV$, differentiate both sides:

$$I(t) = C_{\mathrm{m}} \frac{dV_{\mathrm{m}}}{dt} \tag{3.71}$$

where C_m denotes the membrane capacitance, V_m the membrane potential, and I the membrane current.

If some current is injected into the neuron, then the membrane voltage will increase with time until a threshold V_{th} is reached and then a spike is fired, which could be represented by a delta function and the membrane potential restored to the original resting potential. In this model, the refractory period has not been considered; thus, the firing rate would increase linearly with the stimulating current without bound. Obviously, this is not true for a living neuron; to overcome this deficit, a refractory period t_{ref} must be introduced and thus the frequency f obeys the following formula:

$$f(I) = \frac{I}{C_{\mathrm{m}} V_{\mathrm{th}} + t_{\mathrm{ref}} I} \tag{3.72}$$

Even after such modification, the above model is obviously not realistic enough, as the membrane is not an ideal capacitor; some leak current can pass the membrane, and in the simplest condition, if the voltage-dependent channel and transmitter-dependent channel are not considered, the membrane can be considered as a parallel circuit with a capacitor and a resistor. Thus the modified model is a leaky integrate-and-fire model discussed in the next subsection.

3.4.2 *Leaky Integrate-and-Fire Model (LIF Model)*

Based on the consideration mentioned in the above subsection, a parallel resistor is added to the integrate-and-fire model and then the first equation for the new model should be rewritten as follows:

$$I(t) - \frac{V_{\mathrm{m}}(t)}{R_{\mathrm{m}}} = C_{\mathrm{m}} \frac{dV_{\mathrm{m}}(t)}{dt} \tag{3.73}$$

where R_m is the leaky membrane resistance. Thus the model will fire only if the input current exceeds a certain threshold $I_{\mathrm{th}} = V_{\mathrm{th}}/R_{\mathrm{m}}$; otherwise, owing to the leak current, the firing threshold V_{th} can never be reached to fire a spike. Suppose that the input is constant; then the solution of the above equation is

$$V_m(t) = IR_m(1 - exp(-t/\tau)) \tag{3.74}$$

where $\tau = RC$ is the time constant of the membrane. The neuron will fire a spike, only if the membrane potential reaches a threshold V_{th}, which needs a period of time t_{th} to charge the membrane capacitor; hence

$$t_{th} = -\tau log(1 - V_{th}/(IR)) \tag{3.75}$$

In addition, if an absolute refractory period is considered, then the second equation of the integrate-and-fire model (3.72) should also be modified as follows:

$$f(I) = \begin{cases} 0, & I \leq I_{\text{th}} \\ \left[t_{\text{ref}} - R_{\text{m}}C_{\text{m}}\log\left(1 - \frac{V_{\text{th}}}{IR_{\text{m}}}\right)\right]^{-1} & I > I_{\text{th}} \end{cases} \tag{3.76}$$

The activities of the leaky integrate-and-fire model can be divided into two phases: the integration of the subthreshold potential phase and the firing and resetting phase. For the first phase, Eq. (3.73) could be rewritten as follows:

$$\tau_m \frac{du_i}{dt} = -u_i + RI_i(t) \tag{3.77}$$

where u_i is the membrane potential at the initial segment of the i-th neuron, $\tau_m = RC$ is the time constant of the membrane, and $I_i(t)$ is the membrane current of the i-th neuron. The resting potential is taken as the basis of the potential, i.e., $u_i = 0$, when the neuron is at its resting state. When u_i reaches certain threshold θ at a moment $t = t_i^f$, the second phase starts; the neuron fires a spike and the membrane potential u_i is reset at a value u_{reset} below the resting value.

Like the above models, the neuron model, which does not consider the spatial structure of the neuron, is called a point model.

If a model neuron receives its input from other neurons, then its input current can be written as follows:

$$I_i = \sum_{j,f} w_{ij}\alpha\left(t - t_j^f\right) \tag{3.78}$$

where α is the postsynaptic current induced by a spike from a presynaptic neuron and w_{ij} is the weight coefficient of the synapse between the i-th neuron and the j-th neuron; the function $\alpha(\cdot)$ can be written as follows:

$$\alpha(s) = \delta(s - \Delta^{ax}) \tag{3.79}$$

or more realistically

$$\alpha(s) = \frac{1}{\tau_1 - \tau_2}\left[e^{-(s-\Delta^{ax})/\tau_1} - e^{-(s-\Delta^{ax})/\tau_2}\right] \tag{3.80}$$

where Δ^{ax} is the delay owing to the transmission along the axon and τ_1 and τ_2 are time constants of the synapse.

Assume that the moment of the last firing at the i-th neuron is $\hat{t}_i$; then the solution of the previous set of equations after that moment but before the next firing has the following form:

$$u_i(t) = \eta(t - \hat{t}_i) + \sum_{j,f} w_{ij}\varepsilon\left(t - \hat{t}_i, t - t_j^f\right) \tag{3.81}$$

where function η is the action potential and its after-potential spiking at the moment $\hat{t}_i$ and function ε is the postsynaptic potential of the i-th neuron to the presynaptic impulse of the j-th neuron arriving at the moment t_j^f. If the last spike of the i-th neuron was fired long long ago, i.e., $t - \hat{t}_i \rightarrow \infty$, then $\varepsilon(\infty, s)$ is the postsynaptic potential induced by a single presynaptic spike. When $u_i(t)$ reaches the threshold θ, a new spike will be fired. The above equation itself could be considered as a model—the spike response model (Gerstner 2002). If the forms of the functions η and ε are selected properly, then the timing of the spike train calculated with the H–H model could be predicted by this model with an accuracy of 90 % and the error is less than ± 2 ms (Kistler et al. 1997).

The output of the above neuron model is a spike train and no central clock synchronous control is needed; thus, it is more realistic. In addition, its calculation is rather simple and many authors have been interested in such model. Theoretical analysis has hinted that the neural network model with the above neuron model can carry out almost all the computations that a typical neural network model can do, and it needs much less energy when it is implemented by hardware and it is also more suitable for simulation of biological neural networks.

3.4.3 Hopfield Model

American biophysicist John Hopfield is famous for his neural network model, which gave some hints about the distributive memory mechanism in the living brain and will be discussed in detail later in this book. In this section, we will only introduce how he induced his neuron model as the unit of his network from known biological facts as an example of a realistic simplified neuron model, of which the output is its firing rate.

His basic assumptions are:

1. There are two classes of synapse: excitatory synapse and inhibitory synapse.
2. The neuronal membrane is equivalent to a paralleled electrical circuit with capacitors and resistors.
3. The postsynaptic current can be summed up at the axon hillock.
4. The relationship between the firing rate and the membrane current at the initial segment obeys the law of a sigmoid function.
5. The transmitter releasing by a presynaptic spike has quantum property; thus, only a specific quantity of the postsynaptic current can be induced.

6. The sensory neuron can transduce the outside stimulus into change of the membrane current.
7. An all-or-none spike will be fired if the membrane potential at the axon hillock reaches a threshold; after that, the membrane potential restores to its resting state.

Thus, the membrane current is the input variable and the firing rate of the spike is the output variable in his model. And a frequency coding assumption is implied.

Owing to the all-or-none law of the spike, this can be represented by a Dirac δ function. As every spike will induce a specific quantity of the change of the postsynaptic membrane current and the membrane is equivalent to a paralleled capacitor and resistor, thus the temporal course of the postsynaptic membrane current could be described by the following equations:

$$i_{kj}(t) = 0, \qquad if\, t < t_0 \tag{3.82}$$

$$i_{kj}(t) = S_{kj} exp(-(t - t_0)/\tau), \qquad if\, t > t_0 \tag{3.83}$$

where $i_{kj}(t)$ is the postsynaptic membrane current of the k-th neuron which forms a synapse with the terminal of the j-th neuron, S_{kj} is the peak value of this membrane current, t_0 is the moment at which there reaches a spike at the terminal of the j-th neuron, and τ is the time constant of the postsynaptic membrane. S_{kj} is positive when the synapse is excitatory and negative when the synapse is inhibitory.

If the k-th neuron fires at the moment t_n^k $(n = 1, 2, 3, \ldots)$, then its transient firing rate will be

$$f_k(t) = \sum_n \delta\left(t - t_n^k\right) \tag{3.84}$$

It is easy to see that the integration of $f_k(t)$ in some period is just the number of the spikes fired within this interval; thus, it has the meaning of the transient firing rate.

Differentiate $i_{kj}(t)$ with t:

$$\frac{di_{kj}}{dt} = -\frac{i_{kj}}{\tau} + S_{kj}\delta\left(t - t_n^k\right) \tag{3.85}$$

Sum up all the equations in Eq. (3.85) with different values of j and n; then the membrane current of the k-th neuron at its axon hillock is

$$\frac{di_k}{dt} = -\frac{i_k}{\tau} + \sum_j S_{kj} f_j(t) + I_k(t) \tag{3.86}$$

where I is the injected current from the outside.

As it is assumed that the firing rate–membrane current curve is sigmoid, thus another equation of the neuron can be listed as follows:

$$f_j(t) = V\left(i_j(t)\right) \tag{3.87}$$

where $V(\cdot)$ is a sigmoid function.

Thus the Hopfield model can be described with Eqs. (3.86) and (3.87) or combine them into the following equation:

$$\frac{di_k}{dt} = -\frac{i_k}{\tau} + \sum_j S_{kj}V\big(i_j(t)\big) + I_k(t) \tag{3.88}$$

Here, Eq. (3.88) is only an example of how people develop a neuron model based on some properties of a living neuron under an assumption that the neural coding is rate coding. As for the detailed discussion of Hopfield's neural network, it will be given in Sect. 6.1.

Most of the available simplified neuron models are based on part of the following properties of the Golgi type I neuron with chemical synapses:

1. All the synapses are axodendritic synapse or axosomatic synapse, where they play the role of inputs.
2. There are two types of synapses: excitatory synapse and inhibitory synapse.
3. There is some delay at the synapse.
4. The neuronal membrane is equivalent to a paralleled circuit with capacitors and resistors, if the membrane potential is below a threshold.
5. There is spatial summation and temporal summation for the subthreshold potentials.
6. There is a threshold; the neuron will fire only when the membrane potential at the trigger zone reaches the threshold.
7. The action potential obeys the all-or-none law.
8. There is a refractory period after firing, which includes two phases: absolute refractory period and relative refractory period.
9. There is an after-potential component of the action potential.
10. There is a sigmoid function or another nonlinear monotonic increasing function relationship between the firing rate and the membrane current which induces the firing.

Majority of the available simplified neural network models adopt some of the above properties as their basic assumptions. Some take more and the others take less. For example, the classical McCulloch–Pitts model adopted the properties 1, 2, 3, 5, 6, 7, and 8 as its basic assumption.

According to the fact if the tenth property is adopted as one of its basic assumptions, the simplified neuron model can be categorized into two types: In the first category, the tenth property is adopted as the basic assumption, as a sigmoid function is continuous and smooth enough; it is suitable for mathematical analysis and most of the artificial neural networks adopted such a neuron model as its unit; the Hopfield model is a typical example of such category. However, for these models, their outputs are firing rates, which imply that the neural coding is rate coding. As we will show later in this book, this is not universal. Koch called it a "rate neuron" (Koch and Segev 2000). The second category just omits the tenth property and considers the spike train as its outputs, which implies that the information is carried by the timing of the spike train; such a neuron is called a "spiking neuron," and the integrate-and-fire model is a typical example of such neuron model.

However, it should be noted that even if all the above properties are considered as the basic assumptions, the neuron model discussed in this section only partly simulates the Golgi type I neuron with chemical synapse, but not all kinds of neurons, especially Golgi type II neurons without an axon, which may play some role in the information processing in the core of local neural circuits, which we don't understand much. Within the knowledge of the author, until now there is a rare, if there is any, neuron model of Golgi type II neuron without an axon. If an artificial brain model is to be built using a simplified neuron model as its element, then the neuron model for Golgi type II must be developed. Models of glia cells should also be considered, even the physiological functions of glia cells are still known poorly.

In addition, even only for the Golgi type I neuron, the above ten properties cannot cover everything for these neurons on one hand and some of them are not kept rigorously under some conditions, such as the so-called all-or-none law and a fixed threshold. Moreover, some of the above properties may even not be true in some case. For example, according to the fifth property, all the inputs to a neuron seem to be addable; however, there exists such a case, in which some inputs are not addable. If there is an excitatory synapse at the distal end of a dendrite and, at the proximal part of the dendrite, there is an inhibitory synapse with a $GABA_a$ receptor, the current induced by the activation of the excitatory synapse flows to the activated inhibitory synapse, where it may be shunted (shunting inhibition), a nonlinear interaction; besides, in the case of presynaptic inhibition, two axon terminals from different neurons form an axoaxonic synapse, where the presynaptic axon terminal inhibits the postsynaptic axon terminal; the total operation might be multiplication. In the above, all the dendrites are considered as passive; their effects seem to be only to transmit currents or at most make some delay and low-pass filtering. However, dendrites may have active properties; they may amplify or multiply inputs. All the above problems are largely still unexplored in neural network studies.

In addition, the plasticity of the synapse has not been discussed in the above; we will discuss this problem in detail in the sections about learning and memory later.

3.5 Dynamics of a Single Neuron

3.5.1 Monostability vs. Multistability and Integrator vs. Resonator

In the last section, it has been shown that in many available neuron models, it is assumed that there is a sigmoid function relationship between the firing rate and the stimulus intensity, i.e., the firing rate will increase gradually with the increase of the stimulus intensity until approaching some saturation level, once the stimulus is above some threshold. Historically, people thought that this is a basic property of the neuron. However, it is Hodgkin (1948) himself who found that there is some neuron which only fires when the stimulus intensity is strong enough; however, if such threshold is reached, the neuron fires with a rather higher firing rate and the rate increases very slowly with the stimulus intensity within a narrow range. It is obvious that for such a neuron there is no sigmoid function relationship between the firing rate and stimulus intensity. Such distinct difference hadn't been paid enough attention until Rinzel and Ermentrout (1989) pointed out that the different behaviors could be explained by different bifurcation[4] mechanisms of excitability within a unified theoretical framework from a dynamical system viewpoint. And only after that, people begin to realize: "Information processing depends not only on the electrophysiological properties of neurons but also on their *dynamical properties*. Even if two neurons in the same region of the nervous system possess similar electrophysiological features, they may respond to the same synaptic input in very different manners because of each cell's bifurcation dynamics." "From a dynamical systems point of view, neurons are excitable because they are near a transition, called bifurcation, from resting to sustained spiking activity. While there are a huge number of possible ionic mechanisms of excitability and spike generation, there are only four bifurcation mechanisms that can result in such a transition" (Izhikevich 2007).

As what we summarized in the last section, for most of the simplified neuron models, their prerequisite hypotheses include the following two properties: there is some well-defined threshold; the neuron will fire only if the weighted sum of the postsynaptic potential is above this threshold; otherwise, there is only a non-propagating local subthreshold graded potential; if the weighted sum of a postsynaptic potential at axonal hillock reaches a threshold, an all-or-none action potential or spike will be produced and propagate along the axon. All the formal neuron model, integrate-and-fire model, and leaky integrate-and-fire model accept the above hypotheses. However, as what we indicated in Sect. 3.3, such statements stand only approximately for the F–N model and so do for the H–H model and real neuron. Therefore, Izhikevich said in his excellent book (Izhikevich 2007): "If you find one (well-defined threshold), let the author know!"

In addition, both in real neuron, the F–N model, and the H–H model, people can find an anodal break excitation (also called as a rebound spike or postinhibitory spike), i.e., a neuron fire in response to an inhibitory input after the input stops. In some case, a neuron is stimulated with brief pulses of current; the neuron only fires when the repetitive rate of the pulse series is proper, i.e., a resonant frequency; the neuron will not fire when the stimulus rate is either too slow or too fast.

Besides, neurons can show huge varieties of rhythms which may make laymen confused and feeling dizzy. However, from the dynamic viewpoint, the underlying

[4] The so-called bifurcation is a transition of one kind of attractor to another kind of attractor, i.e., the system's behavior takes place dramatic change qualitatively, such as from an equilibrium to limit cycle (periodic oscillation).

mechanisms are not so complicated if the basic modes of bifurcation are understood.

All the above diversities in the neuron behavior cannot be replicated in simplified neuron models. Although they may be simulated by the H–H model or its simplified versions with proper parameter combinations, it would be difficult to show the essence of such diversities, if there is no certain unified theoretical framework to give a few basic rules to summarize all these behaviors. And the nonlinear dynamical theory just provides such an ideal theoretical framework.

Hodgkin's Class 1 Neural Excitability and Class 2 Neural Excitability

As what we mentioned above, Hodgkin found that there were different classes of responses to pulse stimuli of various amplitudes in squid axons. Mainly, they could be classified into the following two classes: being able to fire with arbitrarily low frequency is the so-called class 1 neural excitability (Fig. 3.19a), while only firing with frequency in a certain narrower band, which is relatively insensitive to stimulus strength changes, is the so-called class 2 neural excitability (Fig. 3.19b).

Hodgkin's finding was long before the bifurcation concept was proposed; thus, it was not paid enough attention for a long time, and most of neuron modelers just neglected class 2 excitability. Only after the mathematical theory of bifurcations was developed, people realized that "…there could be millions of different electrophysiological mechanisms of excitability and spiking, but there are only four – yes, four – different types of bifurcations of equilibrium that a system can undergo without any additional constraints, such as symmetry" (Izhikevich 2007). The four bifurcations are summarized in Fig. 3.20. In these figures, only one parameter (the injected DC current magnitude I) is varied; the membrane potential with activation Na^+ current is the abscissa variable, while the recovery current such as K^+ is the ordinate variable.

It has been pointed out that class 1 excitability corresponds to saddle node on invariant circle bifurcation (Fig. 3.20 upper right) and class 2 excitability corresponds to the other three bifurcation modes (Izhikevich 2007). To simplify analysis, assume that neurons only receive impulse input, so that the phase portraits can keep unchanged, while the neuron state is pushed away to other places in the portrait. From Fig. 3.20, it is easy to know that only if an input can push the state of the system into the shaded area, the neuron can initiate spiking, while spiking stops if the state is pushed into the white area. Thus, for the right half cases in Fig. 3.20, in which there are no stable limit cycle, the trajectory starting from shaded area will make a large-amplitude rotation before returning to the equilibrium, which represents a transient spike. In such two cases, the system is monostable. In contrast, for the left half cases in Fig. 3.20, in which there are both a stable equilibrium and a stable limit cycle, the system is multistable.

If we compare the portraits shown in the upper half and lower half in Fig. 3.20, it is easy to find that inhibition impedes spiking for the former but promotes it for the

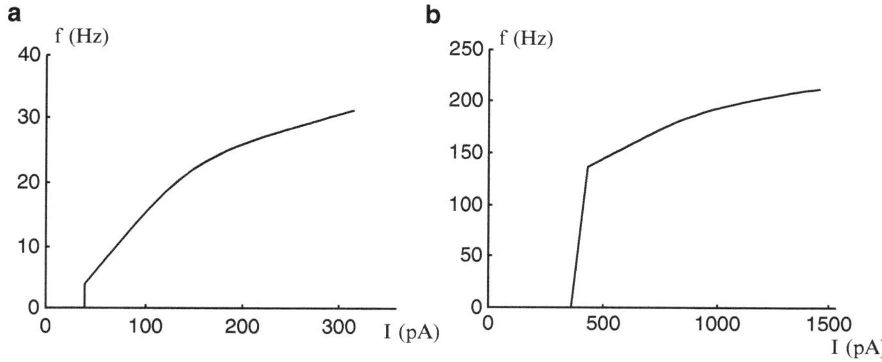

Fig. 3.19 *Frequency–current (F-I) curves of neurons with class 1 excitability* (**a**) *and class 2 excitability* (**b**). *f* firing frequency, *I* injected current

latter. This is owing to the fact that the shaded region is to the right of the equilibrium for the former; thus, excitatory inputs push the state of the system toward the shaded region, while inhibitory inputs push it away. For the latter, there is a shaded area in any direction around the equilibrium; thus, both excitation and inhibition may push the state toward the shaded area. This is why there is rebound spiking phenomenon for the latter. Izhikevich called the former "integrator" and the latter "resonator."

Izhikevich (2007) pointed out the main difference between integrators and resonators as follows:

1. "Integrators have well-defined thresholds; resonators may not." As the white circles near the resting states of integrators are saddles, they are stable along the vertical direction and unstable along the horizontal direction. The vertical lines play the role of threshold since only a perturbation that push the state beyond this line can result in spiking, although it is not a fixed value but a range.
2. "Integrators integrate, resonators resonate." If there is a series of impulse input to integrators, every impulse pushes the state near the vertical line by one step, and the more the impulse and the denser the series, the faster the state reaches the threshold line and the effect is integrated. As for the resonator, whether the next pulse impedes or promotes spiking depends on its timing as shown in Fig. 3.20. For example, the timing marked "2" pushes the state to the equilibrium, while the timing marked "3" pushes the state toward a shaded area and may initiate a spike if the impulse is strong enough.

Thus, the diversity of neuronal behavior could be explained using a unified point of view with the dynamic system theory, as exemplified above.

In summary, neurons can be classified into four different kinds as shown in Fig. 3.20, according to two criteria: monostable vs. multistable and integrator vs. resonator. Great diversities of neuronal behavior now can be explained with such a theoretical framework.

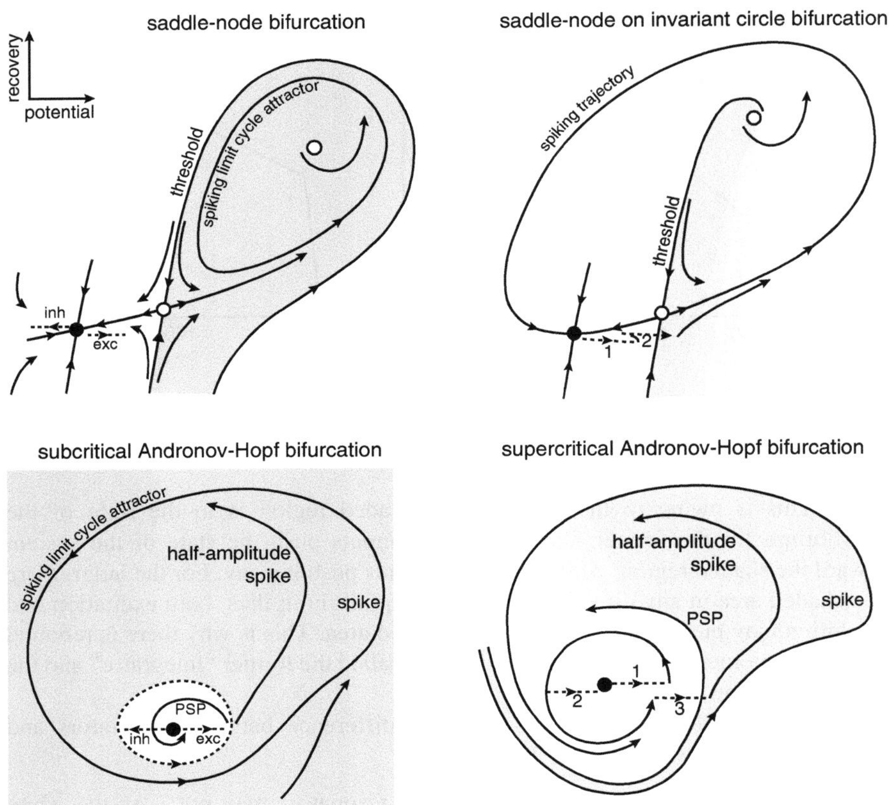

Fig. 3.20 *The geometry of phase portraits of excitable systems near four bifurcations can explain the diversity of neuronal properties with a unified dynamic framework.* In this figure, *black circles* denote stable equilibriums; *white open circles* represent unstable equilibriums and limit cycles mean sustained spiking. *Shaded regions* denote attraction basin of limit cycles, while *white regions* denote attraction basin of equilibriums (Reproduced with permission from Izhikevich (2007). Fig. 1.14)

3.5.2 Neuronal Excitation Rhythms

As we mentioned above, the information carried by a single neuron is represented with the temporal pattern of its spike train or, in other words, with its rhythm. There are a variety of complicated patterns, representing various rhythms. When the neuron is stimulated differently, its firing rhythm is different; its rhythm also changes when the inside or outside environment (such as the various ionic concentrations in the extracellular fluid or the intracellular calcium concentration, etc.) changes. The theoretical studies on nonlinear dynamics tell us that the behavior of a nonlinear dynamical system can experience systematical changes when there are changes of a stimulus or its parameters in a wide range, such as from a resting state to periodic rhythmic activities with different periods until a unpredictable chaotic pattern for its long run or even close to a stochastic pattern. Its behavior pattern may take place as qualitative and abrupt changes at some critical points, i.e., some bifurcation may happen. Thus, the nonlinear dynamics gives us a theoretical framework to study the above complicated phenomena systematically: Under which conditions, a single neuron will stay at its resting state or varies with various periodic or even chaotic oscillations. After the above basic rhythms are understood thoroughly, how the real spike train is constructed by these basic rhythms may be understood from the point of view of bifurcation. Reversely, the stimuli which a neuron receives or the inner- or outer-environment changes might be deduced from the spike train of that neuron based on the above understanding.

In this section, we will discuss the above problem. The term "rhythm" means some elementary event that takes place again and again; it may be periodic, bursting, chaotic, or even stochastic. In our case, the elementary event is the generation of a spike, which is a fast process and was discussed in most of the above sections. In addition, producing a series of spikes with different rhythms needs some slower processes to modulate the generation of spikes, as what the Hindmarsh–Rose model or Ekeberg model suggested. The factors which modulate rhythms can be classified into two categories: (1) some autonomic, intrinsic factors such as intracellular calcium-dependent potassium conductance and (2) external driving inputs, such as synaptic inputs or injected electrical currents (Ren 2004).

Autonomic, Intrinsic Factors

Now it is known that there are more types of ion channels besides voltage-gated potassium channels and voltage-gated sodium channels in the membrane; the ionic currents passing through such channels may be involved in the slow process. For example, changes in intracellular calcium concentration or calcium-dependent potassium conductance may be such factors.

Thus, to simulate autonomic rhythmic behavior, the model must be composed of both a fast subsystem such as the H–H model to generate spikes and a slow subsystem to modulate its rhythms. The two subsystems may interact with each other; for example, repetitive firing may increase intracellular calcium concentration and the latter may change the intervals between successive spikes generated by the fast process, thus forming a variety of different spiking patterns.

Li Li et al. (2004) reported their observation of a transition of a variety of spiking patterns on a nervous pacemaker formed at an injured site of a rat sciatic nerve subjected to chronic ligatures by decreasing extracellular calcium concentration in the bathing fluid systematically. In addition, they also found that such transition might show different scenarios owing to slight difference between different nerve preparations.

They classified the bifurcation scenarios into the following three types: (1) type 1 was a bifurcation scenario from period-1 bursting to period-1 spiking via period-2 bursting; (2) type 2 was a bifurcation scenario form period-1 bursting to period-1

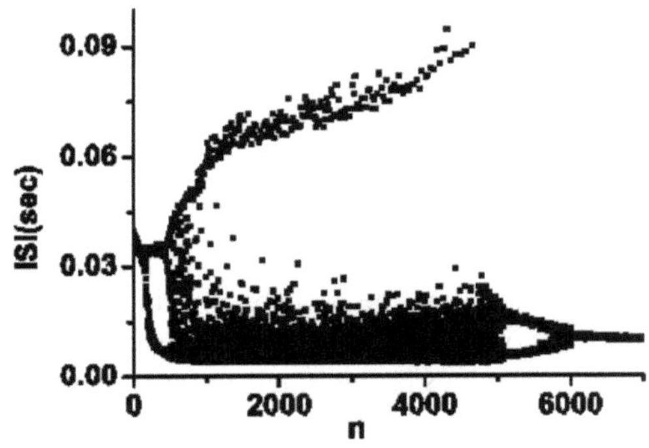

Fig. 3.21 Bifurcation scenario from period-1 bursting to period-1 spiking via a complex procedure including period-doubling bifurcation cascade to chaos. (Reproduced with permission from Li et al. (2004). Fig. 1)

spiking via a period-doubling cascade of bursting and an inverse period-doubling cascade of spiking (Fig. 3.21); (3) type 3 was a bifurcation scenario of period-adding bifurcation without chaos.

The above different scenarios seem very complicated and confusing; however, simulation with the Chay model (Chay 1990; Chay et al. 1995)—a variation of the H–H model with modulation of spike firing by the intracellular calcium concentration—will give insight of the underlying mechanism.

The Chay model can be listed as follows:

$$\frac{dV}{dt} = g_I {m_\infty}^3 h_\infty (v_I - V) + g_{Kv} n^4 (v_K - V) + g_{KC} \frac{C}{1+C} (v_K - V) + g_l (v_l - V)$$
$$\frac{dn}{dt} = \frac{n_\infty - n}{\tau_n}$$
$$\frac{dC}{dt} = \frac{m_\infty^3 h_\infty (v_C - V) - k_C C}{\tau_C}$$

where g_I, g_{KV}, g_{KC}, g_l is the maximal conductance of the Na^+–Ca^{++} channel, K^+ channel, intracellular calcium concentration (C)-dependent K^+ channel, and Cl^- channel, respectively; v_I, v_K, v_L is the equilibrium potential of Na^+–Ca^{++} channel, K^+ channel, and Cl^- channel, respectively; V is the membrane potential; m_∞ and h_∞ are the probability of opening the two gates for Na^+–Ca^{++} channels, respectively; n_∞ is the probability of opening the gate of K^+ channel; τ_n is a constant and $\tau_n = \frac{1}{\lambda_n(\alpha_n+\beta_n)}$; v_C is the equilibrium potential of Ca^{++}; and τ_C is a constant. To be compared with the experimental data, they took λ_n as the conditional parameter; different λ_n values represent different nerves; while they took v_C as a bifurcation parameter, the change of v_C value corresponds to the change of extracellular calcium concentrations. Figure 3.22 shows the result of their simulation for type 2. To save space, only type 2 are shown both in Fig. 3.21 and Fig. 3.22 respectively. The figures for experimental data and simulation for type 1 and type 3 are omitted. Comparing the simulation results with their corresponding experimental data is so well that they concluded: "Using a single model and adjusting two

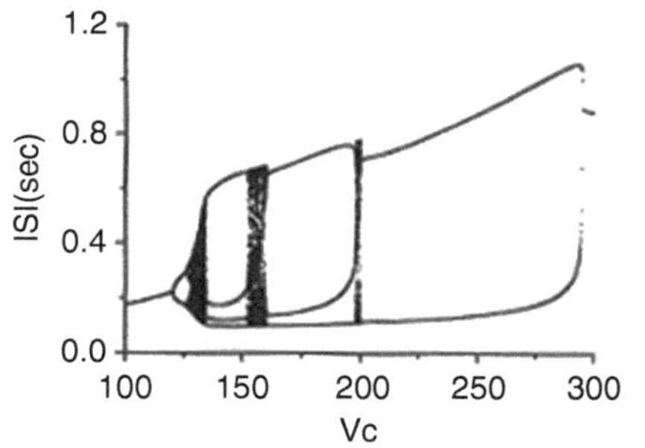

Fig. 3.22 *Type 2 of bifurcation scenarios in the firing pattern transitions simulated with the Chay model, where* $\lambda_n = 225.8$ (Reproduced with permission from Li et al. (2004). Fig. 3)

experimentally related parameters, we realized that the three very different bifurcation scenarios were related and caused by the differences in the configuration of parameters."

External Driving Factors

Giving external driving inputs corresponds to introduction of some slow subsystem to modulate the fast subsystems generating action potentials. Aihara and Matsumoto (1986) found that the giant axon of the squid could show chaotic behavior with sinusoidal driving current, and they also found different types of bifurcation scenarios in the firing pattern transitions in such axon. Solving H–H equations with sinusoidal input, the above phenomena had also been simulated by them. After their pioneering work, a series of works on dynamic behavior of a single neuron or single nerve have been made (Faure and Korn 2001; Korn and Faure 2003).

3.6 Summary and Advanced Readings

Owing to their biological importance and being well studied and modeled, models of single neurons have been picked up at the very beginning of the parts discussing neural computation in this book as an example to show how computational neuroscientists use mathematical and informatic ideas and approaches to solve biological problems.

In the first section, the biological problem which Hodgkin and Huxley faced was explained and how they proposed that there were active potassium channels and sodium channels at the neuronal membrane based on the contradiction in Bernstein's theory to the experimental data and how they used mathematical and physical thinking to developed a circuit model of neuronal membrane, combining with new biological experiments to test their hypothesis with mathematical modeling, especially to predict new biological phenomena, are discussed.

In the second section, considering that the H–H model is only a model of a patch of neuronal membrane with voltage-gated ion channels, a neuron is considered as a multi-compartmental system, the compartments of which can be modeled by H–H-like models. Thus, an example of a realistic neuron model was introduced in this section. And at the same time, owing to the variability of neurons, the model introduced in this section is only an example of such models. In addition, owing to the higher dimensionality of the model and since no analytic solution can be got, only a numerical solution can be calculated for a very specific condition.

In the third section, to overcome the difficulties listed in the above section, a simplified version of the H–H model with only two dimensions was introduced, so that its global behavior can be viewed in a phase plane. The FitzHugh–Nagumo model is introduced as an example of such models. In addition, the H–H model and F–N model only solve the genesis and propagation problem of action potentials with only potassium channels and sodium channels, which cannot elucidate the essence of the rhythm of the spike train. More channels with much slower kinetics have to be introduced; in this way, the new model not only is more realistic but also can elucidate the underlying mechanism of spike rhythms. The Hindmarsh–Rose model is introduced as an example of such models.

In the fourth section, further simplification was made, to make the model easier to be treated as an element of a network. In such models, the underlying subcellular mechanisms were neglected; only the input–output relationship is considered. Two kinds of such models, the spiking neuron model and rate neuron model, are considered. Although some basic properties of a Golgi type I neuron were summarized as the presumption of such models, as a matter of fact, some of the assumptions are not rigorously valid such as the all-or-none law and fixed threshold of action potentials and the variability of neuronal behaviors could not be elucidated by such models.

In the fifth section, neurons are considered as some dynamic systems; the variability of their behaviors can be explained with nonlinear dynamical approaches. Thus a problem is raised, which are the basic functional units of nervous systems? Neurons or ion channels? Even if ion channels were considered as the units owing to the variability of neurons, then a similar question can also be raised due to the variability of ion channels. Thus, to discuss the universal basic functional units of the nervous system might be meaningless. Only at a specific level, and for a concrete problem, some kind of neurons or ion channels or something else such as a neuronal assembly or an even higher module could be selected as the basic units and only for that problem or similar problems.

3.6.1 Advanced Reading

[1]. Izhikevich (2007) Dynamical Systems in Neuroscience: The Geometry of Excitability and Bursting. The MIT Press.
[2]. Sterratt et al. (2011) Principles of Computational Modelling in Neuroscience. Cambridge University Press.

[3]. Plonsey and Barr (2007) Bioelectricity - A Quantitative Approach (3rd Edition), Springer
[4]. Lytton WW (2002) From Computer to Brain – Foundations of Computational Neuroscience. Springer.
[5]. Trappenberg (2002) Fundamentals of Computational Neuroscience. Oxford University Press
[6]. Dayan P and Abbott LF (2001) Theoretical Neuroscience - Computational and Mathematical Modeling of Neural Systems, The MIT Press
[7]. Koch (1999) Biophysics of Computation: Information Processing in Single Neurons. Oxford University Press.

References

Aihara K, Matsumoto G. Chaotic oscillations and bifurcations in squid giant axons. In: Holden AV, editor. Chaos. Princeton: Princeton University Press; 1986. p. 257–69.

Bahill AT. Bioengineering – biomedical, medical and clinical engineering. Englewood Cliffs: Prentice-Hall Inc; 1981.

Bernstein J. Untersuchungen zur Thermodynamic der bioelektrischen Ströme. Pflugers Arch. 1902;92:521–62.

Chay TR. Electrical bursting and intracellular Ca2+ oscillations in excitable cell models. Biol Cybern. 1990;63:15–23.

Chay TR, et al. Bursting, spiking, chaos, fractals and universality in biological rhythms. Int J Bifurcat Chaos. 1995;5:595–635.

Ekeberg O, Wallen P, Lansner A, et al. A computer based model for realistic simulations of neural networks – I. The single neuron and synaptic interaction. Biol Cybern. 1991;65:81–90.

Ermentrout GB, Terman DH. Mathematical foundations of neuroscience. New York: Springer; 2010.

Faure P, Korn H. Is there chaos in the brain? I. Concepts of non-linear dynamics and methods of investigation. Life Sci. 2001;324:773–93.

Finger S. Origins of neuroscience – a history of explorations into brain function. New York: Oxford University Press; 1994.

FitzHugh R. Impulses and physiological states in theoretical models of nerve membranes. Biophys J. 1961;1:445–66.

Gerstner W. Integrate-and-fire neurons and networks. In: Arbib MA, editor. The handbook of brain theory and neural networks. 2nd ed. Cambridge, MA: MIT Press; 2002.

Hindmarsh JL, Rose RM. A model of the nerve impulse using two first-order differential equations. Nature. 1982;296:162–4.

Hindmarsh JL, Rose RM. A model of neuronal bursting using three coupled first-order differential equations. Proc R Soc London, Ser B. 1984;221:87–103.

Hodgkin AL. The local electrical changes associated with repetitive action in a non-modulated axon. J Physiol. 1948;107:165–81.

Hodgkin AL. The ionic basis of nervous conduction. Nobel Lecture, December 11, 1963.

Hodgkin AL, Huxley AF. Action potentials recorded from inside a nerve fibre. Nature. 1939;144 (3651):710.

Hodgkin A, Huxley A. A quantitative description of membrane current and its application to conduction and excitation in nerve. J Physiol. 1952;117:500–44.

Hopfield J. Neurons, dynamics and computation. Phys Today. 1994;47:40–6.

Izhikevich EM. Dynamical systems in neuroscience: the geometry of excitability and bursting. Cambridge, MA: The MIT Press; 2007.

Izhikevich EM, FitzHugh R. FitzHugh-Nagumo model. Scholarpedia. 2006;1(9):1349.
Kandel ER, Schwartz JH, Jessell TM. Principles of neural science. 4th ed. New York: McGraw-Hill; 2000.
Keating SM, Novère NL. Encoding neuronal models in SBML. In: Novère NL, editor. Computational systems neurobiology. Dordrecht: Springer; 2012. p. 459–88.
Kistler WM, Gerstner W, Van Hemmen JL. Reduction of Hodgkin-Huxley equations to a single-variable threshold model. Neural Comput. 1997;9:1015–45.
Koch C. Computation and the single neuron. Nature. 1997;385:207–10.
Koch C. Biophysics of computation: information processing in single neurons. New York: Oxford University Press; 1999.
Koch C, Segev I. The role of single neurons in information processing. Nat Neurosci. 2000;3:1171–7.
Korn H, Faure P. Is there chaos in the brain? II. Experimental evidence and related models. C R Biol. 2003;326:787–840.
Lapicque L. Recherches quantitatives sur l'excitation électrique des nerfs traitée comme une polarisation. J Physiol Pathol Gen. 1907;9:620–35.
Li L, et al. A series of bifurcation scenarios in the firing pattern transitions in an experimental neural pacemaker. Int J Bifurcation Chaos. 2004;14:1813–7.
McCulloch W, Pitts W. A logical calculus of the ideas immanent in nervous activity. B Math Biophys. 1943;7:115–33.
Nagumo J, Arimoto S, Yoshizawa S. An active pulse transmission line simulating nerve axon. Proc IRE. 1962;50:2061–70.
Plonsey R, Barr RC. Bioelectricity: a quantitative approach. 3rd ed. New York: Springer; 2007.
Rall W. Perspective on neuron model complexity. In: Arbib MA, editor. The handbook of brain theory and neural networks. 2nd ed. Cambridge: MIT Press; 2002.
Ramachandran VS. The astonishing Francis Crick. Perception. 2004;33:1151–4.
Ren W. Neurodynamic studies on neuronal excitation rhythms. Fudan Lectures in Neurobiology. 2004;XX:109–24 (in Chinese).
Rinzel J, Ermentrout GB. Analysis of neural excitability and oscillations. In: Koch C, Segev I, editors. Methods in neuronal modeling. Cambridge, MA: The MIT Press; 1989.
Seyfarth E-A. Julius Bernstein (1839–1917): pioneer neurobiologist and biophysicist. Biol Cybern. 2006;94:2–8.
Skinner FK. Conductance-based models. Scholarpedia. 2006;1(11):1408.
Sterratt D, Graham B, Gillies A, Willshaw D. Principles of computational modelling in neuroscience. New York: Cambridge University Press; 2011.
Trappenberg TP. Fundamentals of computational neuroscience. Oxford: Oxford University Press; 2002.
Van der Pol B. A theory of the amplitude of free and forced triode vibrations. Radio Rev. 1920;1:701–10, 754–63.
Van der Pol B, Van der Mark J. Frequency demultiplication. Nature. 1927;120:363–4.

Chapter 4
Neural Coding

Abstract In traditional studies, neuronal response was normally characterized by measuring the neuron's firing rate changes in exposure to particular stimulation. But recent works provided evidence that some other aspects of neuronal activities greatly contribute to neural information coding. This chapter will be devoted to coding properties of neuronal activities, including rate coding and precise time coding, the basic ideas about dynamic population coding based on spatiotemporal activities of multiple neurons, the theory of population coding, and Bayesian inference. Relevant analytical approaches will be introduced, together with some typical research examples.

Keywords Neural coding • Single neuron coding • Population coding • Bayesian inference • Reinforcement learning

Through the preceding chapters, we have got to know that the nervous system transmits information via action potentials. In other words, action potentials are the basic elements of the neural "language" and provide the fundamental process for neural information transmission. Although the generation and propagations of action potential, as well as the relevant membrane mechanism, are well understood, we still know relatively little about the following issues, i.e., in which form is information carried by the action potentials and how to quantitatively measure the information amount transferred, etc. In traditional studies, neuronal response was normally characterized by measuring the neuron's firing rate changes in exposure to particular stimulation via neuroelectrophysiological approaches. But recent works provided evidence that some other aspects of neuronal activities, such as the precise timing of firing sequence, as well as the spatiotemporal characters of population activities, greatly contribute to neural information coding. In the meantime, neural information can also be stored via the adjustment of synaptic strength. Therefore, to study the information coding property of the nervous system, the abovementioned aspects of the neuronal activity should be quantitatively measured, and it will be helpful to incorporate mathematical tools to analyze the results obtained from biological experiments which are properly designed and conducted.

The researches incorporating neurobiological experiments and informatics approaches can be categorized into two kinds. While one is to examine the properties of neurons' activity in response to particular stimulations, the other is

P. Liang et al., *An Introduction to Neural Information Processing*,
DOI 10.1007/978-94-017-7393-5_4

to judge the information content that the neurons' activity carries according to the neuronal activation properties, which actually reflects the working process of the genuine nervous system, although it is more difficult to carry out this kind of work.

This chapter will be devoted to coding properties of neuronal activities, including rate coding and precise time coding, the basic ideas about dynamic population coding based on spatiotemporal activities of multiple neurons, the theory of population coding, and Bayesian inference. Relevant analytical approaches will be introduced, together with some typical research examples.

4.1 Rate Coding

One of the fundamental issues in neuroscience is how the information is encoded by the neural activity. It was forwarded by Lord Adrian in early 20s century that the firing rate of sensory neuron is related to the stimulation intensity. It is thus well accepted that firing rate encodes information about the stimulation parameters. Up to recent days, most of the knowledge about sensory neural coding comes from similar observations, which show that the neuron's firing rate is related to various stimulation parameters.

From basic knowledge about sensory systems, we know that such stimulation-property-dependent firing rate changes are ubiquitous in sensory neurons. One example in visual system is retinal ganglion cell. As has been introduced in Sect. 2.5, for a ganglion cell with on-center receptive field property, if a light spot is projected onto its receptive field center, its firing rate will be increased, and more increment is related to higher luminance level of the light; in the meantime, when light is projected on the peripheral part of its receptive field, the cell's firing activity will be reduced. A full-field light which covers both receptive field center and surround causes no obvious change in the cell's firing activity, due to the counter-balance between the excitatory and inhibitory signals. The firing rate changes observed from off-center ganglion cell is exactly the opposite way.

As for a particular kind of ganglion cells which is specified as motion–direction sensitive, their firing rate is modulated by the motion direction of moving target. When visual stimulation enters the cell's receptive field with a particular direction, its firing activity will be most intensively elicited, and such a direction is identified as the cell's "preferred direction," but when the visual target is moving into its receptive field along the direction opposite to the preferred direction, the cell will not be activated, and this direction is identified as the cell's "null direction." The cell's activity shows a monotonic reduction when the stimulus direction changes from the preferred direction to the null direction. Figure 4.1 illustrates the response properties of such ganglion cells. In this example, the neuron's activity was presented by the spike count in response to visual stimulation. Given that firing rate refers to spike count within a certain time period, these two measures are

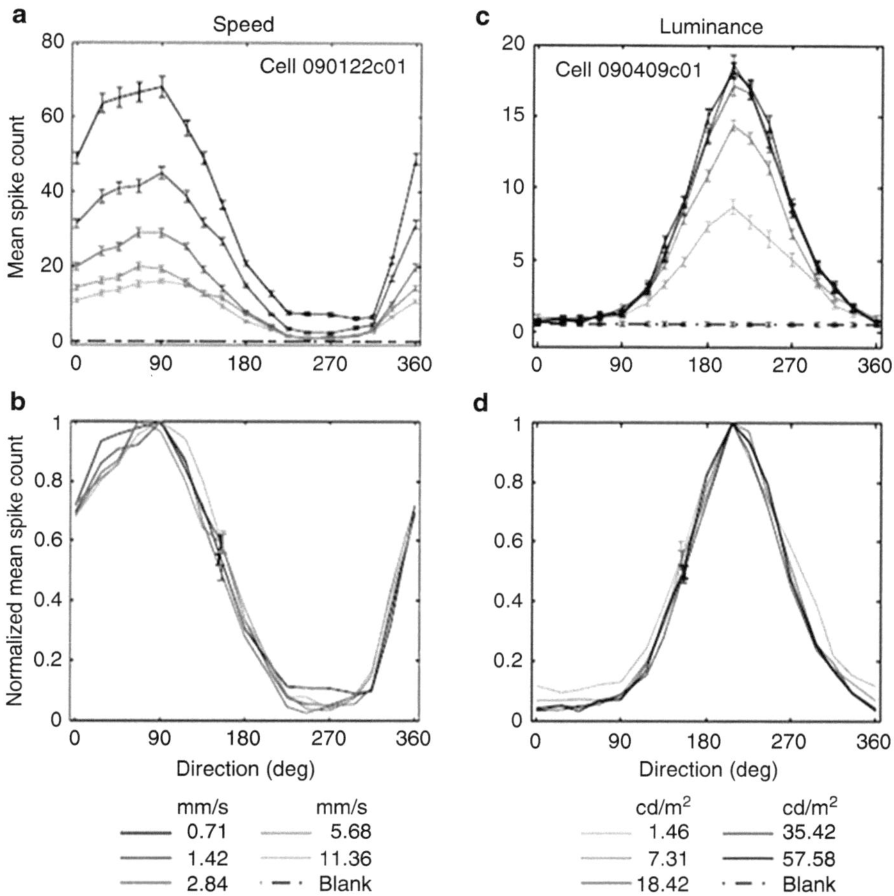

Fig. 4.1 Response properties of rabbit direction-selective ganglion cells. Direction tuning curves were measured either for different speeds or for different luminance levels. (**a**) Direction tuning curves of an example cell measured at 16 directions separated by ~22.5° for a bar with luminance of 57.58 cd/m^2 under 5 different speeds. (**b**) Direction tuning curves same as plotted in (**a**) after being normalized against the peak level of each curve. (**c**) Direction tuning curves of another example cell measured at 16 directions separated by ~22.5° for a bar swept at the speed of 2.84 mm/s for 5 luminance levels. (**d**) Direction tuning curves same as plotted in (**c**) after being normalized against the peak level of each curve (Modified from Nowak et al. 2011, Fig. 3)

closely related to each other. The examples illustrated in Fig. 4.1a, c show that, in addition to motion direction, the cell's spike count was also dependent on other stimulation parameters, such as the moving speed of the visual target, as well as the luminance level of the stimulation. When the moving speed was increased from 0.71 mm/s to 11.36 mm/s, the firing activity was decreased monotonically for the example cell whose "preferred direction" was 90°, and such firing activity reduction was consistent for all the tested motion directions. When the direction tuning curves measured under different moving speeds are normalized against the cell's response to its preferred direction, it is clear that the tuning curves are almost identical (Fig. 4.1b). In the meantime, when the luminance of the stimulation was increased from 1.46 cd/m^2 to 57.58 cd/m^2, firing activity measured from another example cell, whose "preferred direction" was 202.5°, was monotonically increased, such firing activity change was also independent on the motion direction and can be demonstrated by identical normalized direction tuning property under different luminance level.

Such stimulation-feature-dependent firing rate modulation is also a general feature for central visual neurons. In primary visual cortex, the neurons are with elongated receptive field and are thus sensitive to orientation of visual target. When the orientation of visual stimulation fits the long axis of the receptive field, the neuron will be greatly activated due to the activation of the excitatory subareas in its receptive field, but when a visual target being orthogonal to the long axis is presented, the neuron will not be activated because the excitatory signals are cancelled out by the inhibitory signals (Fig. 4.2).

Furthermore, some neurons in the inferior temporal lobe, which is part of the ventral pathway in visual information flow, are selective to complex spatial patterns. A group of neurons are particularly sensitive to faces or face-like patterns, and are named as "face cell" (Fig. 4.3).

In the auditory system, the neuron's firing rate is basically dependent on two parameters: the frequency and intensity of the sound stimulation. Cochlear hair cells, in which sound stimulation is transferred into neuronal signals, are frequency selective. Each cochlear cell is specially sensitive to a particular frequency which is identified as the characteristic frequency of the cell. While the frequency selectivity of cochlear hair cell is determined by its position on the basal membrane, the frequency selectivity of central auditory neuron is dependent on the hair cell from which the neural signal is sent. In exposure to sound stimulation with a certain frequency, a louder sound can elicit more action potentials from the auditory neurons, as has been explained in Sect. 2.6 (see Fig. 2.52).

Taste neurons and olfactory neurons are chemo-sensitive and their activation is dependent on the stimulation properties. While tastant can be roughly classified into four classes, i.e., sweet, salt, sour, and bitter, a taste neuron can generally respond to one or more tastants (but insensitive to others), with the activation degree being concentration dependent. For example, some primary taste fibers are sensitive to more than one sodium salt, as well as HCl, but are insensitive to sweet and bitter stimulations (Fig. 4.4a); in the meantime, some other taste fibers can be most sensitive to HCl and can also respond to NaCl and NH_4Cl (Fig. 4.4b). A similar situation can be found in olfactory neurons such that a certain kind of odorant can activate a small group of olfactory neurons but not the others.

Such general principle also applies to somatic sensory neurons. For example, cold receptor is sensitive to the downregulation of skin temperature. When a cold

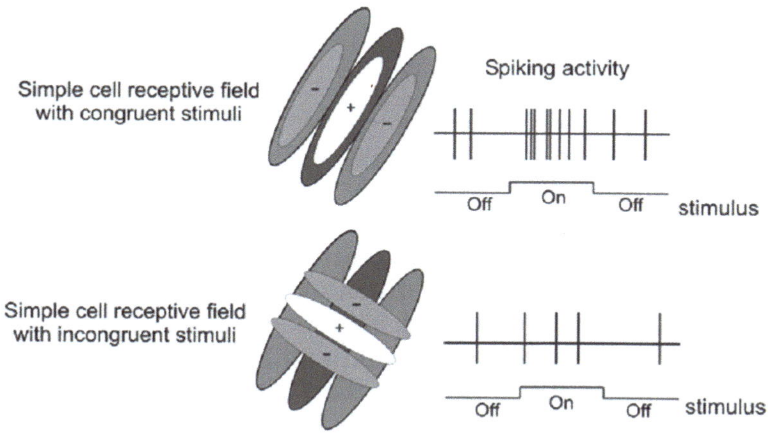

Fig. 4.2 Receptive field property and orientation selectivity of simple cell in the primary visual cortex (Adapted with permission from Pfaff 2013, Fig. 20.10)

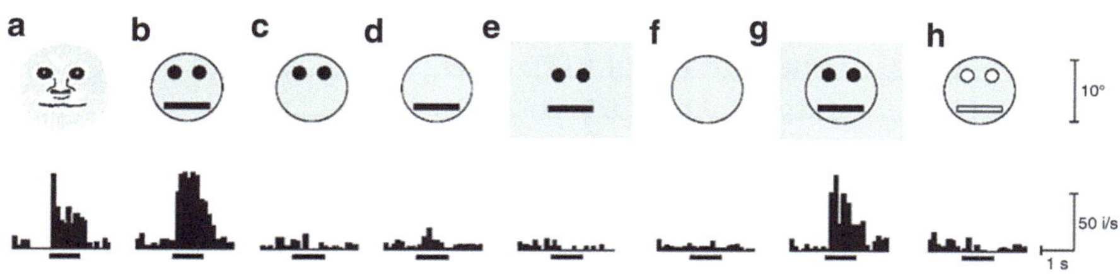

Fig. 4.3 Response of a neuron in the inferior temporal cortex is particularly sensitive to face (Adapted with permission from Kandel et al. 2000, Fig. 28–18)

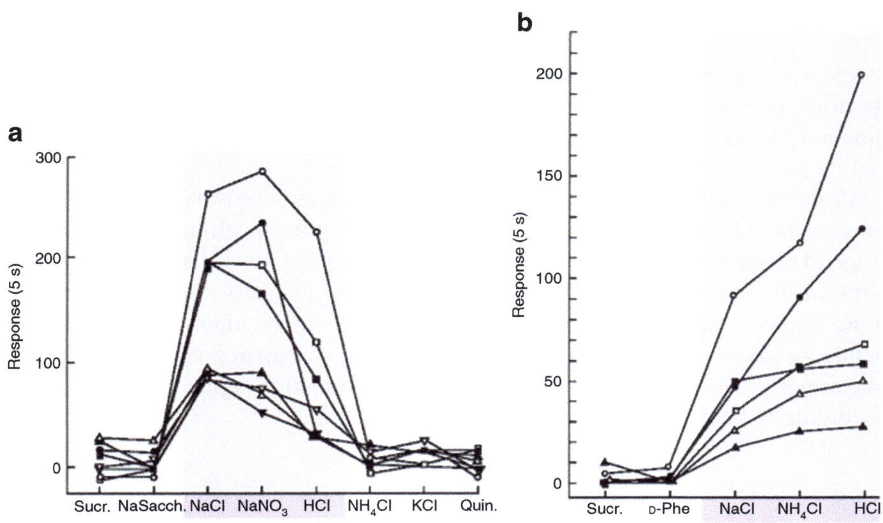

Fig. 4.4 Response profiles of hamster chorda tympani fibers. (**a**) Fibers that respond to sodium; (**b**) fibers sensitive to chloride (Adapted with permission from Kandel et al. 2000, Fig. 32.18)

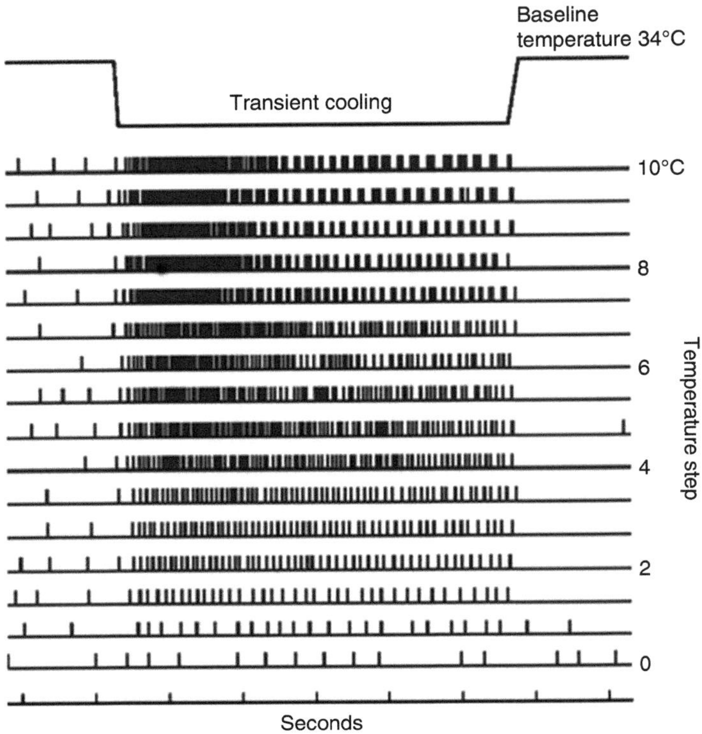

Fig. 4.5 Cooling stimulation elicits temperature-dependent responses in a cold receptor. While mild changes in temperature caused modest increments in the cell's firing activities, severe cold stimulations induced more active firing activities (Adapted with permission from Galizia and Lledo 2013, Fig. 15.7)

receptor is exposed to a mild reduction of skin temperature, a modest increment in its firing rate can be induced, but if it is in exposure to a transient cooling with dramatic temperature change, the cell's firing rate will be greatly increased (Fig. 4.5).

From these cases, one can tell that a common feature of sensory neurons is that a sensory neuron can be activated by a particular type of stimulations (and can respond to more than one type of stimulations for the neurons in chemical systems), with the degree of activation being dependent on the intensity of the stimulus, i.e., more densely fired action potentials are related to more intensive stimulations. While we say that sensory neurons can encode input stimulation via spike activities, one important issue in neuroscience is whether stimulation signal can be identified according to neuron's firing activity.

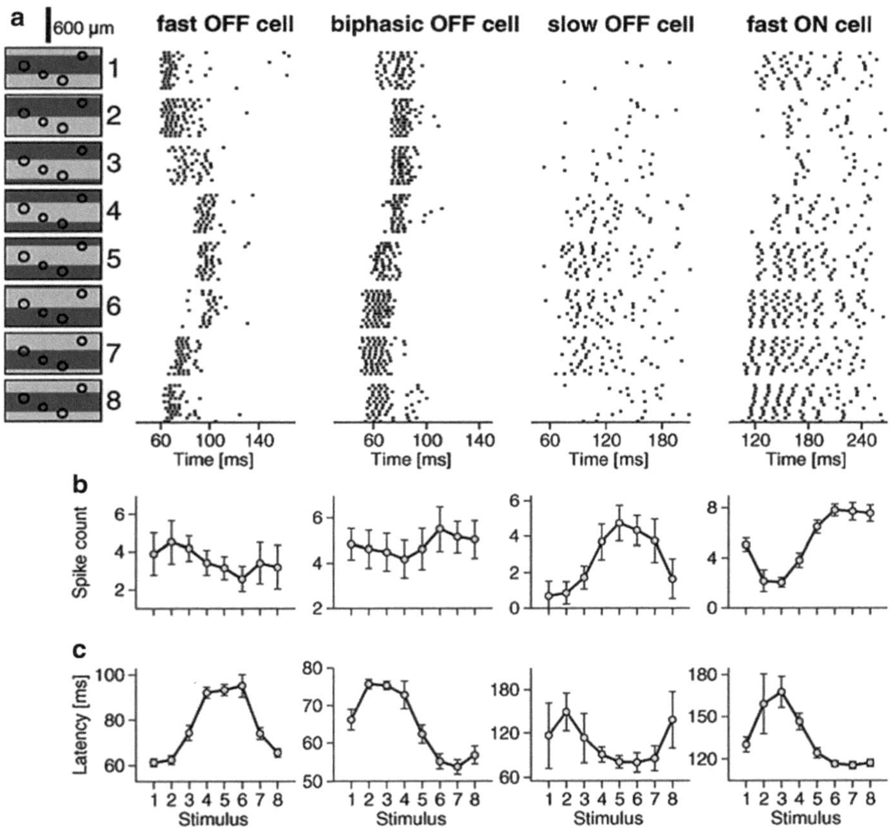

Fig. 4.6 Ganglion cell responses to gratings with different spatial phases. (**a**) Raster plots of spike responses from four ganglion cells (*right*) elicited by 150-ms presentations of eight gratings with different spatial phases (*left*). Time is measured from stimulus onset. The *circles* show the receptive fields of each cell, correspondingly from *left to right*. (**b**) Tuning curves of the elicited spike count. Error bars show the standard deviation across trials with the same stimulus. (**c**) Tuning curves of the first-spike latency. Error bars show the standard deviation across trials with the same stimulus (Adapted with permission from Gollisch and Meister 2008, Fig. 1)

In one recent experiment performed on salamander retinal ganglion cells, the relevance between the neuron's firing rate and the visual stimulation was tested. It is shown that different subtypes of retinal ganglion cells responded to visual stimulation with different properties (Fig. 4.6a). For the ON cell, it fired strongly when a light stimulation was projected in its receptive field center, whereas dark stimulation suppressed its firing activity, and the cell fired at intermediate levels when its receptive field center was partly luminated; the responses of the OFF cells were to the opposite (Fig. 4.6b). It is apparent that the neuronal response strength was closely related to the stimulation intensity—ON cell's response was positively related to the light intensity, and OFF cells' responses were negatively related to the light intensity.

Now, what we are seriously concerned is whether the stimulation intensity can be inferred according to the neuron's firing rate or spike count. For a certain ganglion cell whose response feature can be measured, with its receptive field property being examined as indicated in Fig. 4.7a, a natural image containing 1000 pixels arranged in a 40×25 matrix was given as visual stimulation. Each pixel was presented sequentially to the receptive field center of the ganglion cell, and the neuron's responses to these pixel stimulations were recorded, part of which are plotted in Fig. 4.7b. Given that the cell was an OFF cell and having known the cell's receptive field property, the stimulation intensity of each pixel (darkness of each pixel in this case) can be inferred according to the cell's firing activity as illustrated in Fig. 4.7d. It is shown that the original picture can be roughly restored according to the cell's spike count in response to each pixel, although with some estimation error for a certain portion of pixels. Such result actually confirms that spike count (or firing rate) contributes to visual information coding in the retinal ganglion cells. However, firing frequency is not the sole source for information coding. Actually in this case, the firing latency contributes more to information coding (Fig. 4.7c), and we will give more detailed discussion on this issue in the next section.

4.2 Precise Time Coding

From the example presented in Figs. 4.6 and 4.7, one can tell that for salamander retinal neuron, the spike count, or firing rate, is dependent on visual stimulation intensity. On the other hand, visual information can be retrieved from the cell's spike count or firing rate, and such phenomenon is termed as "rate coding."

However, it is also illustrated in Fig. 4.6 that in addition to stimulation-dependent spike count changes, there was also close relationship between the response latency (time delay of the first spike in respect to stimulation onset) and the stimulation intensity—the cells responded to stronger stimulation with shorter response latency (Fig. 4.6a, c), which suggests that the response latency of the retinal neuron may also carry stimulation information. Therefore, the image restoration task illustrated in Fig. 4.7 could also be performed using the cell's response latency properties (Fig. 4.7c), and the image restored based on the cell's response latency yielded better result as compared to that restored based on the cell's spike count as presented in Fig. 4.7d.

Similar stimulation-dependent response patterns were also observed from primate retina. Figure 4.8 illustrates the firing activities recorded from on-center and off-center ganglion cells in response to binary pseudorandom full-field light flickering. When in exposure to repeated sequence of stimulation, the neurons' response

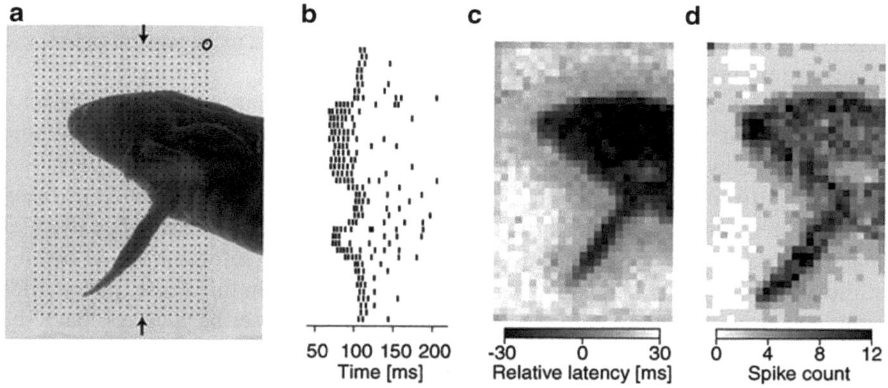

Fig. 4.7 Responses of an off-center ganglion cell of salamander retina in exposure to a natural image. (**a**) Photograph of a salamander larva projected on the retina, which contains 1000 pixels, with each pixel presented sequentially to the receptive field center of the retinal ganglion cell. The ellipse in the upper right corner shows the receptive field of the cell. (**b**) Spike trains of the ganglion cell when its receptive field was shifted along the column marked by the *arrows* in (**a**). (**c**) Gray-scale plot of stimulation image restored according to the differential spike latency on single-trial presentations at the locations marked with dots in (**a**). The reference latency was chosen as the average value at all locations. (**d**) Corresponding gray-scale plot restored for the original stimulation image according to the cell's spike counts (Adapted with permission from Gollisch and Meister 2008, Fig. 4)

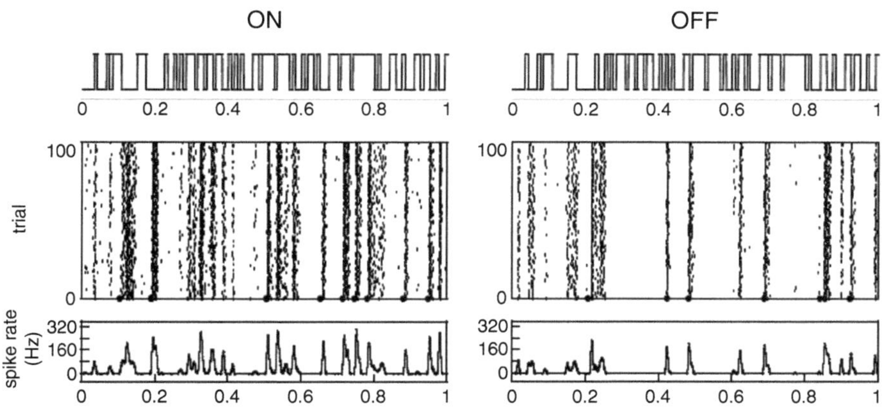

Fig. 4.8 Primate retinal ganglion cell firing patterns. Rasters and firing rates over time are shown for representative ON and OFF parasol cells from one retina. The stimulus was spatially uniform binary noise with time course shown above the rasters. The stimulus refresh interval was 8.33 ms. The *dots* on the time axis represent identified times of firing onset; periods with unreliable firing were excluded. Beneath each raster, the time-varying firing rate is shown, calculated in 0.1-ms bins, and smoothed by a Gaussian filter with a SD of 2 ms (Modified From Uzzell and Chichilnisky 2004, Fig. 1)

pattern, including firing rate and precise timing of the action potentials, was also repeatable. But on the other hand, when the stimulation intensity was changed, the firing pattern had a stimulation-dependent modulation.

From these examples, we get an impression that stimulation information is not solely encoded by firing rate; the spiking time of the first action potential (the response latency), as well as precise timing of the following firing activities, also contributes to information coding. When we say that the time sequence of firing activity represents some properties of the stimulation signal through particular manners or in other words the firing sequence encodes "information" about the stimulation, a certain "language" is required to express it. In this sense, informatics approach provides an effective tool to quantitatively describe and compare the amount of information carried and transmitted by the neuronal firing activity.

The basic property of information is related to the uncertainty of events. The information content is related to the reduction of the unawareness or uncertainty of events; thus, any determined event carries no information. While a stochastic event x can be characterized by its probability of occurrence $p(x)$, the uncertainty of a stochastic event is related to its probability. The uncertainty of a determined event (whose probability is 1) is 0. To the contrary, a stochastic event with a low probability has a relatively high uncertainty, because it is difficult to predict its occurrence in a certain time. Quantitatively, the probability-dependent information quantity $I(x) = -\log p(x)$ is dependent on the uncertainty of each event in the signal assembly. When the probability $p(x)$ is small, the occurrence of x is rare; therefore, once such event occurs, the information quantity it provides is large. Given that x is a stochastic variable, $I(x)$, as a function of x, is also a stochastic variable.

The unit of information quantity is related to the base of the logarithm. While the information quantity is normally calculated using base 2, it has a unit of bit. For a signal X, which is composed of several components $x_1, x_2, \cdots, x_K$, the information entropy $S(X) = \sum_i p(x_i)I(x_i) = -\sum_i p(x_i)\log p(x_i)$ is determined by the general property of the signal sources. When the signal source only contains one component, i.e., $p(x) = 1$, the information entropy is $S(X) = 0$; while the probability of a certain component x_i is equal to zero, i.e., $p(x_i) = 0$, the relevant information entropy $p_i \log p_i$ is meaningless, and in this case, its value is arbitrarily set to 0.

For a signal X with K signals having equal probability $p_i = 1/K$, we have:

$$
\begin{aligned}
S &= -\sum_{i=1}^{K} p_i \log_2 p_i \\
&= -\sum_{i=1}^{K} \left(\frac{1}{K}\right) \log_2 \left(\frac{1}{K}\right) \\
&= \log_2 K
\end{aligned}
\tag{4.1}
$$

The quantitative calculation employing "entropy" to measure the amount of information carried by action potential sequence was first proposed by MacKay and McCulloch (1952), which might be the earliest exploration on applying the information theory on nervous system after the establishment of information theory by

Fig. 4.9 Action potential sequence and the corresponding "0-1" presentation

Shannon. In the work of MacKay and McCulloch, they examined the action potential sequence using short time windows (time bin) $\Delta\tau$. If the time bin is short enough, each time bin should contain at most one spike. A time bin is assigned a value "1" when action potential occurs, or otherwise a value of "0" is assigned, as illustrated in Fig. 4.9. Then in a certain time period T, the action potential sequence can be described by a series of symbols with length of $L = T/\Delta\tau$. In the meantime, the temporal patterns of the subsequences are not evenly distributed. For instance, when T is big but $\Delta\tau$ is very small, a subsequence "1111...11111" is not likely to occur, because a neuron cannot maintain a high firing frequency for a long time. Therefore, when measuring the changing degree of the temporal pattern of neuron's firing sequence in response to certain stimulation using information entropy, the calculation result is heavily influenced by the sampling time window $\Delta\tau(S(\Delta\tau))$.

Since the neuronal firing pattern is dependent on the stimulation condition, therefore the temporal structure of action potential sequence carries information about the stimulus. In this sense, the temporal structure of neuron's firing sequence should be repeatable in response to identical stimulation, to ensure the reliability of information transmission.

On the other hand, the time sequence of neuronal firing activity is often irregular and full of noise. This is due to the existence of stimulation-independent changes in the neuron's firing activity. By examining the firing sequence, one can find that in response to repeated stimulation, the temporal pattern of a neuron's firing activity is pretty similar to each other but not entirely identical (Fig. 4.8), and the difference among the firing sequences can be attributed to noise, and accordingly the measured variability can be defined as noise entropy $N(\Delta\tau)$. The stimulation information carried by the firing sequence is then defined as the difference between the total entropy and the noise entropy $I = S(\Delta\tau) - N(\Delta\tau)$. Because the noise entropy is (semi-)positive, the real information amount I will never exceed the total information amount $S(\Delta\tau)$, and thus the information transmission efficiency can be calculated as $\eta(\Delta\tau) = I/S(\Delta\tau)$.

For calculating information entropy, the sequence constructed by neuronal firing activity is first transformed into "words" containing $L = T/\Delta\tau$ letters; thus, the estimation of information entropy is dependent on two important parameters: the length of the firing sequence T and the time bin for sampling $\Delta\tau$. If the probability of the i-th "word" is p_i, the information entropy is:

$$S_0(T, \Delta\tau; \text{size}) = -\sum p_i \log_2 p_i \tag{4.2}$$

This value is dependent on the length of the data set. When the length of the data sequence tends to be infinitive, the entropy approaches:

$$S(T, \Delta\tau) = \lim_{\text{size}\to\infty} S_0(T, \Delta\tau; \text{size}) \tag{4.3}$$

A classic work examining the activity of sensory neuron by information theory approach is from H1 neuron of fly visual system, which is sensitive to horizontal movement and provides visual feedback for flying control.

From the experimental observations illustrated in Fig. 4.10, one can tell that when the stimulation velocity was maintained constant (0.022°/s in this case) (Fig. 4.10a), the activities of a single H1 neuron showed irregular temporal patterns (each trace in Fig. 4.10b represents a single record which was lasted for 1 s, with the intervals between successive recordings being 20 s). PSTH (calculated using time bin of 3 ms) reveals fluctuating firing rates around the mean level which was stimulation velocity dependent (approximately 50 spikes/s) (Fig. 4.10c). On the other hand, stimulation with fluctuating velocities which was similar with natural stimulation induced the neuronal firing activities with more consistent temporal patterns. From the plot we can see that in response to such changing velocity stimulation (Fig. 4.10d), the temporal pattern of the neuron's firing activity in different trials tended to be consistent with each other (Fig. 4.10e), while the firing rate changed in a larger range and showed apparent time dependence (Fig. 4.10f).

As compared to the results illustrated in Fig. 4.10b, c, the firing patterns presented in Fig. 4.10e, f are more complex—the neuron was more likely to burst a series of action potentials, instead of fired single solitary action potentials. Using information theory algorithms, quantitative analysis can be performed to compare the variability and repeatability of the neuron's activities in these two conditions and examine the contribution of the temporal patterns of the neuron's activity to information coding.

As has been discussed, the information entropy can be calculated as the difference between the total entropy of the firing sequence and the noise entropy. While the total entropy can be directly calculated according to the recorded firing sequences, the noise entropy is estimated from the difference between different trials.

The algorithm is illustrated by Fig. 4.11, in which 100 experimental trials were recorded from one cell (a and b). To obtain the full range of temporal variability, the stimulus was generated following the same statistical property but not identical one with another. Then the "events" were constructed according to the time length T and sampling bin $\Delta\tau$, and a series of "words" w containing L "letters" ($L = T/\Delta\tau$) were obtained. If the time bin is sufficiently short, no more than one action potential should be contained in each bin; therefore, the words w can be "translated" into "0-1" sequences (b). By calculating the probability of each single event $p(w)$ in the whole experimental recording, the total entropy can be obtained:

$$S(\Delta\tau) = -\sum_{w} p(w) \log_2 p(w) \text{ bits} \tag{4.4}$$

This value reflects the total amount of changes in the response sequence when the stimulation condition is changed, which reflects the ability of neuronal activity in carrying stimulation information.

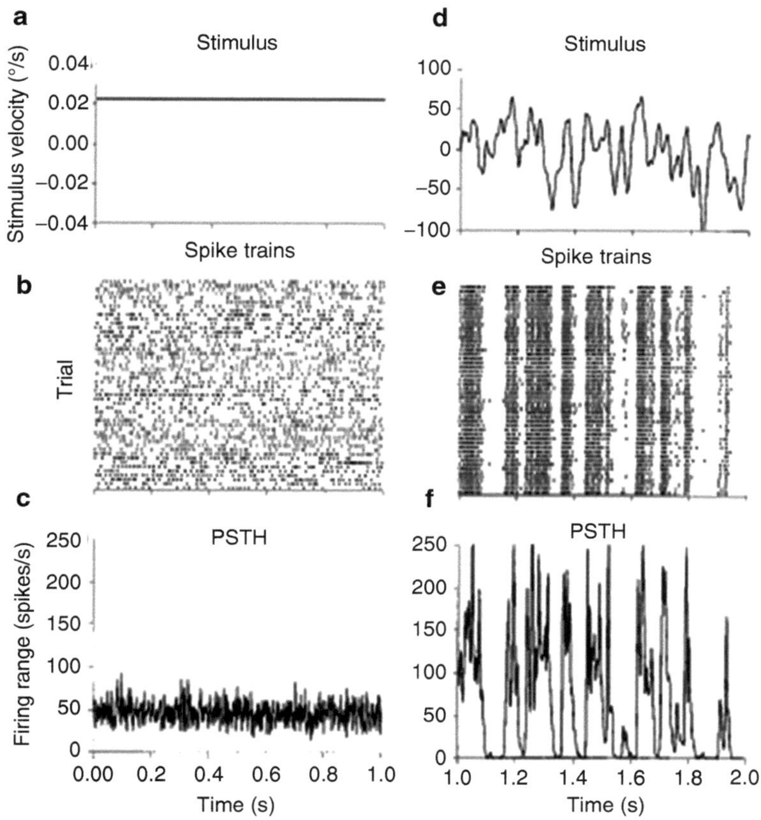

Fig. 4.10 The temporal pattern of an H1 neuron's firing activity in response to moving stimulation with different velocity patterns. (**a**) A visual target moving along with the H1 neuron's preferred direction, with a constant speed (0.022°/s). (**b**) A set of 50 response traces elicited by the stimulus in (**a**), with each dot representing a spike and each row representing the firing sequence recorded in a single trial. (**c**) PSTH calculated based on the cell's response presented in (**b**). (**d**) The visual stimulation moving with time-dependent velocity, mimicking motion during random walk. (**e**) Fifty response traces elicited by the stimulus in (**d**). (**f**) PSTH calculated based on data shown in (**e**) (Modified From de Ruyter van Steveninck et al. 1997, Figs. 1 and 2)

By applying an entirely identical stimulation sequence repeatedly and calculating the probability of event w at time t ($p(w|t)$), the entropy for each time t can be obtained. Because such variation in firing sequences is not caused by the stimulation but rather reflects the noise property of the firing activity, therefore, the noise entropy can be obtained by taking its average across time:

$$N(\Delta\tau) = \left\langle -\sum_{w} p(w|t)\log_2 p(w|t) \right\rangle_t \text{ bits} \tag{4.5}$$

where $\langle \ldots \rangle_t$ is the average across time t.

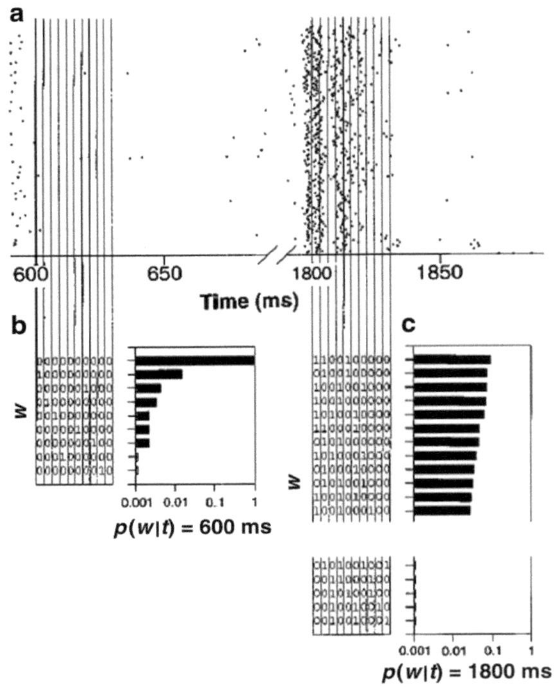

Fig. 4.11 The calculation of information amount carried by action potential sequence. (**a**) Two segments from 100 response traces, starting at about 600 and 1800 ms, respectively. (**b**) Responses starting from about 600 ms were elicited by stimulus with same probability distributions and were applied for calculating the total entropy, with each 30-ms response sequence being divided into ten 3-ms time bins to form local "words." (**c**) Response starting from about 1800 ms were elicited by identical stimulus and were applied for calculating the noise entropy (Adapted with permission from de Ryter van Steveninck et al. 1997, Fig. 3)

The difference between the total entropy and noise entropy $I = S(\Delta\tau) - N(\Delta\tau)$ is the estimated information amount carried by the action potential sequence. It can be inferred that the information quantity carried by the neuron's firing sequence reaches its upper limitation when there is no noise.

It is worth noting that such calculation is based on statistical data; therefore, insufficient data will cause bias in estimation. For the example illustrated in Fig. 4.11, when T and $\Delta\tau$ were taken as 30 ms and 3 ms, respectively, the calculation results were that in exposure to moving target with fluctuating velocities, the total information and the noise information were $S(\Delta\tau) = 5.05 \pm 0.01$ bits and $N(\Delta\tau) = 2.62 \pm 0.02$ bits, respectively; this gives that the neuron carried the stimulation information of 2.43 ± 0.03 bits during the 30-ms time period.

Another example about neural information encoded by action potential sequence comes from cat LGN neuron which responds to intensity changes of light within its receptive field. The experimental results (Fig. 4.12) show that when the cell's receptive field was elicited by a full-field light with its intensity varying with time in a random manner, the cell responded with time-varying activity. When different stimulation sequences were applied, firing sequences with different temporal patterns were elicited. But when the stimulation sequences were fully identical, the neuron's responses were also highly repeatable and showed high temporal precision. The occurrences of firings were well aligned, which results in sharp peaks in the PSTH. This implies that the occurrence of firing activities encodes information about stimulation, which can be quantified by information entropy.

We have explained that while estimating the information entropy, the time length T and time bin width $\Delta\tau$ are two crucial parameters. Given "words" with length of $L = T/\Delta\tau$ and time bin of $\Delta\tau$ and their occurrence probabilities, information entropy can be estimated by calculating the total entropy $S(\Delta\tau)$ and noise entropy $N(\Delta\tau)$. When the entropy (S) and information (I) are estimated for unit time and yield entropy rate and information rate, it should take the time duration into account:

$$S(L, \Delta\tau) = -\frac{1}{L\Delta\tau} \sum_{w \in W(L, \Delta\tau)} p(w) \log_2 p(w) \tag{4.6}$$

Figure 4.13 shows the probability distribution of "words" for the calculation of total entropy rate $S(L, \Delta\tau)$ (Fig. 4.13a) and noise entropy rate $N(L, \Delta\tau)$ (Fig. 4.13b) based on the experimental data presented in Fig. 4.12 when taking $L = 8$, $\Delta\tau = 1$ ms. But according to (4.6), the estimated values for $S(L, \Delta\tau)$ and $N(L, \Delta\tau)$ are both dependent on L value (Fig. 4.13c). The estimated entropy rate is approaching its real value when $L \to \infty$. One can tell from the plot that when L is changing between 2 and 8 (with $1/L$ being changing between 0.125 and 0.5), the entropy rate is linearly related to $1/L$; therefore, the value of entropy rate can be obtained by linear extrapolation. On the other hand, the time bin size $\Delta\tau$ is also crucial (Fig. 4.13d). When $L \to \infty$ and $\Delta\tau$ are sufficiently small, the estimated entropy rate is approaching its real value.

From the abovementioned examples, one can tell that the precise timing of the firing activity carries information. This leads to another question, i.e., is the neural information entirely encoded by the precise timing of the firing activity—or in other words, is the temporal structure of firing sequence involved in information coding? An intuitive way of examining this issue is to look into the correlation between time bins. If the temporal structure does not carry information, then the time bins should be independent to each other. In this case, the calculated information entropy should be independent on the length of the "words."

To examine the contribution of the temporal structure to information coding, we can define a new variable:

$$Z(\Delta\tau) \equiv \lim_{L\to\infty} I(L, \Delta\tau) - I(L = 1, \Delta\tau) \tag{4.7}$$

If the information is encoded by the temporal pattern, the Z value should be positive, because in the coding process, the information quantities carried by firing activities in adjacent time bins are synergistic. To the contrary, when the temporal

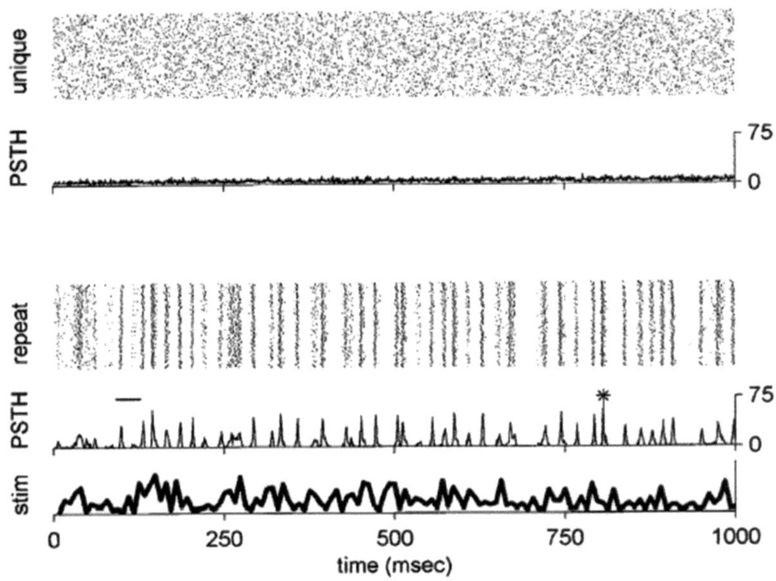

Fig. 4.12 LGN neuron in response to white-noise stimulation. The *upper panels* show the activity an on-center LGN X-cell from anesthetized cat in response to non-repeated white-noise stimulation; the raster plot shows the cell's response to 128 unique 1-s stimulation segments with each *dot* representing the occurrence of an action potential; the *lower panels* show the cell's activity in response to repeated identical white-noise sequences (128 repeats) (Adapted with permission from Reinagel and Reid 2000, Fig. 1)

pattern is redundant for coding, Z value should be negative, because information contained in one time bin is partly redundant with that carried by its adjacent bin(s). From the calculation results shown in Fig. 4.13, it is shown that the information quantity is increased when the word length is increased, which is to say that the Z value is positive; this shows that in addition to the precise time of each individual spikes, the information is also partly carried by the temporal correlation of the spikes.

According to the coding theory, "temporal coding" of stimulation information requests that the temporal pattern of firing sequence is dependent on the stimulation condition. In other words, for a single neuron, its firing sequences elicited by the same stimulation pattern should be similar with each other, while the firing sequences generated during responses to different stimulation conditions can be distinguished from each other.

Following such an idea, the temporal precision of spike trains can be analyzed using a "metric-space" method based on the measurement of "similarity" and "dissimilarity" of spike trains was developed. To measure the "distance" between spike trains, one spike train can be transformed into any other train by some elementary manipulations (Fig. 4.14) with certain "cost." While adding or deleting

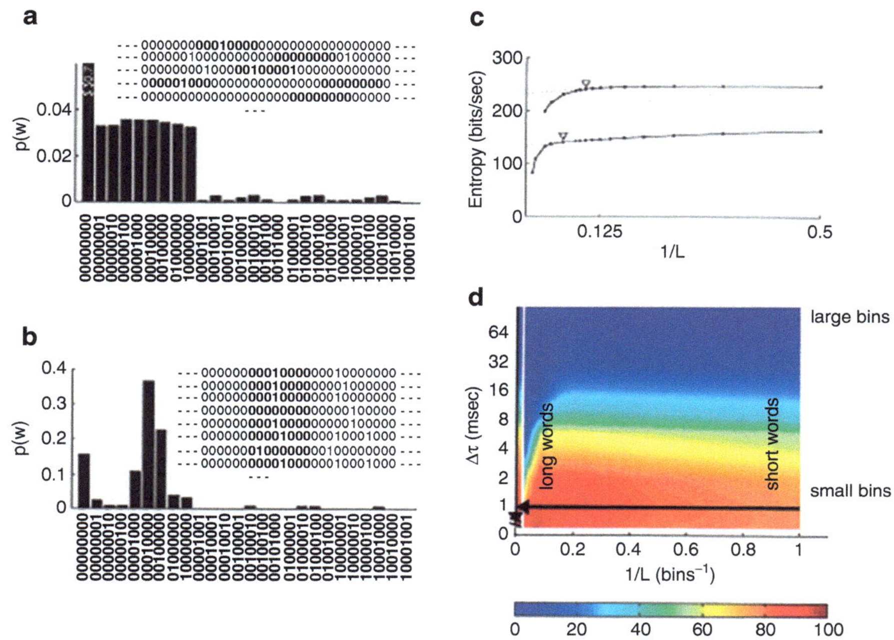

Fig. 4.13 Calculation of information entropy based on experimental data shown in Fig. 4.12. (**a**) The probability distribution of "words" for total entropy rate $S(L, \Delta\tau)$ estimation, with $L = 8$, $\Delta\tau = 1$. (**b**) The probability distribution of "words" for noise entropy rate $N(L, \Delta\tau)$ estimation, with $L = 8$, $\Delta\tau = 1$. (**c**) The estimations for the total entropy rate (the upper trace) and the noise entropy rate (the lower trace) are both dependent on the word length L, with the information entropy rate being equal to the difference between the two $I = S(L, \Delta\tau) - N(L, \Delta\tau)$. (**d**) The changes of estimated information entropy rate I with the word length L and time bin $\Delta\tau$ (Adapted with permission from Reinagel and Reid 2000, Fig. 2)

a spike has a cost of unity, moving a spike costs q (in units of 1/s) per unit time of moving: $c = q|\Delta t|$, where the value of q expresses the relative sensitivity to the precise timing of the spikes.

Actually, there are two possible manners to transform a spike train A into spike train B. One is by pure deletion and insertion: to delete all spikes from train A, then insert all spikes according to train B; another is in combination with shifting single spikes in time. Then the distance between two spike trains is defined as the minimum total cost of a set of elementary steps that transforms one spike into the other. A small q value reflects little influence of spike timing on the "distance," and $q = 0$ means that spike can be shifted for "free cost," and the "distance" between two sequences is entirely presented by spike counts and is insensitive to spike timing. Increasing values of q correspond to a growing dependence of the metric on the spike timing.

Given that deleting a spike from sequence A and reinserting it according to sequence B has a cost of $1 + 1 = 2$, while shifting this spike has a cost of $q|t_a - t_b|$,

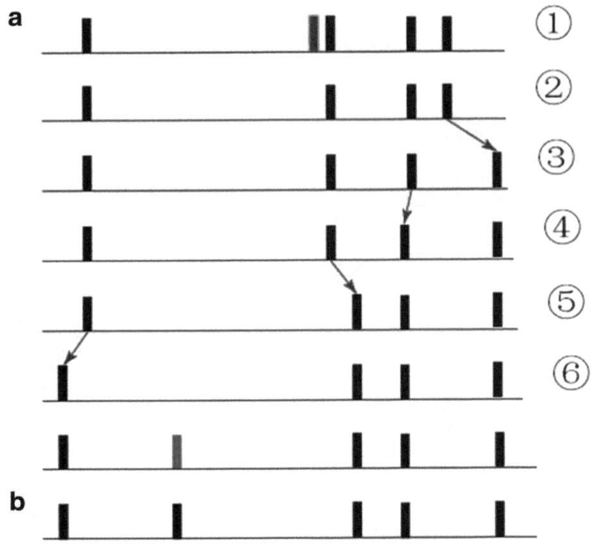

Fig. 4.14 A diagram for transforming spike train *A* into spike train *B* via several elementary steps. Each *vertical bar* represents one spike, and the length of the horizontal line denotes time, ① deleting a spike, ②–⑤ shifting a spike in each step, ⑥ inserting a spike (Adapted with permission from Victor 2005, Fig. 1)

therefore, the minimum cost for treating a single spike equals to $q|t_a - t_b|$ when $|t_a - t_b| \leq 2/q$, and otherwise it should be equal to "2." This is also to say that spikes in two different sequences are only considered as comparable if they occur with a time interval of $2/q$.

If a neuron's activities elicited by different stimuli show different response properties including spike count and temporal structure in spike sequence, the neuron's response to different stimuli can be classified according to the distance between the spike trains. For a neuron with K_{tot} spike trains elicited by L_{sti} different stimuli, the shortest distance ($D[q](r_i, r_j)$) between one spike train (r_i) and other spike train (r_j) of this neuron ($i, j = 1, 2, \ldots, K_{\text{tot}}$. and $i \neq j$) can be computed based on the abovementioned three elementary manipulations. If one spike train (r) is elicited by a stimulus s_α, the average distance ($d(r, s_\gamma)$) from the spike train (r) to each of the spike trains elicited by stimuli s_γ ($\gamma = 1, 2, \ldots, L_{\text{sti}}$) is defined as:

$$d(r, s_\gamma) = \left[\left\langle D[q]\left(r, r'\right)^z \right\rangle_{r' \text{ elicited by } s_\gamma} \right]^{1/z}, \tag{4.8}$$

where $\langle \cdot \rangle$ denotes the average over all the spike trains (r') elicited by stimulus s_γ, and z is arbitrarily set as -2. After computing the average distance for every spike train, K_{tot} spike trains can be classified into M_{sti} response classes.

The result of classification can be summarized by a matrix N (s_α, r_β), whose entries indicate the number of times that spike trains elicited by the stimulus s_α are classified into response class r_β. If a spike train r is elicited by stimulus class s_α ($\alpha = 1, 2, \ldots, M_{\text{sti}}$), it is classified into the response class r_β ($\beta = 1, 2, \ldots, M_{\text{sti}}$) when

$d(r, s_\beta)$ is the minimum of all the average distances and increment $N(s_\alpha, r_\beta)$ by 1. If there are k average distances sharing the minimum, elements of the matrix N corresponding to these k average distances are incremented by $1/k$. Then, the clustering performance can be quantified by the transmitted information H:

$$H = \frac{1}{N_{tot}} \sum_{\alpha,\beta} N(s_\alpha, r_\beta) \left[\log_2 N(s_\alpha, r_\beta) - \log_2 \sum_\alpha N(s_\alpha, r_\beta) \right. \\ \left. - \log_2 \sum_\beta N(s_\alpha, r_\beta) + \log_2 N_{tot} \right], \quad (4.9)$$

If the data set includes spike sequences elicited by C equally distributed stimulus classes, perfect clustering ($N(s_\alpha, r_\beta) = K_{tot}/C$ for $\alpha = \beta$ and others are zero) corresponds to a maximal value of the transmitted information $H_{max} = \log_2 C$ bits, while random clustering ($N(s_\alpha, r_\beta) = K_{tot}/C^2$) leads to $H = 0$. In the meantime, H value changes with the cost parameter q. When $q = 0\ s^{-1}$, H_0 represents the amount of information contained in the spike count or firing rate. If the peak value of H (H_{peak}) occurs at $q > 0\ s^{-1}$, it implies that part of information is contained in the temporal structure of spike train. The information contributed by response latency of the first spike is obtained by selecting the first spike in each sequence only, and those trials in which no spike fired are excluded.

Such spike metric-space method has been employed to analyze data obtained from a variety of neural systems (including visual, auditory, chemical senses, and motor system) to characterize neural coding property.

One example analyzing the coding property of monkey visual cortical neuron in exposure to gratings of different orientations is illustrated in Fig. 4.15. In the experiment, gratings with a spatial frequency of 21 c/deg were presented to the monkey at eight orientations (0–157.5 deg with steps of 22.5 deg). The raster plots of responses recorded from one V1 neuron is plotted in Fig. 4.15a, and the metric-space analysis was performed as shown in Fig. 4.15b. To test whether the transmitted information H (thick line in Fig. 4.15b) was dependent on the firing sequence, the clustering matrix $N(s_\alpha, r_\beta)$ was reconstructed using surrogate data in which the response sequences were randomly assigned to the stimulus categories. In this example, H values expected by chance was estimated via 10 resamplings and yielded confidence intervals of mean $\pm$ 2 SE (thin lines in Fig. 4.15b). The results show that when $q = 0$ (which refers to clustering merely according to firing account), the H value was even lower than the chance level, while the H value reaches a maximum value of 0.49 bits (which significantly exceeds the chance level) at $q_{max} = 64$. These results suggest that in this neuron, the visual information about orientation is not entirely encoded by spike count, and precise spike timing is also involved in information coding. Given that a shift in time by an amount $2/q$ is equivalent in cost to spike deleting and reinserting, the peak H value at $q_{max} = 64$ means that within a precision of ~31 ms ($2/64 \approx 0.031$ s), the arrival of spikes is systematically dependent on stimulus orientation.

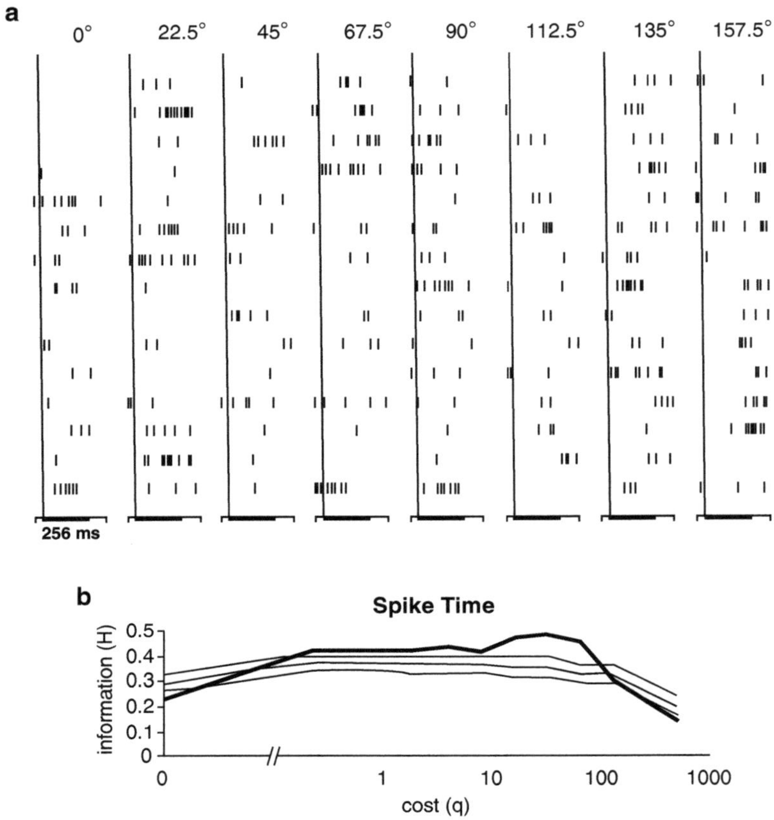

Fig. 4.15 Coding property analysis based on metric-space method. (**a**) Raster plots of a monkey V1 neuron's responses elicited by grating presented at eight orientations (0–157 deg in 22.5-deg steps). (**b**) Analysis of stimulus-dependent clustering induced by D(q). *Thick line*: the transmitted information H measured based on the clustering matrix derived from the response sequences in (**a**); *thin lines*: mean $\pm$ 2 SE of distribution of H values calculated using surrogate data sets in which observed response sequences were randomly assigned to the stimulus categories (Adapted with permission from Victor and Purpura 1996, Fig. 3)

4.3 Spatiotemporal Properties of Neuronal Activity

The brain functions mainly involve information processing and transmission in the nervous system. Traditionally, knowledge related to this issue was mainly accumulated based on the understanding of the activity and signal processing mechanism at single neuron level. But actually, functions of the central nervous system are organized based on the ensemble behavior emerged from the interactions among single units, with the connection among single neurons changing according to the environmental conditions.

As we have discussed, the neuron's firing count within a certain period of time or the temporal structure of the firing sequence both contain useful information about the input stimulation. But these indices are normally used to describe the activity properties of single neurons. Although single neuron has long been considered as independent unit for information processing, which means that the activities observed from single neurons can provide insight for neural information coding and processing, such hypothesis is with many deficits.

First of all, the environmental stimulation events greatly overnumber the total amount of neurons in the brain, which challenges the possibility that stimulation information can be precisely encoded by single neuron's activity. On the other hand, when the brain is partly injured, its function can be restored in the remaining parts. Such restoration of brain function can hardly be explained by single neuron-coding theory. Given these considerations, it is less likely that a single neuron can work as an independent encoder.

Besides, many lines of evidence have revealed that while in response to environmental stimulation, the neuron's activity is not solely dependent on the stimulation input but instead largely related to the activities of other neurons in the circuit. In fact, as compared to single neuron's activity, the spatiotemporal pattern of population neurons activity normally carries more information and is more accurate in reflecting the stimulation properties.

Furthermore, the abundant dendrite structure in neuronal circuitry allows a postsynaptic neuron to receive multiple inputs from its presynaptic neurons and respond to the integrated input signals. Due to the spatial distribution of synapses and nonlinearity of synaptic process, there exists intensive interaction between adjacent synapses, which makes that dendritic inputs can be synergized when they are synchronized. Therefore, the spatial correlation of presynaptic neurons' activity is crucial for the postsynaptic neuron, because as compared to independent input, synchronized inputs are more likely to elicit the postsynaptic neuron's activity.

Actually, connections among neurons in the network can be modulated by stimulation conditions, and a single neuron can form various neuron populations with its adjacent neurons under different stimulation conditions. Therefore, although single neuron's activity contributes to the organization of brain function, the information coding process in the nervous system is largely dependent on the concerted activity of population neurons. The spatiotemporal pattern of population neurons' activity is crucial for information processing and transmission.

Upon these facts, it has been gradually established that in the neuronal network, stimulation information is processed and transmitted via population cells activities in addition to single neuron's activity. It is also noticed that population coding process is generally dynamic. In response to environmental stimulation, adjacent neurons can form cell groups dynamically via functional connections, so as to ensure the information coding and transmission in the neural network.

Population coding thus has the following two aspects.

One is characterized by clustering of neurons with similar properties, which is to enhance the reliability of signal transmission. There are many examples for this kind of population activity in sensory systems. For example, in visual system, including retina, LGN, and even in primary visual cortex, nearby neurons often have similar response selectivity and normally have similar firing behavior in response to the same stimulation. Neurons can thus have coordinated activities, so as to encode stimulation information efficiently.

The coordinated activities of this type can be measured using cross-correlation function calculated for pair-wise neurons' firing activities:

$$c_{xy}(m) = \begin{cases} \dfrac{\sum_{n=0}^{N-|m|-1} x_n y_{n+m}}{\sqrt{\sum_{i=1}^{N} x_i^2 \sum_{i=1}^{N} y_i^2}}, & m \geq 0 \\ c_{yx}(-m), & m < 0 \end{cases} \tag{4.10}$$

where y_{n+m} denotes the value of sequence y at time $n+m$; x_n denotes the value of sequence x at time n; and $c_{xy}(m)$ denotes the correlation between sequence x and y with a time lag of m. The peak value of the normalized cross-correlation function can be taken as the correlation strength between neurons. The cross-correlation calculation is normalized using the firing activities of both neurons, to reduce the influence of the firing rates of the cells.

But on the other hand, adjacent neurons of the same subtype normally receive similar inputs from their presynaptic neurons due to the network structure, so that they are likely to have correlated activities. One may thus argue that when two adjacent neurons of the same subtype are exposed to the same stimulation, they will be activated at similar level, with similar manner, and thus their activities might be synchronized by chance. To rule out such possibility, the cross-correlation function can be "corrected" using a "shift predictor." The calculation of the "shift predictor" is similar to the cross-correlation function, except that the two sequences are taken from different experimental trials (Fig. 4.16a, insets). The "corrected" cross-correlation function is defined as the difference between the original cross-correlation function and the "shift predictor."

Although it is easily understandable that the correlation pattern between the neurons' firing activities is related to the input pattern and is thus dependent on the stimulation pattern, one can still question that whether stimulation information is encoded and transmitted by population activity pattern. Experimental evidence collected from different sensory systems in different species confirmed that population activity pattern contributed to the animals' perceptual behaviors.

One example is from frog visual system. Dimming detector, a certain subtype of frog retinal ganglion cells, could be activated by dark stimulation. Experimental results showed that nearby dimming detectors have correlated activities, with oscillatory pattern presented in cross correlogram between pair-wise neurons (Fig. 4.16a). However, when GABAergic signal pathway, which provides inhibitory modulation in the retina, was blocked by bicuculline, the synchronized oscillation was eliminated, although the cells' response to dark stimulation was even increased in the presence of bicuculline (Fig. 4.16b). In the meantime, the

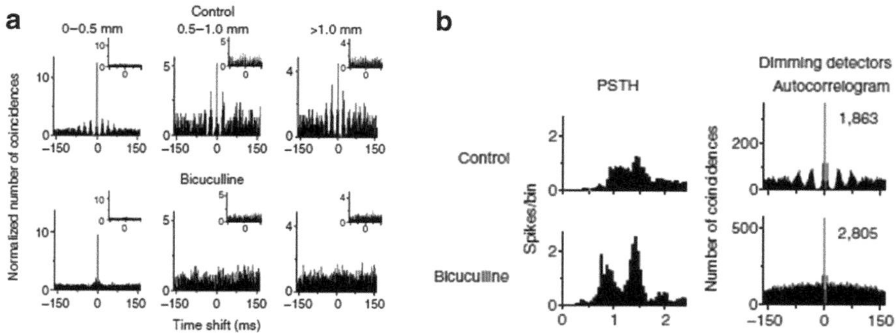

Fig. 4.16 Bicuculline effects on synchronized oscillation between neuron pairs and single cell's activity. (**a**) Examples of cross-correlograms and shift predictors (*inset*) obtained from three pairs of dimming detectors before (*top*) and during (*bottom*) application of bicuculline. Bicuculline application suppressed the synchronized activity and eliminated the synchronized oscillation. Distance between receptive field centers of the cells was categorized into three groups (*top*). (**b**) Effects of bicuculline on single dimming detector was to increase its firing activity, although the oscillation in the auto-correlogram was also eliminated (Adapted with permission from Ishikane et al. 2005, Figs. 5 and 2)

application of bicuculline also destroyed the frog's dark perception, which was confirmed by behavioral experiments that in control condition, the dark stimulation was able to elicit the animal's escaping behavior, but when $GABA_A$ receptor was blocked by bicuculline, the animal's dark-elicited escape behavior was very much suppressed, as shown in Fig. 4.17.

Correlated activity can be measured from various parts of neuron networks and can have various forms which are related to the functional connectivity of the network. In neuronal network, two postsynaptic neurons can receive common presynaptic input. In this case, the activity of presynaptic neuron can induce synchronized activation of the two postsynaptic neurons, and the postsynaptic neurons' activity will thus show synchronized property. But if the synaptic connection is formed between two neurons, then the activation of presynaptic neuron will have it influence on the activity of the postsynaptic neuron, and the two neurons' firing activity will be correlated with a certain time delay.

As presented by the above example, correlated activity can result in one or more peaks in the cross-correlation function. A peak at 0-lag means that the two neurons might have synchronized activities due to common presynaptic input, while a peak at time lag of 1–2 ms means that there is electrical coupling between the two cells, and a time delay of several tens of milliseconds means that the signal is transmitted from one neuron to the other via chemical synapses. To some extent, while correlated activity between pair-wise neurons with certain delay reflects the synaptic connection between the neurons, oscillations shown in the cross-correlation function may reflect specific circuitry structure.

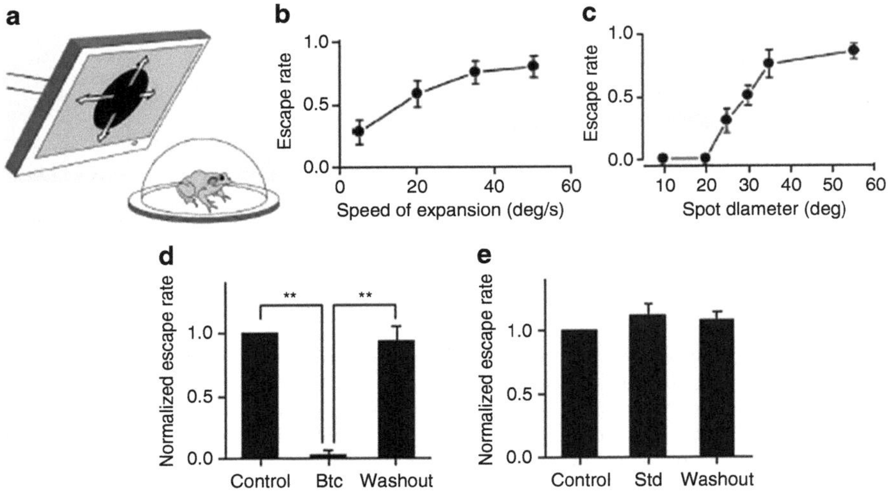

Fig. 4.17 Bicuculline suppressed escape behavior in response to the expanding dark spot. (**a**) Escape behavior of a frog was elicited by an expanding dark spot presented on a computer-controlled display. (**b**) The escape rate was positively related to the expansion speed of a dark spot (mean ± SE, $n = 6$ frogs). (**c**) The escape rate was also increased when the final diameter of an expanding *dark spot* was increased (mean ± SE, $n = 5$ frogs). (**d**) Intraocular injection of bicuculline suppressed the escape behavior. Behavioral experiments were performed before (*Control*), 2 h after (*Bic*) and 1 day after (*Washout*) injection. The escape rate was normalized against the control level. Mean ± SE, $n = 5$ frogs. $^{**}P < 0.01$. (**e**) Intraocular injection of the standard saline did not cause significant changes in the escape behavior (mean ± SE, n = 5 frogs) (Adapted with permission from Ishikane et al. 2005, Fig. 1)

Another important aspect of correlated activity is that the dynamic temporal properties of neuronal interaction are sensitive to environmental stimulation, which can be accomplished within several milliseconds and therefore has swift changing property. Such time-dependent dynamic pattern changes are not necessarily accompanied by single neuron's firing rate change.

While function-related neuronal population is characterized by correlated activities among neurons, dynamically formed neuronal population can contain only a few cells but can also have up to several dozens of members. Due to the complexity of network wiring, a single neuron may join cell groups formed by different sets of neurons. The grouping manner is not only related to the stimulation property but also dependent on the recent history of relevant neurons' activation status.

Such dynamic property of population activity allows population neurons to process more information as compared to single neurons, while single neurons can participate in different information processing via such dynamic adjustment according to various needs. This enhances the coding efficiency, as well as increases the reliability of information transmission.

In neuronal network, large-scale synchronous activities can be represented by pair-wise synchrony. However, while cross-correlation function is a useful tool for measuring concerted activities between pair-wise neurons, it is not realistic to use such a tool to describe population activity pattern formed by a large amount of neurons. To identify the synchronous activity neuron group, other calculation methods were developed.

One algorithm to identify the synchronization among large groups of neurons was proposed based on information theory, the purpose of which was to find groups of cells that fired together more often than expected by chance.

In this algorithm, the firing sequences were firstly digitized into "0-1" sequences:

$$r_j^{(A)} = \begin{cases} 1, \text{ if cell } A \text{ fired } \geq 1 \text{ spikes in time bin } j \\ 0, \text{ otherwise} \end{cases} \tag{4.11}$$

When two cells A and B fire in synchrony, they should exhibit "1" in the same time bin. Thus, synchronous firings of the two real cells A and B can be represented by a new symbolic unit AB (Fig. 4.18):

$$r_j^{(AB)} = r_j^{(A)} r_j^{(B)} = \begin{cases} 1, \text{ if cell } A \text{ and } B \text{ fired in time bin } j \\ 0, \text{ otherwise} \end{cases} \tag{4.12}$$

Then two new symbolic units $A'B$ and AB' are built with those events where A and B fires alone, respectively. Such recoding may induce information compression, which can be tested by comparing the information entropy:

$$H = -\sum_{i=1}^{M} p_i \log p_i + (1 - p_i)\log(1 - p_i) \tag{4.13}$$

where M is the number of spike trains, p_i is the firing probability of the i-th spike train:

$$p_i = \frac{1}{N}\sum_{j=1}^{N} r_j^{(i)} \tag{4.14}$$

N is the number of time bins in the data set, and *log* denotes the logarithm to base 2, denoting the probability that A and B fire together by:

$$p_{AB} = \frac{1}{N}\sum_{j=1}^{N} r_j^{(AB)} \tag{4.15}$$

Then the net reduction in entropy from the recoding is:

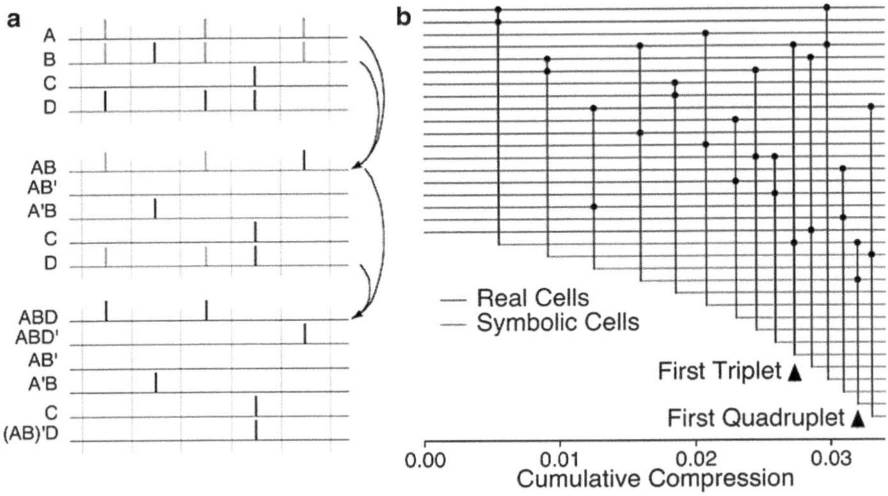

Fig. 4.18 Multineuronal firing pattern identification. (**a**) Procedure for "recoding." *Top*: schematic spike trains from four neurons (*A*, *B*, *C*, and *D*). *Middle*: synchronized firings of cells *A* and *B* are represented by a symbolic cell *AB*. The events where either *A* or *B* fires alone are represented by symbols *AB'* and *A'B*. *Bottom*: synchronized activities of symbols *AB* and *D* can be further recoded by *ABD*. This procedure continues until no more combinations can be found according to the preset criterion (**b**) The iterative search procedure on a group of salamander retinal ganglion cell spike trains. *Horizontal lines* correspond to real or symbolic cells. *Vertical lines* indicate the combination of two cells into a new symbol, with its horizontal offset indicating the compressing ratio as a fraction of its original size (Adapted with permission from Schnitzer and Meister 2003, Fig. 2)

$$\begin{aligned}\Delta H_{AB} = \log&\left(\frac{(1-p_{AB})(1-p_A+p_{AB})(1-p_B+p_{AB})}{(1-p_A)(1-p_B)}\right)\\ &+ p_A\log\left(\frac{(p_A-p_{AB})(1-p_A)}{(1-p_A+p_{AB})p_A}\right)\\ &+ p_B\log\left(\frac{(p_B-p_{AB})(1-p_B)}{(1-p_B+p_{AB})p_B}\right)\\ &+ p_{AB}\log\left(\frac{p_{AB}(1-p_B+p_{AB})(1-p_A+p_{AB})}{(1-p_{AB})(p_A-p_{AB})(p_B-p_{AB})}\right)\end{aligned} \tag{4.16}$$

In this algorithm, a positive ΔH_{AB} reflects the information storage (bits) saved for each time bin when representing the joint firing of cells *A* and *B* by a new symbolic unit *AB*. In real cases, it is generally true that $P_{AB} << P_A, P_B$, and then Eq. (4.16) can be simplified as

$$\Delta H_{AB} \approx p_{AB}\log(p_{AB}/p_A p_B) \tag{4.17}$$

It is apparent that the factor $\log(p_{AB}/p_A p_B)$ is large when joint firing occurs more often than predicted from the product of the individual firing rates. Thus, the identification of groups begins by computing ΔH_{AB} for all cell pairs, with the pair yielding the largest value being chosen and a new symbolic unit AB is defined. Then this process is repeated, with the identified symbolic cells being treated same as the real cells. The iterations stop when the largest available ΔH_{AB} falls below a predetermined threshold of significance.

A suitable threshold can be calculated as follows: first to shift each spike train with a different random number of time bins and then to compute the ΔH_{AB} values for the shuffled data set, and then a threshold can be chosen just above the largest ΔH_{AB} obtained from the calculation.

To compute the correlation index among M cells, their frequency of synchronous firing is

$$p_{1\ldots M} = \frac{1}{N}\sum_{j=1}^{N}\prod_{i=1}^{M} r_j^{(i)} \tag{4.18}$$

whereas the synchronous firing frequency expected by chance is

$$p_1 \ldots p_M = \prod_{i=1}^{M}\frac{1}{N}\sum_{j=1}^{N} r_j^{(i)} \tag{4.19}$$

Therefore, the correlation index can be calculated as

$$C_{1\ldots M} = \frac{p_{1\ldots M}}{\prod_{i=1}^{M} p_i} \tag{4.20}$$

In experiments measuring retinal ganglion cells' synchronized activity via planar multielectrode system, it was found that adjacent neurons often show tendency of concerted activity. Using the information–theory-based algorithm introduced above, it was identified that a single neuron can participate different synchronous firing groups dynamically according to different visual tasks, whereas a synchronous group may only contain 1–2 neurons but can also accommodate up to 6–7 neurons (Fig. 4.19).

During visual information coding, one of the advantages of population activity is to provide better details. Figure 4.20 gives an example about spatial information encoded by retinal ganglion cells' population activities estimated using "stimulus-triggered average" method which is based on the relationship between the stimulation pattern and the neuronal response (the method will be introduced in details in Chap. 5). While each single neuron's firing activity is related to a particular receptive field, ganglion cells with anatomically adjacent position often have overlapped receptive fields and have a tendency of firing in synchrony. As compared to single ganglion cell's activity, the synchronized activity of adjacent ganglion cells normally encode finer receptive field.

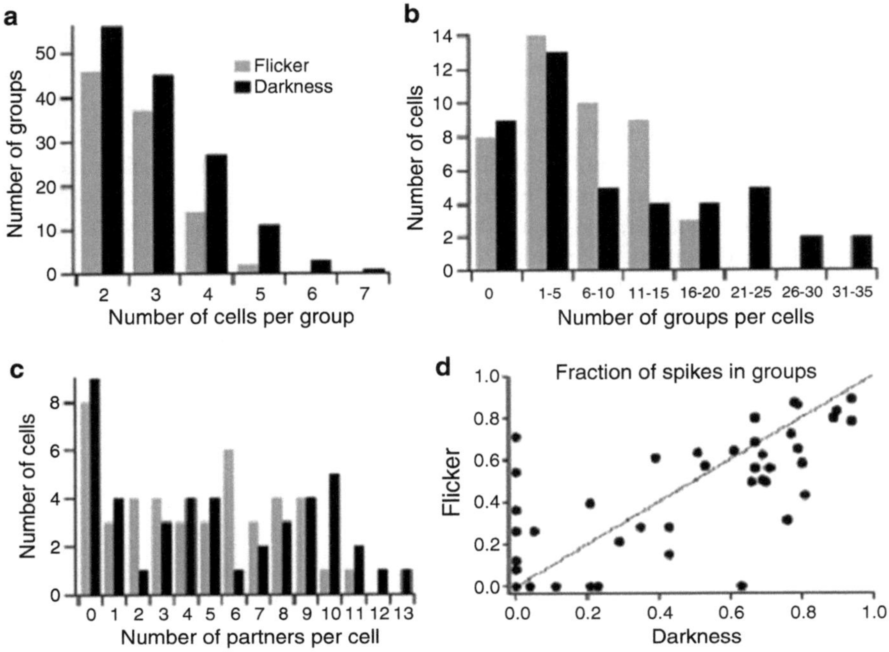

Fig. 4.19 Population activity properties of salamander retinal ganglion cells. (**a**) Histogram of cell number in single neuron group, observed in dark or with flickering light. (**b**) Histogram of group number that a single neuron participates. (**c**) Histogram of "partner neurons" of single neuron while participating different groups. (**d**) Comparison of each neuron's activity in "population activity" in different conditions (Adapted with permission from Schnitzer and Meister 2003, Fig. 4)

The synchronous firings among adjacent ganglion cells reflect that multiple neurons may carry similar information about the environmental stimulation and may thus result in some redundancy in information. This seems to be conflict with the efficient principal for information processing. But in fact, neurons can form different activation populations dynamically. In neuronal network, information encoded by such dynamical population activity may greatly exceed the total information encoded by each individual neuron's firing pattern.

The combination of multineuronal activities results in abundant population firing patterns, which carry more information. To describe the spatiotemporal pattern of concerted activities among many neurons, a measurement of subsequence distribution discrepancy (MSDD), termed "information discrepancy," was proposed to handle a group of spike train sequences ($n > 2$).

In MSDD calculation, the spike trains are symbolized into "0-1" sequences. Reasonable bin size is chosen to allow for that each time bin contains at most one spike. Therefore, in the "0-1" sequences, "1" represents that there is one spike in the time bin of interest and "0" represents that there is no spike. When the MSDD

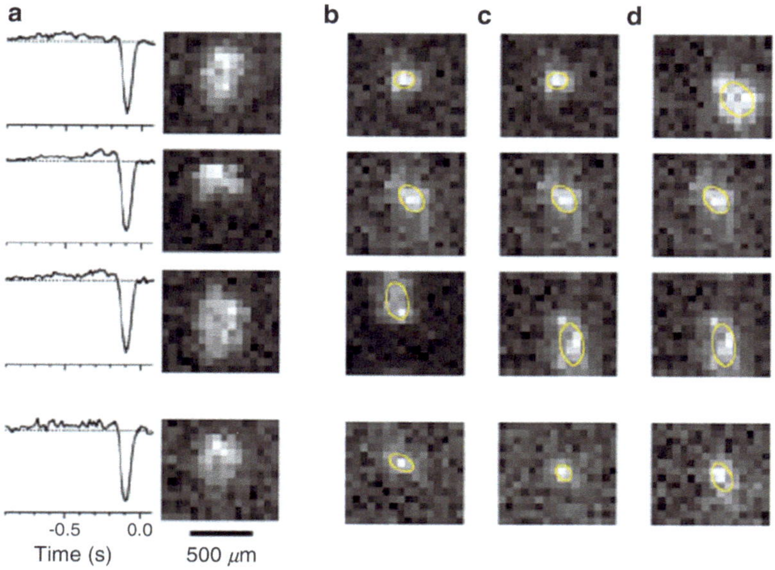

Fig. 4.20 Spatial information encoded by single neuron's activity and ganglion cells' population activity. (**a**) The reverse correlation estimation of three ganglion cells receptive field (three *top panels*) and that of synchronized activity (the *bottom panel*). The time course of the stimulus-triggered average is also plotted in each channel. (**b–d**) Receptive fields for three more triplets of ganglion cells (*circled area* in the *upper panels*) and their synchronous spikes (*circled area* in the *lower panel*). It shows that the receptive field encoded by synchronous activity is finer as compare to that encoded by single neuron's activity (Adapted with permission from Schnitzer and Meister 2003, Fig. 6)

method is applied to measure discrepancy among multiple sequences, the constructive information of a sequence is uniquely transformed into a set of subsequence distributions, which is defined as follows:

Let $G = \{a_1, a_2, \ldots, a_m\}$ be a set of m symbols, and suppose $S = \{S_1, S_2, \ldots, S_s\}$ is a set of s sequences (firing sequences recorded from s neurons, in this case) formed from the symbol set G. Then the total number of all different subsequences formed from G with length l should be equal to m^l. The neuronal firing activities are represented by two symbols "0" and "1" ($m = 2$). In case all the sequences to be analyzed are separated into 6-letter overlapping subsequences ($l = 6$), with moving step being one time bin, $m^l = 2^6 = 64$. Suppose S_k is the k-th sequence of L bins and given the subsequence length of l, the probability distribution of the subsequences in S_k is:

$$U^l_{S_k} = \left\{p^l_{k,1}, p^l_{k,2}, \ldots, p^l_{k,64}\right\} \tag{4.21}$$

where $p_{k,j}$ denotes the probability of subsequence pattern j in sequence k, $\sum_{i=1}^{64} p^l_{k,i} = 1,\ l << L,\ k = 1, 2, \ldots, s.$

Given a set of s sequences, we have:

$$
\begin{aligned}
U_{S_1}^{l} &= \left\{p_{11}^{l}, p_{21}^{l}, \ldots, p_{m(l)1}^{l}\right\}^{T} \\
U_{S_2}^{l} &= \left\{p_{12}^{l}, p_{22}^{l}, \ldots, p_{m(l)2}^{l}\right\}^{T} \\
&\cdots\cdots \\
U_{S_s}^{l} &= \left\{p_{1s}^{l}, p_{2s}^{l}, \ldots, p_{m(l)s}^{l}\right\}^{T}
\end{aligned}
\tag{4.22}
$$

Then the difference between the temporal structure of the k-th sequence and the rest of the group can be calculated as:

$$
B_k\left(U_{S_1}^{l}, U_{S_2}^{l}, \ldots, U_{S_s}^{l}\right) = \sum_{i=1}^{m(l)} p_{ik}^{l} \log \frac{p_{ik}^{l}}{\sum_{k=1}^{s} p_{ik}^{l}/s}
\tag{4.23}
$$

A large B_k value is related to a profound difference between the k-th sequence and the rest of the group, and if two spike trains are completely synchronized, they will have exactly equal values for B_k.

This algorithm was applied for analyzing spatiotemporal pattern of retinal ganglion cells' activities in response to different spatial patterns of visual stimulation as illustrated in Fig. 4.21. Retinal ganglion cells' activities in responses to these visual stimulations were recorded using multielectrode arrays as illustrated in Fig. 4.22, with each trace corresponding to a 200-ms recording of one cell's response.

To describe the concerted pattern among multiple spike trains, MSDD algorithm was applied to perform multidimensional data analysis, and B_k values for each neuron's activities in response to three different stimulation patterns were calculated and plotted in Fig. 4.23. It is shown in Fig. 4.23a that when checkerboard stimulus was applied, the neuronal activities were with larger B_k values as compared to the neuronal responses elicited by the vertical and horizontal gratings, and this is true for all the 34 neurons recorded. According to the principle of MSDD method, larger B_k values reflect that the firing patterns were less coincident among the group cells in response to checkerboard stimulus than to the gratings. On the other hand, the population activity patterns in response to vertical gratings and horizontal gratings were somewhat similar in a sense that the B_k values obtained for these situations were quite similar for most cells as illustrated in Fig. 4.23a. However, the results plotted in Fig. 4.23b also show that there were a small group of neurons whose B_k values were remarkably different in response to the two grating stimulation patterns. Taken together, these results reveal that: (1) Different stimulation patterns can hardly be distinguished according to the changes in single neuronal firing rates. (2) The population activity patterns were different in response to different stimulus patterns.

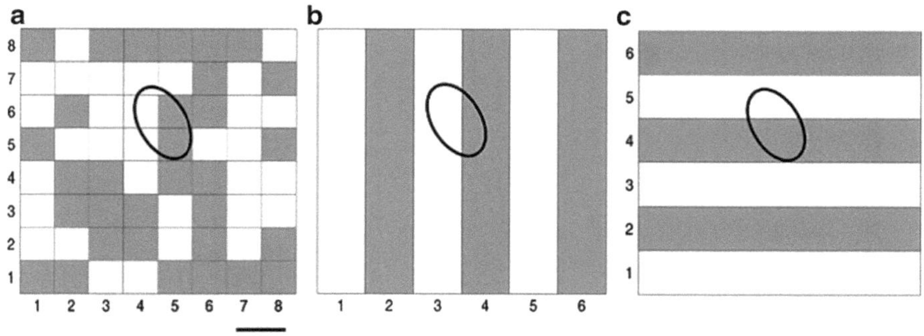

Fig. 4.21 Three stimulation patterns used as visual stimulation and one example RGC's receptive field profile. (**a**) Checkerboard, (**b**) vertical gratings, (**c**) horizontal gratings. The ellipse indicates the 1-s.d. boundary of the neuron's receptive field fitted with a two-dimensional Gaussian distribution (scale bar, 200 μm) (Adapted with permission from Jing et al. 2010, Fig. 2)

In addition to the abovementioned concerted activity among neurons with similar properties, population coding involves another manner in which neurons with different properties have status-dependent coordinated activities, which allows for information coding and transmission for various conditions. One example is taken from primary motor cortex M1 (see Fig. 2.72), in which two major parameters of motion, direction and amplitude, can be encoded by population neurons' activities. In this case, a single neuron's activity related to a certain motion direction can be represented by a vector, with the direction and amplitude of the vector determined by the neuron's optimal movement direction and the condition-related firing rate, respectively. When each single neuron's activity during a particular movement can be represented by a vector, the average population vector can be used to predict the actual movement direction and amplitude, provided sufficient amount of neurons are examined.

Actually, several cortical areas in the frontal and parietal lobes (including dorsal premotor cortex (PMd), primary motor cortex (MI), and posterior parietal cortex (PP)) are involved in motor commands, with distributed representations of motor information. Therefore, according to the simultaneous activity of large populations of neurons distributed in the PM, MI, and PP areas, accurate real-time prediction of movement trajectory can be obtained by properly analyzing the cortical neuronal ensemble activity, as demonstrated by a pioneer work.

In the experiment, a monkey was trained to make one-dimensional hand movements to displace a manipulandum (left or right) or make three-dimensional hand movements to access a target which was randomly placed in a tray. Cortical recordings were obtained while the animal was trained and tested on both tasks (Fig. 4.24a). The raw neuronal data obtained while the animal performed the movement task (Fig. 4.24b) were then analyzed to predict the animal's movement (Fig. 4.24c).

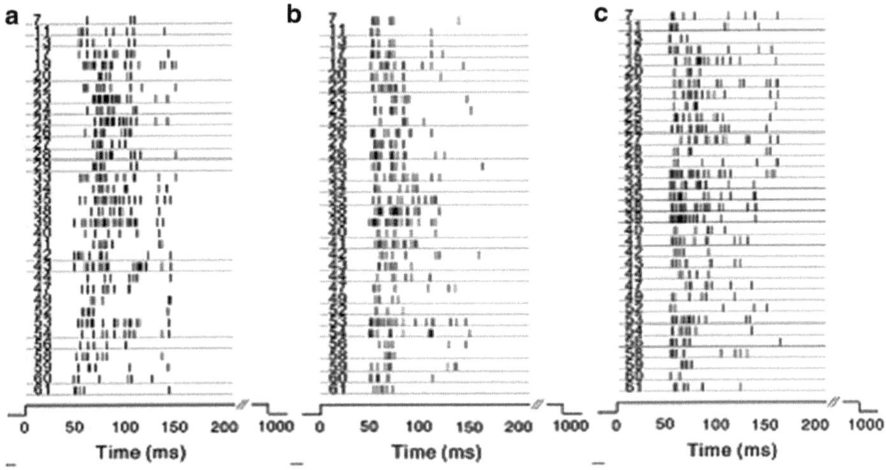

Fig. 4.22 Raster plot of 34 neurons' firing activities (200 ms after the stimulus onset) in response to (**a**) checkerboard, (**b**) vertical gratings, and (**c**) horizontal gratings (Adapted with permission from Jing et al. 2010, Fig. 5)

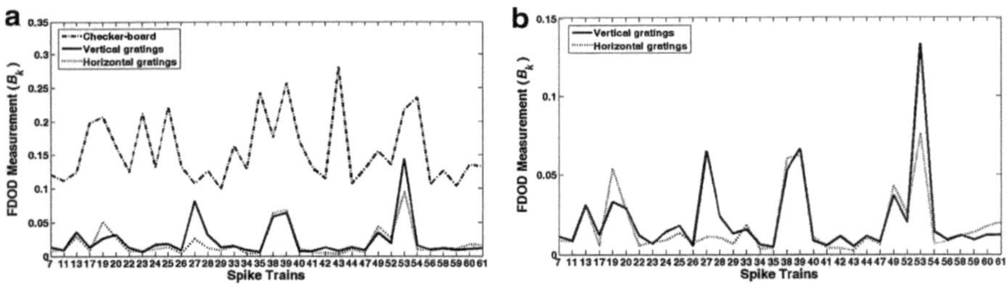

Fig. 4.23 B_k values calculated for the neuronal activities (**a**) in response to checkerboard, vertical gratings, and horizontal gratings; (**b**) enlarged details for difference in neuronal population activity patterns in response to vertical gratings and horizontal gratings (Adapted with permission from Jing et al. 2010, Figs. 6 and 7)

According to the population neurons firing activity, the animal's hand position can be predicted using either a linear model or an artificial neural network (ANN) model. In the linear model, the hand position $Y(t)$ is dependent on the neuronal discharges $X(t)$:

$$Y(t) = b + \sum_{u=-m}^{n} a(u)X(t-u) + \varepsilon(t) \tag{4.24}$$

where b are the Y-intercepts in the regression, a is a set of impulse response functions, and $\varepsilon(t)$ is the residual error. Therefore, the neuronal firing series $X(t)$ are convolved with the weight functions a, so that the sum of these

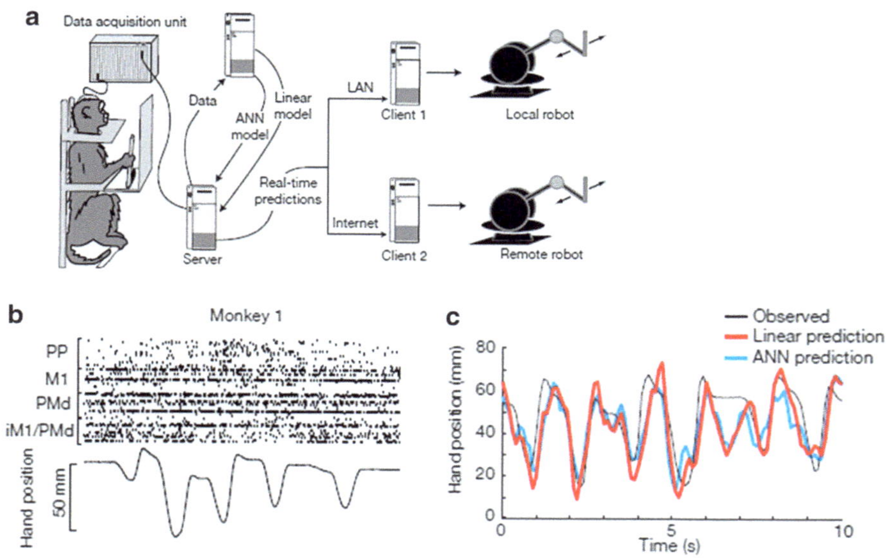

Fig. 4.24 Experimental design. (**a**) Schematic diagram showing the experimental apparatus for motor control of both local and remote robotic devices based on simultaneously recorded cortical ensemble data from primate motor cortices, with the cortical data being analyzed using a linear model or an artificial neural network (ANN) model. (**b**) Simultaneously recorded neuronal activity in five cortical areas in the monkey. PP, posterior parietal cortex; M1, primary motor cortex; PMd, dorsal premotor cortex; iM1/PMd, ipsilateral M1 and PMd. (**c**) Observed (*black*) and real-time predicted one-dimensional hand movements using linear (*red*) and ANN (*blue*) models (Adapted with permission from Wessberg et al. 2000, Figs. 1 and 2)

convolutions plus b provides the prediction of $Y(t)$ that in equation (4.24). Parameters are estimated via off-line analysis. Real-time prediction of the animal's movement and the control of robot arm movement can then be achieved by using a slightly modified linear model, in which parameters are updated according to the most recently recorded data. In animal experiment, this linear model yielded highly significant real-time predictions (Fig. 4.25c). The prediction based on the activity of 27 PMd, 26 MI, 28 PP, and 19 ipsilateral MI/PMd neurons recorded from the monkey cortices achieved an average correlation coefficient of 0.61 (range 0.50–0.71) between the observed and predicted hand position (60-min session), while the ranging was between 0.45 and 0.73 for ANN model.

The performance of the linear-model-based prediction was highly dependent on the number of neurons according to whose activities the model was built. The relationship between ensemble size and the goodness of fitting can be analyzed off line by randomly removing single neurons, one at a time, and fitting the models repeatedly using the progressively shrinking neuron population. The results show that hyperbolic functions could be used to describe the relationship between the goodness of fitting and the ensemble size, for either neurons recorded from single or multiple cortical areas (Fig. 4.25), which confirmed that the movement information was encoded by population neurons' activities.

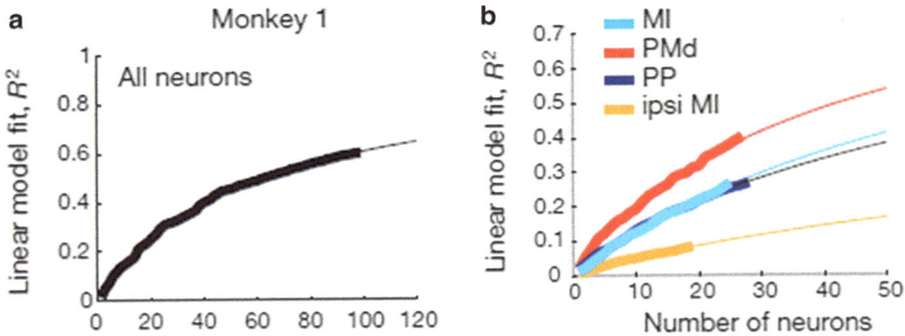

Fig. 4.25 Neuron-dropping analysis for movement prediction. (**a**) Neuron-dropping curve (*thick line*) obtained in a single session for all cortical neurons in the monkey and its hyperbolic fitting (*thin lines*). (**b**) Neuron-dropping curves (*thick lines*) and corresponding hyperbolic functions (*thin lines*) for each cortical area for the monkey (Adapted with permission from Wessberg et al. 2000, Fig. 4)

4.4 Theory of Population Coding

The preceding sections focus on the properties of single neurons in information processing. In reality, however, a neural system typically exploits the joint activity of a large number of neurons to encode the stimulus information, the so-called population code. An advantage of population code is that by taking into account the joint effort of a neuron ensemble, fluctuations in the responses of individual neurons, which are ubiquitously observed in experiments, are averaged out, leading to a much improved decoding accuracy about the stimulus. In this section, the theory of population coding is reviewed.

4.4.1 Tuning Function

Although the response of a neuron to a given stimulus value is very noisy in a single trial, its averaged value over many trials is rather smooth, which is called tuning function (or tuning curve). The tuning function summarizes the response nature of a neuron to all possible stimulus values. Two forms of tuning function, the bell-shaped and the sigmoid shape, are observed in the experiments. The former is often described as Gaussian or Cosine function depending on the fitting of experiment data.

Figure 4.26a shows the extracellular recordings of a neuron in the primary visual cortex (V1) of a monkey in response to a light bar of different orientations. The firing rates of the neuron can be fitted as Gaussian function in the space of orientation angles (Fig. 4.26b). The Gaussian tuning function can be written as

$$f(s) = r_{\max}\exp\left[-\frac{(s - s_{\max})^2}{2a^2}\right], \tag{4.25}$$

where s denotes the stimulus value, i.e., the orientation angle is the orientation angle evoking the maximum average response rate $r_{\max}$ and the parameter a determines the width of the tuning curve. The neuron responds most vigorously when a stimulus $s_{\max}$ is presented, so $s_{\max}$ is called the preferred stimulus of the neuron.

Sigmoid function is another important tuning curve observed in the experiments. Figure 4.27 shows an example of sigmoid tuning function observed in V1 neurons of a cat in response to retinal disparity of different degrees, which can be written as

$$f(s) = \frac{r_{\max}}{1 + \exp\left[\left(s_{1/2} - s\right)/\Delta_s\right]}, \tag{4.26}$$

where s denotes the retinal disparity, and $s_{1/2}$ is the disparity degree that produces a firing rate half as big as the maximum value $r_{\max}$. The firing rate monotonically increases with s, with Δ_s controlling the increasing speed.

4.4.2 *Encoding Model*

The tuning function only reflects the averaged response property of a neuron. In a single trail, the response of a neuron is very noisy; this is where the challenge of neural decoding arises and why population coding becomes necessary. Incorporating the noise contribution, the activity of the i-th neuron in a single trail is expressed as

$$r_i = f(s - s_i) + \sigma\varepsilon_i, \qquad i = 1, \ldots, N, \tag{4.27}$$

where N is the number of neurons, s_i is the preferred stimulus of the i-th neuron, and the parameter σ denotes the noise intensity. ε_i is a random number and satisfies $\langle\varepsilon_i\rangle = 0, \langle\varepsilon_i\varepsilon_j\rangle = A_{ij}$, with $\langle\cdot\rangle$ representing averaging over many trials. The covariance matrix A_{ij}, for $\mathrm{i,j} = 1, \ldots, \mathrm{N}$, specifies the pair-wise correlation between neurons, with $A_{ij} = \delta_{ij}$ meaning noises of different neurons independent to each other.

Overall, the encoding process of a neuron population is described by the conditional probability of observing $\mathbf{r}$ given the stimulus value s, which is written as:

$$p(\mathbf{r}|s) = \frac{1}{\sqrt{(2\pi\sigma^2)^N \det(A)}} \times \exp\left[-\frac{1}{2\sigma^2}\sum_{ij} A_{ij}^{-1}(r_i - f_i(s))\left(r_j - f_j(s)\right)\right]. \tag{4.28}$$

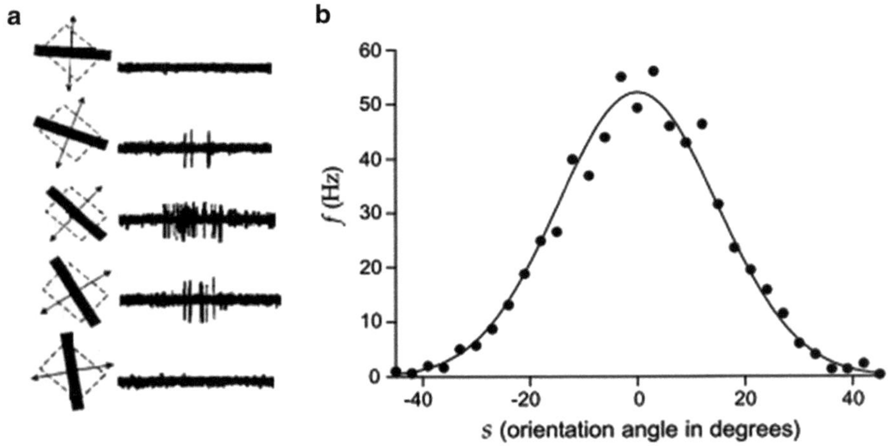

Fig. 4.26 (**a**) Recordings from a neuron in the primary visual cortex of a monkey. A light bar of different orientations was moved across the receptive field of the neuron. (**b**) Average firing rate of a cat V1 neuron plotted as a function of the orientation angle of the light bar stimulus. ((**a**) from Hubel and Wiesel 1968; adapted from Wandell 1995; (**b**) adapted with permission from Dayan and Abbott 2001)

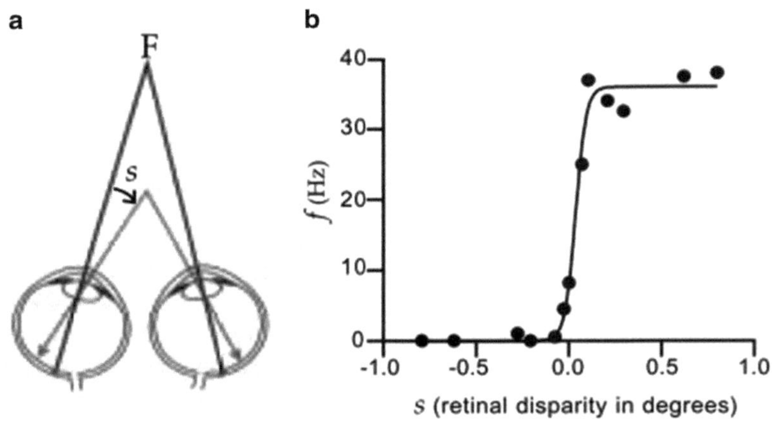

Fig. 4.27 (**a**) Illustration of retinal disparity. (**b**) Average firing rate of a cat neuron as a function of the disparity ((**a**) Adapted with permission from Wandell 1995; (**b**) Adapted with permission from Dayan and Abbott 2001, (**b**) data points from Poggio and Talbot 1981)

4.4.3 Upper Bound of Neural Decoding

The task of neural decoding is to infer the stimulus value based on the observed neural population activity. Depending on the strategies used, different approaches may have different performances. Nevertheless, their accuracies are bounded by the Cram*ér* – Rao bound, which is given by the inverse of the Fisher information.

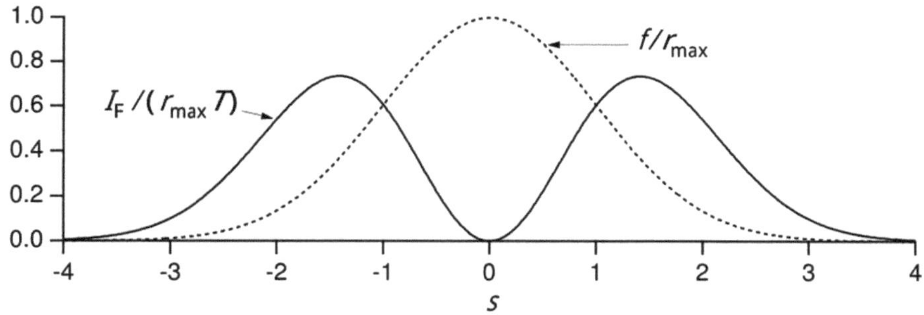

Fig. 4.28 The Fisher information of a single neuron with Gaussian tuning function with s = 0, $a = 1$ and $\sigma A_{ij} = f_i^{\frac{1}{2}}\delta_{ij}$. Note that the Fisher information is greatest where the slope of the tuning curve is highest and vanishes at s = 0 where the tuning curve peaks (Adapted with permission from Dayan and Abbott 2001)

Thus, the Fisher information indicates the amount of stimulus information contained in neural data. Given the conditional probability $p(\mathbf{r}|s)$, the Fisher information is calculated to be:

$$I_F(s) = \left\langle -\frac{\partial^2 \ln p(\mathbf{r}|s)}{\partial s^2} \right\rangle = \int dr\, p(\mathbf{r}|s) \left(-\frac{\partial^2 \ln p(\mathbf{r}|s)}{\partial s^2} \right). \tag{4.29}$$

For any unbiased estimator (i.e., the mean of estimation equals to the true stimulus value), the Cramér – Rao bound states that the variance of estimation satisfies

$$\left\langle (\hat{s} - s)^2 \right\rangle \geq \frac{1}{I_F(s)}, \tag{4.30}$$

where ŝ denotes the estimation.

Figure 4.28 displays an example of Fisher information contributed by a single neuron.

4.4.4 Population Decoding Methods

A simple way to decode the stimulus by using neural population activity is center of mass (COM), whose estimation is expressed as

$$\hat{s} = \frac{\sum_i s_i r_i}{\sum_i r_i}. \tag{4.31}$$

Note that this method is equivalent to population vector when the stimulus is a one-dimensional variable. Population vector was used by Georgopoulos et al. to reconstruct the moving direction of monkey's arm based on neural population activity recorded in the motor cortex, and it provides a strong evidence supporting population coding is used in the brain (Fig. 4.29).

Statistically, a more accurate decoding approach than COM is maximum likelihood, which infers the stimulus value by maximizing the log likelihood function (the conditional probability $p(\mathbf{r}|s)$), i.e.,

$$\hat{s} = \underset{s}{\text{Max}} \ln p(\mathbf{r}|s). \tag{4.32}$$

If the correlation between fluctuations in neuronal responses is weak, ML achieves the Cramér – Rao asymptotically when the number of neurons is sufficiently large.

If the likelihood function is given by Eq. (4.28) and that fluctuations of neuronal responses are independent to each other, i.e., $A_{ij} = \delta_{ij}$, the solution of ML can be written as:

$$\hat{s} = \underset{s}{\text{Max}} \sum_i f(s) r_i. \tag{4.33}$$

In this case, ML can be interpreted as template matching, where the tuning function $f(s)$ is the template, and the decoding is to adjust the position of the template s, such as the overlap between the template and the neural population activity $\mathbf{r}$ has the maximum value. It has been shown that template matching can be implemented by continuous attractor neural networks (Fig. 4.30).

In case there is prior knowledge about the stimulus value, referred to as $p(s)$, maximum a posterior (MAP) can be used to decode the stimulus, which is given by:

$$\begin{aligned}\hat{s} &= \underset{s}{\text{Max}} \ln p(s|\mathbf{r}) \\ &= \underset{s}{\text{Max}} \left[\ln p(\mathbf{r}|s) + \ln p(s)\right].\end{aligned}$$

ML is a special case of MAP when no prior knowledge exist, i.e., $p(s)$ is a constant.

Both ML and MAP assumes that the encoding process $p(\mathbf{r}|s)$ is available to the estimator. However, this is not always true for neural decoding—in reality, the noise structure of external inputs may not be known to the neural estimator. Thus, the brain faces a constant challenge of inferring the stimulus value based on partial knowledge of the encoding process, which is called unfaithful decoding.

Consider an example. The true neural encoding process is given by Eq. (4.28), but the neural estimator does not know about the correlation matrix A_{ij}. Instead, the neural estimator assumes that the neuronal response fluctuations are independent to each other and employs an unfaithful encoding model given by:

$$q(\mathbf{r}|s) = \frac{1}{\sqrt{(2\pi\sigma^2)^N}} \times \exp\left[-\frac{(r_i - f_i(s))^2}{2\sigma^2}\right].$$

The estimate of unfaithful decoding is written as:

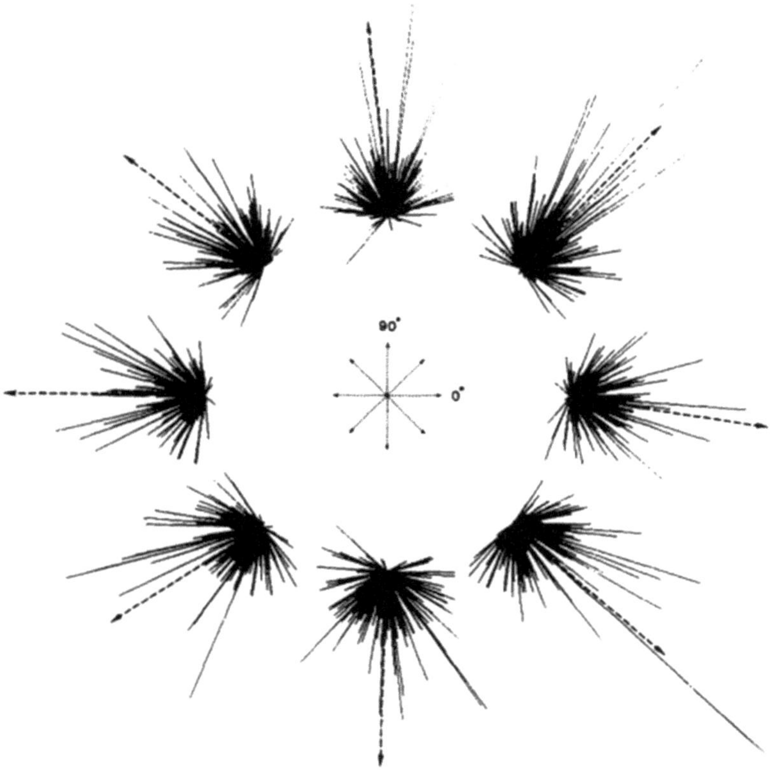

Fig. 4.29 The population vector provides a cortical representation of movement. Vector contributions of 241 directionally tuned motor cortical cells are shown for each of the eight movement directions tested. Notice the spatial congruence between the direction of the vectorial sum (*thick interrupted lines in each plot*) and the direction of movement (*thin interrupted lines at center*) Adapted with permission from Georgopoulos and Pellizzer (1995)

$$\begin{aligned}\hat{s} &= \underset{s}{\text{Max}} \ln q(\mathbf{r}|s) \\ &= \underset{s}{\text{Max}} \sum_i f(s) r_i.\end{aligned}$$

Note that in this case, unfaithful decoding is actually equivalent to template matching. In general, by choosing a proper unfaithful encoding model, the neural estimator can achieve a proper balance between decoding accuracy and computational complexity (e.g., by neglecting neuronal correlation, decoding becomes much simpler).

4.5 Bayesian Inference

Human and animals are living in a world with full of uncertainty, a consequence arising from the noisy nature of external environments and the stochastic character of neural signaling (e.g., the stochastic release of neural transmitters). Information

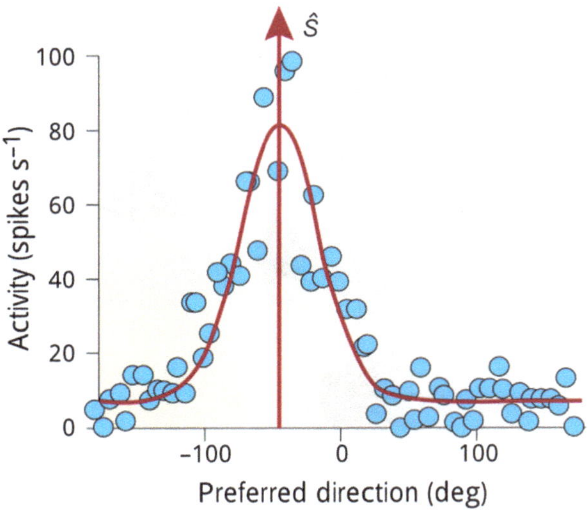

Fig. 4.30 Maximum likelihood works as if implementing template matching. The template is obtained from the noiseless (or average) population activity in response to a stimulus moving in direction s. The peak position of the template with the best fit, $\hat{s}$, corresponds to the maximum likelihood estimate, that is, the value that maximizes $p(\mathbf{r}|s)$ (Adapted with permission from Pouget et al. 2000)

extraction in the brain is essentially a process of probabilistic inference. Thus, it is not surprising that Bayesian inference, a statistically efficient way, is widely used in brain functions, including, for instance, object recognition, motor control, and multisensory information integration. The fundamental idea of Bayesian inference is to utilize the prior knowledge about the stimulus to enhance the extraction of the stimulus information. The population decoding method, MAP, presented in the preceding section is one such example. The present section introduces the basic concept of Bayesian inference and its application to multisensory integration.

4.5.1 The Bayes' Theorem

Consider two stochastic variables, X and Y. The joint probability of them taking a particular pair of values $(X = x, Y = y)$ is written as $p(x, y)$. The conditional probability of observing $X = x$ given $Y = y$ is denoted as $p(x|y)$, and $p(y|x)$ is defined similarly. We have the following relationship,

$$p(x, y) = p(x|y)\,p(y) = p(y|x)\,p(x). \tag{4.34}$$

From the two ways of representing the joint probability in the above equation, the two conditional probabilities can be related as

$$p(x|y) = \frac{p(y|x)\,p(x)}{p(y)}, \tag{4.35}$$

as long as $p(y)$ never becomes exactly zero. The above equation is called the *Bayes' theorem*, which provides a way of converting one conditional probability into the other.

In the framework of inference, the Bayes' theorem can be interpreted as:

$$p(\text{hypothesis}|\text{data}) = \frac{p(\text{data}|\text{hypothesis})\, p(\text{hypothesis})}{p(\text{data})}. \tag{4.36}$$

It tells us how to update the belief on the hypothesis based on the observed data, $p(\text{hypothesis}|\text{data})$, by using the fact of how well the hypothesis predicts the data, $p(\text{data}|\text{hypothesis})$, and the prior knowledge of the hypothesis, $p(\text{hypothesis})$.

4.5.2 Bayesian Inference Applied to Multisensory Integration

Multisensory integration is an important function of the brain, which helps us to improve perception by integrating ambiguous cues about the same object from different sensory organs. Figure 4.31 displays an example of estimating the position s of an object by integrating the visual and auditory cues, denoted as x_1 and x_2, respectively (Knill and Pouget 2004). The is equivalent to computing the conditional probability $p(s|x_1, x_2)$, also called the posterior function. According to the Bayes' theorem, we have,

$$\begin{aligned} p(s|x_1, x_2) &= \frac{p(x_1, x_2|s)\, p(s)}{p(x_1, x_2)} \\ &\propto p(x_1, x_2|s)\, p(s) \end{aligned} \tag{4.37}$$

To get the above result, the condition that $p(x_1, x_2)$ is irrelevant to the position s is used.

In practice, it is often reasonable to assume that the noise sources in the visual and auditory pathways are statistically independent to each other, which gives

$$p(x_1, x_2|s) = p(x_1|s)\, p(x_2|s). \tag{4.38}$$

By taking the above condition, we get

$$p(s|x_1, x_2) \propto p(x_1|s)\, p(x_2|s)\, p(s). \tag{4.39}$$

Furthermore, if we assume $p(s)$ as a flat function, i.e., all values of s occur with equal probability and that each likelihood function $p(x_i|s)$, for $i = 1,2$, has the Gaussian form with mean μ_i and variance σ_i^2, then the posterior distribution $p(s|x_1, x_2)$ is also a Gaussian function with the mean and variance given by:

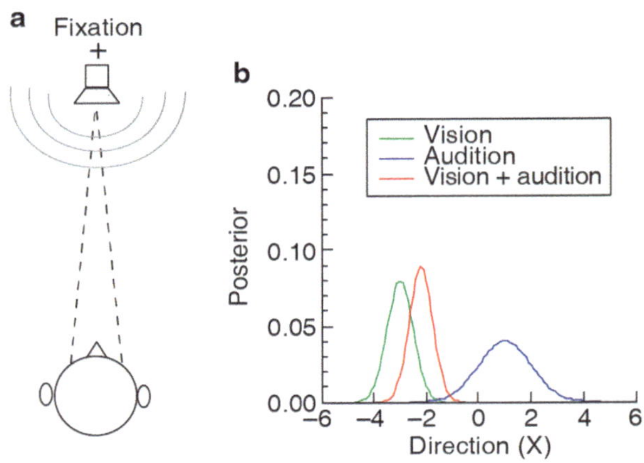

Fig. 4.31 Integration of visual and auditory cues for inferring object location. (**a**) Diagram of the visual and auditory integration. A subject receives the visual and auditory inputs from an object and needs to estimate object direction by integrating visual and auditory cues. (**b**) Posterior functions when presenting either cue alone or both cues simultaneously. The positions of posterior function given visual and auditory cues individually are different due to the corruption of noise (Adapted with permission from Knill and Pouget 2004, Fig. 1A)

$$\mu_3 = \frac{\sigma_2^2}{\sigma_1^2 + \sigma_2^2}\mu_1 + \frac{\sigma_1^2}{\sigma_1^2 + \sigma_2^2}\mu_2, \tag{4.40}$$

$$\sigma_3^2 = \frac{\sigma_1^2 \sigma_2^2}{\sigma_1^2 + \sigma_2^2}. \tag{4.41}$$

These above two equalities are the results of Bayesian integration with two independent cues. Equation (4.40) tells us that the Bayesian integrated mean is a weighted average of the means when only single cues are used, and the values of weights are determined by the reliabilities (inverse of variance) of the cues. When one cue is less reliable than the other, the integrated mean is biased toward the more reliable cue. This result is intuitively understandable. Taking the integration of visual and auditory cues in Fig. 4.31a as the example, if light is very dark, we would rely more on the auditory cue to infer the object location; otherwise we trust more on the visual cue. After integration, the variance of estimate is always smaller than that when a single cue is used (e.g., for $\sigma_1 = \sigma_2$, $\sigma_3^2 = \sigma_1^2/2$), implying that Bayesian integration always improves the decoding performance. Figure 4.31b illustrates an example of the integration behavior, where the visual cue is more reliable than auditory cue, and consequently, the posterior function with combined cues (red line) is closer to the posterior function with only the visual cue (green line).

4.5.3 *The Brain Behaves Like a Bayesian Decoder*

A large volume of psychophysical studies has demonstrated that the brain does implement Bayesian inference in multisensory integration. These studies include the integration of texture and motion cues for depth perception (Jacobs 1999), stereo and texture cues for slant perception (Knill and Saunders 2003; Hillis et al. 2004), visual and vestibular cues for motion detection (Fetsch et al. 2013), and visual and auditory cues for object localization (Alais and Burr 2004). In these experiments, subjects were asked to perform a discrimination task in the presence of either a single cue or combined cues.

Let us look at the integration of visual and haptic cues to infer the height of an object as the example (Fig. 4.32). In the experiment, the subject could touch an object by using fingers and/or looking at the object through an opaque mirror (Ernst and Banks 2002) (Fig. 4.31). Four levels of noise were added to the visual cue to vary its reliability. Figure 4.32b presents the result of the experiment, which is displayed in the form of psychometric function showing the proportion of correct choice versus the stimulus parameter. The experiment also artificially introduced a disparity between the visual and haptic cues. When the level of visual noise is decreased, the psychometric function becomes sharper, indicating that the visual cue becomes more reliable; meanwhile, the psychometric function is shifted toward to the visual cue, implying that the human subject automatically adjusted the weights of the cues according to their reliabilities. Thus, Bayesian inference is realized.

4.5.4 *Neural Correlates of Bayesian Multisensory Integration*

Psychophysical studies demonstrate the application of Bayesian inference in brain functions. Neurophysiological experiments further unveil where this happens in the brain. In a series of monkey experiments, Angelaki and her colleagues identified the brain regions involved in the integration of visual and vestibular cues for self-motion detection (Fetsch et al. 2013). In their experiments, monkeys were trained to carry out a discrimination task which infers the heading direction based on either a visual cue or a vestibular cue or both of them (Fig. 4.33a–c). The behavior data demonstrates that the monkeys were performing Bayesian inference (Fig. 4.33d, g). They further recorded neuronal responses in several regions at the posterior parietal cortex (PPC), including the dorsal medial superior temporal area (MSTd) and the ventral intraparietal area (VIP). They found that neuronal activities in these areas also exhibit the optimal Bayesian integration performance, including the improved decoding accuracy when both cues were applied and the shift of estimation toward to the more reliable cue (Fig. 4.33e, f, h and i).

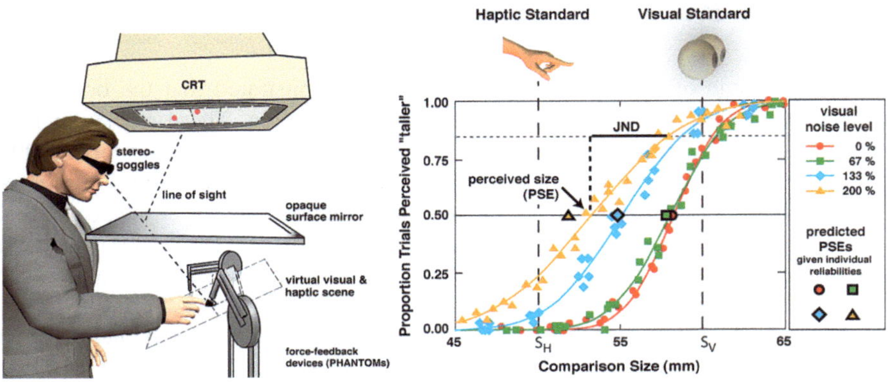

Fig. 4.32 Integration of visual and haptic cues for inferring object height. (**a**) Experimental setup. A subject needs to infer the height of an object through direct touching it with fingers and/or looking the object through an opaque mirror. (**b**) Psychometric function of integration of visual and haptic cues for inferring object height. When decreasing visual noise, the psychometric function becomes sharper and moves toward the position of visual cue (Adapted with permission from Ernst and Banks 2002, A Bayesian view on multimodal cue Integration, Figs. 3 and 4)

4.5.5 Network Architecture for Implementing Bayesian Multisensory Integration

The psychophysical and physiological experiments have confirmed the application of Bayesian multisensory integration in the brain. It prompts a question: how does a neural circuit implement this optimal integration?

In engineering applications, two principled architectures have been proposed to integrate observations from different information sources, namely, the centralized and the decentralized architectures (Durrant-Whyte and Henderson 2008). The former considers that raw observations from sensors are sent directly into a central fusion center, which estimates the underlying state from the raw observations of all of the modalities. Although simple in structure, the centralized architecture suffers from the computational burden of the fusion center, the high communication load (because all raw observations of all modalities must be delivered into a single center), and the susceptibility of being paralyzed once the fusion center fails. On the other hand, the decentralized architecture considers that all of the processors communicate with the others directly, so that a central fusion center becomes obsolete. Each processor first makes a local estimate according to its own observation and then corrects it by integrating the local estimates from the other processors to obtain a global estimate. Decentralized computing has been favored for engineering applications due to its robustness, computational efficiency, and modularity and has recently become the state-of-the-art system architecture for diverse applications such as cloud computing and data sharing.

The available anatomical and neurophysiological data tends to suggest that the decentralized architecture is applied for multisensory information integration. One

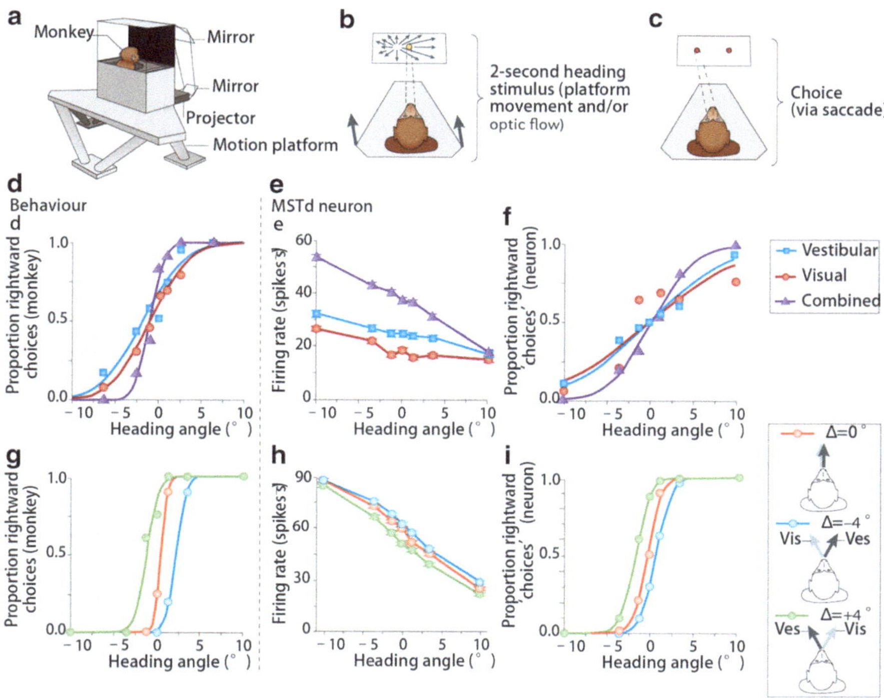

Fig. 4.33 Experiments of visual and vestibular integration for perception of self-motion direction. (**a**) Experiment setup. A monkey is sitting on a motion platform and watching the screen playing optic flow. (**b** and **c**) The stimulus lasts for 2 s, and then the monkey needs to judge whether the motion direction is left- or rightward by making a saccade. (**d** and **g**) Psychometric functions when visual and vestibular cues are congruent (**d**) or conflict (**g**). (**e** and **h**) Tuning curves of a neuron under three stimulus conditions around 0° when two cues are congruent (**d**) or conflict (**h**). (**f** and **i**) Neurometric functions under three stimulus conditions when two cues are congruent (**f**) or conflict (**i**) (Adapted with permission from Fetsch et al. 2013, Fig. 3)

supporting observation is that there exist abundant reciprocal connections between multisensory areas. This reciprocal connectivity is essential for a decentralized integration system, whereas, in a centralized system, reciprocal connectivity is not needed and thus hard to justify. Another supporting observation is that in the integration of visual and vestibular cues, not one single central area but instead many areas, including MSTd, VIP, the frontal eye field (FEF), and the visual posterior Sylvian area (VPS), individually display integrative responses to combine visual and vestibular inputs. This distributed fashion of information integration seems to contradict with the view of having a single central multisensory integration area and instead favors a decentralized view of individual processors that all individually integrate information by communicating with one another.

Consider a decentralized integration system formed by two local processors, MSTd and VIP (Zhang and Wu 2013) (Fig. 4.34a). They receive, respectively, the

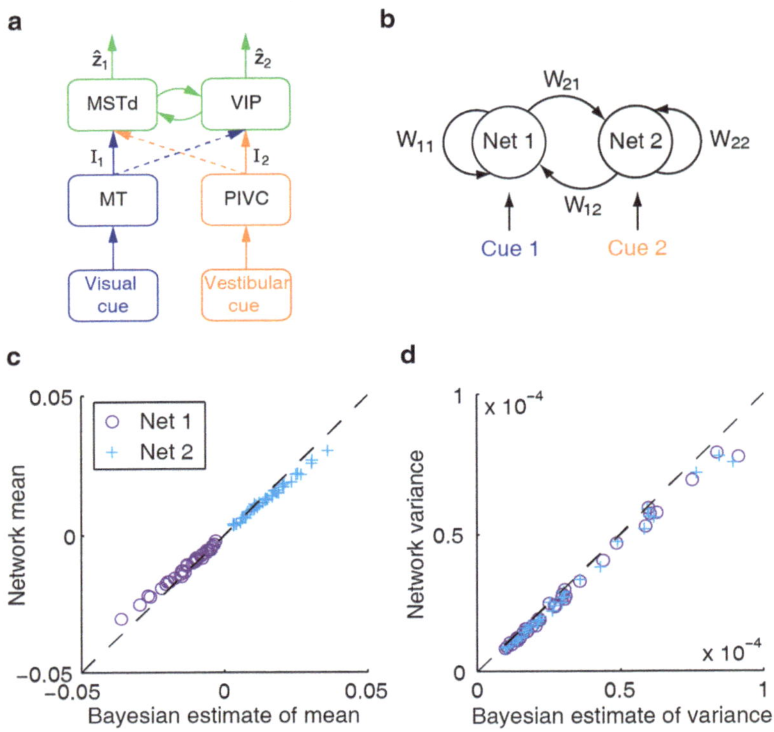

Fig. 4.34 Optimal information integration with the decentralized architecture. (**a**) The structure of a decentralized system formed by MSTd and VIP. (**b**) The network model consists of two reciprocally connected local processors, and each of them receives feedforward inputs from the associated cue. (**c** and **d**) Each network can optimally integrate both cues. Comparison between the network estimation with the Bayesian prediction: (**c**) for the mean of estimation and (**d**) for the variance of estimation (Adapted with permission from Zhang and Wu 2013)

visual and the vestibular cues but are connected with each other reciprocally, implying that each local processor also accesses the information from the other processor indirectly (Fig. 4.34b). Such a decentralized system achieves Bayesian information integration at each local processor (Fig. 4.34c, d).

4.6 Summary and Advanced Readings

One of the central issues in neuroscience is how information is carried and transferred in the nervous system. To address this, it is necessary to understand the properties of neuronal activity in response to stimulation or in delivering motor command.

Traditionally, neuronal activity was characterized by the neuron's firing rate changes. But recent evidence showed that the precise timing of firing sequence, as well as the spatiotemporal characters of population activities, also contributes to neural information coding. Therefore, to reveal the information coding property of the nervous system, the abovementioned aspects of the neuronal activity should be quantitatively measured, and mathematical tools should be incorporated.

This chapter is focused on coding properties and coding theories.

Section 4.1 gives an introduction about rate coding by giving a group of examples showing that the firing rate of sensory neuron is stimulation-feature dependent, which reveal that information about stimulation parameters can be encoded by firing rate. It is also shown that stimulation signal can be identified according to neuron's firing rate.

Section 4.2 gives some evidence that in addition to firing rate, precise timing of the spike activities, as well as the temporal structure of neurons firing sequence, also carry stimulation information. Following this temporal coding idea, informatics approaches for calculating the amount of information carried and transmitted by the neuronal firing activities are introduced.

Section 4.3 introduces the idea about population coding. Although single neuron acts as functional unit of the nervous system, the spatiotemporal pattern of population neurons' activity is crucial for information processing and transmission. In neuronal network, information carried by dynamical population activity may greatly exceed the total information encoded by each individual neuron's activity. To describe the spatiotemporal pattern of population neurons activity, mathematical approaches should be adopted, and algorithms for identifying concerted activities between pair-wise neurons and that among large groups of neurons are introduced.

Section 4.4 introduces the mathematical theory that has developed for neural population code. The theory addresses the issues, including the lower bound of decoding error given the neural encoding process and a number of decoding methods including center of mass, maximum likelihood inference, and unfaithful decoding.

Section 4.5 introduces the theory of Bayesian inference, a statistically efficient method to infer stimulus information with prior knowledge. As an example, we present the application of Bayesian inference to multisensory integration, a phenomenon that has been widely observed in human psychophysical experiments.

4.7 Advanced Readings

Dayan P and Abbott LF (2001) Theoretical Neuroscience – Computational and Mathematical Modeling of Neural Systems, The MIT Press

Mallot HA (2013) Computational Neuroscience – A First Course, Springer

References

Alais D, Burr D. The ventriloquist effect results from near-optimal bimodal integration. Curr Biol. 2004;14:257–62.

Dayan P, Abbott LF. Theoretical neuroscience: computational and mathematical modeling of neural systems. London: The MIT Press; 2001.

de Ruyter van Steveninck RR, et al. Reproducibility and variability in neural spike trains. Science. 1997;275:1805–8.

Durrant-Whyte H, Henderson TC. Multisensor data fusion. In: Springer handbook of robotics. Berlin/Heidelberg. Springer; 2008. pp. 585–610.

Ernst MO, Banks MS. Humans integrate visual and haptic information in a statistically optimal fashion. Nature. 2002;415:429–33.

Fetsch CR, DeAngelis GC, Angelaki DE. Bridging the gap between theories of sensory cue integration and the physiology of multisensory neurons. Nat Rev Neurosci. 2013;14:429–42.

Galizia CG, Lledo PM, editors. Neuroscience – from molecule to behavior: a university textbook. Berlin: Springer Spectrum; 2013.

Georgopoulos AP, Pellizzer G. The mental and the neural: psychological and neural studies of mental rotation and memory scanning. Neuropsychologia. 1995;33(11):1531–47.

Gollisch T, Meister M. Rapid neural coding in the retina with relative spike latencies. Science. 2008;319:1108–11.

Hillis JM, Watt SJ, Landy MS, Banks MS. Slant from texture and disparity cues: optimal cue combination. J Vis. 2004;4:967–92.

Hubel DH, Wiesel TN. Receptive fields and functional architecture of monkey striate cortex. J Physiol. 1968;195(1):215–43.

Ishikane H, et al. Synchronized retinal oscillations encode essential information for escape behavior in frogs. Nat Neurosci. 2005;8:1087–95.

Jacobs RA. Optimal integration of texture and motion cues to depth. Vis Res. 1999;39:3621–9.

Jing W, et al. Visual pattern recognition based on spatio-temporal patterns of retinal ganglion cells' activities. Cogn Neurodyn. 2010;4:179–88.

Kandel ER, et al., editors. Principles of neural science. New York: McGraw-Hill; 2000.

Knill DC, Pouget A. The Bayesian brain: the role of uncertainty in neural coding and computation. Trends Neurosci. 2004;27:712–9.

Knill DC, Saunders JA. Do humans optimally integrate stereo and texture information for judgments of surface slant? Vis Res. 2003;43:2539–58.

MacKay D, McCulloch WS. The limiting information capacity of a neuronal link. Bull Math Biophys. 1952;14:127–35.

Nowak P, et al. Separability of stimulus parameter encoding by on-off directionally selective rabbit retinal ganglion cells. J Neurophysiol. 2011;105:2083–99.

Pfaff DW, editor. Neuroscience in the 21st century: from basic to clinical. New York: Springer; 2013.

Poggio GF, Talbot WH. Mechanisms of static and dynamic stereopsis in fovea cortex of the rhesus monkey. J Physiol. 1981;315(1):469–92.

Pouget A, Dayan P, Zemel R. Information processing with population codes. Nat Rev Neurosci. 2000;1(2):125–32.

Reinagel P, Reid RC. Temporal coding of visual information in the thalamus. J Neurosci. 2000;20:5392–400.

Schnitzer MJ, Meister M. Multineuronal firing patterns in the signal from eye to brain. Neuron. 2003;37:499–511.

Uzzell VJ, Chichilnisky EJ. Precision of spike trains in primate retinal ganglion cells. J Neurophysiol. 2004;92:780–9.

Victor JD. Spike train metrics. Curr Opin Neurobiol. 2005;15:585–92.

Victor JD, Purpura KP. Nature and precision of temporal coding in visual cortex: a metric-space analysis. Sunderland, MA, USA. J Neurophysiol. 1996;76:1310–26.

Wandell BA. Foundations of vision. Sunderland, MA, USA: Sinauer Associates; 1995.
Wessberg J, et al. Real-time prediction of hand trajectory by ensembles of cortical neurons in primates. Nature. 2000;408:361–5.
Wu S, Nakahara H, Amari SI. Population coding with correlation and an unfaithful model. Neural Comput. 2001;13(4):775–97.
Zhang W, Wu S. Reciprocally coupled local estimators implement Bayesian information integration distributively. Paper presented at: Advances in Neural Information Processing Systems; 2013.

Chapter 5
Neural Information Processing in Different Brain Areas

Abstract Information processing in different sensory systems might have different characteristics. On the other hand, different systems might share common rules for information coding and transmission which can be analyzed using similar computational tools. This chapter will be devoted to introduce some aspects of information coding in various nervous systems, focusing on visual and olfactory systems.

Keywords Neural information processing • Visual information processing • Visual adaptation • Olfactory information processing

Through neuroelectrophysiology, the function of sensory systems can be measured as the neural activity in response to stimulation, and computational neuroscience helps to analyze the stimulation–response relationship quantitatively at activity during relevant movements. In Chap. 3, we have introduced several aspects of neural activity and coding property, as well as relevant algorithms for quantitative computation.

Information processing in different sensory systems might have different characteristics; e.g., "receptive field" is specified for visual neurons and somatosensory neurons, while auditory neurons are with "tonotopic" character. On the other hand, different systems might share common rules for information coding and transmission which can be analyzed using similar computational tools.

This chapter will be devoted to introduce some aspects of information coding in various nervous systems, including visual and olfactory systems. Information processing property in the visual system will be introduced, which involves the spatial and temporal property of neural activity during visual information processing, including receptive field property, gain control in response to stimulation with different intensities, as well as coding property during adaptation. Information processing in olfactory system is essentially to discriminate odors whose neural representations are largely overlapped. The basic structure of the olfactory system and its role in odor discrimination are introduced.

P. Liang et al., *An Introduction to Neural Information Processing*,
DOI 10.1007/978-94-017-7393-5_5

5.1 Models of Retinal Ganglion Cell's Receptive Field

One of the central issues in research about retinal information coding is about the relationship between the visual stimulation and firing activity of the retinal ganglion cells, the output neurons of retina. During early experiments, it was observed that the retinal ganglion cell's firing activity was closely dependent on the spatial distribution of the stimulation luminance. Quite different response patterns can be observed from a retinal ganglion cell while the cell is stimulated using a small light spot or full-field light with identical luminance intensity. When a small light spot is projected onto a small patch of retina, it will induce changes in a ganglion cell's firing activity—the assembly of such small patches on the retina is defined as the ganglion cell's "receptive field."

The original concept of "visual receptive field" came from H. K. Hartline's research on frog retinal ganglion cells. In his experiments, Hartline found that the frog retinal ganglion cells were activated when light stimulation was applied in a small region close to the cell body. Hartline defined such small area as the cell's "receptive field." The definition made by Hartline was later modified by S. W. Kuffler who defined "receptive field" as "those areas of the retina which must be illuminated to cause a discharge in a ganglion cell." By this definition, Kuffler first reported that the cat retinal ganglion cells' receptive field had the "center–surround antagonist" property: the light stimulation applied in the receptive field center and surround-induced opposite changes in the cell's firing activity.

The extension of a ganglion cell's receptive field is generally between 10 and 100 μm. Receptive field property differs according to the cell subtype. In some ganglion cells, a small light spot projected within the receptive field center increases the cell's firing activity, while the cell's firing activity is reduced when the light is switched off. These cells are termed "on-center cell." But in some other ganglion cells, light projected in the receptive field center reduces the cell's firing activity, whereas the cell's firing activity can be increased at the light-off transient. Such cells are called "off-center cell." For these two kinds of cells, light applied to the central or peripheral part of the receptive field will cause opposite effects. While the central area and the peripheral area are lit up simultaneously, the neuron's response elicited by light stimulating the receptive field center will be substantially attenuated by the peripheral stimulation (Fig. 5.1). Apart from these two subtypes of ganglion cells, some other cells respond to both light-on and light-off transients; these cells are called "on–off cell."

Generally speaking, the activity changes in a ganglion cell are dependent on the intensity of light projected onto its receptive field. In addition, the relationship between the ganglion cell's firing activity and the stimulation intensity is also time dependent. To quantitatively describe the relationship between the visual input and the neuronal response, one should consider both spatial and temporal properties of the stimulation. The stimulation can be defined as the light intensity distribution on the retina $I(x, y, t, \lambda)$, where (x, y) stands for the stimulus position in the space, t is the timing, and λ represents the wavelength of the stimulation; in the meantime, the

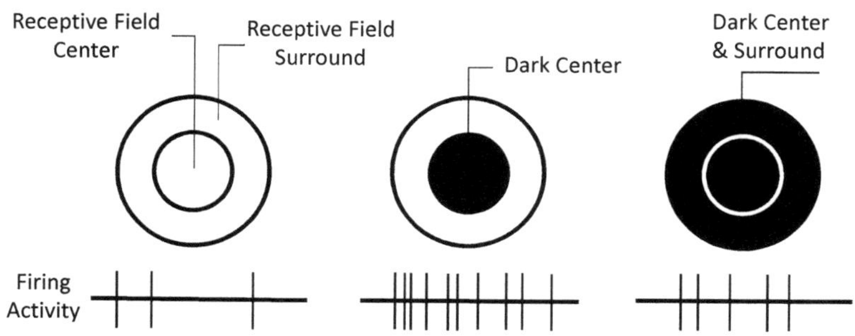

Fig. 5.1 A center–surround ganglion cell receptive field (Adapted with permission from Bear et al. 2007, Fig. 9.23)

retinal ganglion cell's firing rate is denoted as $R(t)$. The mathematical description of the relationship between $I(x, y, t, \lambda)$ and $R(t)$ is thus the receptive field model we are discussing here.

Early in 1965, R.W. Rodieck forwarded a quantitative description about the receptive field properties of cat retinal ganglion cell:

$$f(x, y, t) = f(x, y) \cdot h(t) \tag{5.1}$$

where $f(x, y)$ refers to the spatial "shape" of the receptive field and $h(t)$ is the temporal filter.

In retinal ganglion cells, the neuronal responsiveness is dependent on the spatial location of the light spot. Take the on-center ganglion cell as an example; when the light spot is projected on the receptive field center, the cell's firing rate is increased with relatively large amplitude; when the light spot is located on the peripheral area of its receptive field, it induces a reduction in the cell's firing activity, with relatively small amplitude. Therefore, the spatial structure of the cell's receptive field $f(x, y)$ can be described as the spatial summation of two Gaussian distributions:

$$f(x, y) = A_C \cdot \exp\left(-\frac{x^2 + y^2}{2r_C^2}\right) - A_S \cdot \exp\left(-\frac{x^2 + y^2}{2r_S^2}\right) \tag{5.2}$$

where r_C and r_S are the radius of the receptive field center and surround, respectively, whereas A_C and A_S are the amplitudes of the central and peripheral Gaussian components, respectively.

The temporal component $h(t)$ can be described as

$$h(t) = \left(\alpha + ke^{-t/\tau}\right)u(t) \tag{5.3}$$

where α and k are, respectively, the magnitude of the maintained and transient components and $u(t)$ is a step function:

$$u(t) = \begin{cases} 0 \text{ for } t < 0 \\ 1 \text{ for } t \geq 0 \end{cases} \tag{5.4}$$

Given light stimulation with certain wavelength, the Rodieck model deduces the firing rate elicited from a ganglion cell by summing up the light stimulation effect spatially and temporally:

$$R(t) = R_0 + \iint I\left(x, y, t'\right) \cdot f(x, y) \cdot h\left(t - t'\right) dxdydt' \tag{5.5}$$

where R_0 denotes the cell's background activity which is measured as the cell's firing rate when no stimulation is given, while stimulation $I(x, y, t)$ is the spatial assembly of light.

This model provides such an input–output relationship: a small patch of light projected in the receptive field center of an on-center ganglion cell causes the transient firing activity increase, together with a small undershoot, and then returns to the base level.

This receptive field model can be presented using the cascade shown in Fig. 5.2a. The light intensity on the retina is a weighted sum of the stimulation in the cells' receptive field (the upper panel, thick line), which is the algebraic summation of the two concentric opposite Gaussian functions (the upper panel, thin lines). This input signal is convolved with the retinal impulsive response (the middle panel) and forms a positive impulse which is followed by a negative undershoot. The estimation of firing rate is the additive result of this convolution output and the base-level firing rate (negative values are rectified as zero). This model nicely describes the retinal ganglion cell's light response (Fig. 5.2b).

The retinal ganglion cell's light response depends on both temporal and spatial characteristics of the stimulation, and in the model, the time-dependent component and the spatial-dependent component appear to be separable to each other (Eq. 5.5). This seems to suggest that the temporal process can be separated from the spatial distribution, which is termed as "spatiotemporal separability." But the real process is not that simple; the spatial and temporal components in the ganglion cell's response are not entirely separable. In fact, due to the reason that horizontal cells and amacrine cells participate in the formation of the peripheral area of ganglion cells' receptive field, whereas the receptive field center only receives the direct input from photoreceptors via bipolar cells, the time delay of the ganglion cells' response to light spots located in the central or peripheral subregions of the receptive field is different. Based on such a fact, a "modified Gaussian difference model" was forwarded (Fig. 5.2c). In this model, the signals in the central and peripheral parts of the receptive field are processed in parallel. The Gaussian distributions for the central and peripheral parts are not necessarily concentric, and the impulse responses could also be different. The signals from these two pathways are summed up at a later stage to generate the firing rate of the cell.

Another important aspect is that, in the Rodieck model, the signal processing in the retinal ganglion cells is with linear characteristics. But in fact, the visual

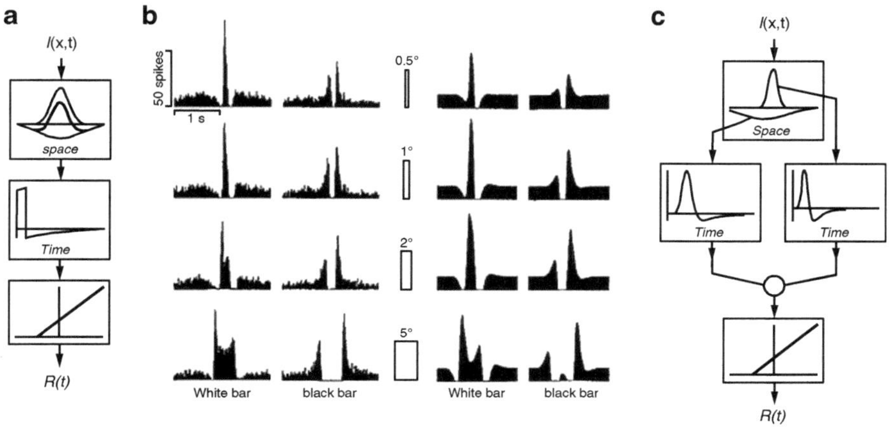

Fig. 5.2 The "difference-of-Gaussian" model proposed for retinal ganglion cell's firing rate prediction. (**a**) The signal cascade includes three parts: the weighted spatial summation of light stimulation (*upper panel*); a temporal filter (*middle panel*); the input–output transition (*lower panel*). (**b**) An on-center ganglion cell of cat retina in response to a light bar (*white* or *black* on a *gray* background) swept across it receptive field (*left panels*) and the model output (*right panels*). (**c**) The "modified difference-of-Gaussian" model for ganglion cell's light response in which the center and peripheral parts are opposite but not necessarily concentric and the impulse responses are also different (Adapted with permission from Meister and Berry 1999, Fig. 1)

information processing in the retina is regulated by the overall properties of the visual environment. In general, retinal ganglion cells' sensitivity is decreased when the ambient mean luminance level is increased. Such modification allows the system to dynamically adjust its activity range according to the environmental changes and approach the optimal coding for the visual stimulation. For the natural environment, the luminance intensity change in a circadian day is around 10^8–10^9, whereas the ganglion cells' firing activity change is limited to a range of ~10^2. Such response adjustment will result in some nonlinearity in the ganglion cells' activity: when the system output is driven to be increased, the system's gain will be reduced accordingly. The receptive field model can thus be modified such that the spatial response property of the cell follows that of the classical receptive field model, but the gain of the temporal filter is modified by its output:

Intuitively, the most important function of the retina is to measure the light intensity changes. Retinal ganglion cells receive input signals from many photoreceptors via bipolar cells, and detect the light intensity changes by integrating the input signals. Such integration favors the sensitive light intensity detection, at the same time, it allows for filtering the random signals generated by photoreceptors and enhances the signal-to-noise ratio of the retinal circuitry.

In addition to light detection, another distinct function of the visual system is moving target detection, and such motion detection starts with retinal ganglion cell. When the visual target is moving across the retina, the light intensity is increased in some sub-region of the retina, while decreased in others. Actually, cat retinal ganglion cells can be classified as X and Y subtypes according to their spatial integration property. X-cell can perform simple summation for light stimulation located in different sub-regions of its receptive field, this kind of cells have relatively high spatial precision, and can carry information about the location of light stimulation. Y-cell has relatively big receptive field as compared to X-cell, its receptive field is composed of many "sub-units", and these sub-units can process and integrate the input stimulation signals, including linear filtering, nonlinear gain-control, and further linear filtering, these processes are usually termed as LNL cascade, and has antagonistic receptive field and biphasic impulse responses.

The Y cell is therefore sensitive to a moving target. It is activated when the visual target is moving, but it is insensitive to motion direction and spatial pattern of the visual target. Such Y cell can be activated by visual pattern with high spatial frequency. It even responds to moving gratings which is finer than its receptive field (Fig. 5.3). In this case, if the neuron only makes simple summation for the input signals within its receptive field, the positive inputs provided by the moving gratings will be canceled out by the negative ones. But in fact, the neuronal circuits work in a way that the presynaptic bipolar cells rectify the inputs in a parallel manner before sending their outputs to the ganglion cell to form the final nonlinear output. Such circuitry mechanism explains why the Y cell is sensitive to a moving target without direction selectivity or spatial pattern sensitivity. When the visual target is moving across the receptive field, the subregion in the moving target activates the subregion of the receptive field. The subregions' activities are then rectified individually, which allows for the overall positive output and prevents the cancelation between the positive and negative signals.

In real life, what is more frequently encountered is "object motion", i.e., the visual target moving across a static background. Some retinal ganglion cells are sensitive to such object motion, and are called "object-motion-sensitive ganglion cells" (OMS GS). Such neuron is not sensitive to global moving, but is sensitive to local moving in the receptive field center against the global background (Fig. 5.4). Similar to the Y-cell, such OMS cell rectifies the excitatory input from bipolar cells in the receptive field center, while the bipolar signal from the peripheral part are sent to amacrine cells before forming inhibitory signal to the ganglion cell. The underlying mechanism of the OMS cell can be explained as follows: the object motion in the central area provides the ganglion cell with excitatory input via bipolar cells, whilst the motion in the peripheral area sends inhibitory input via amacrine cells. When motion in the central area is identical with that in the peripheral area (i.e., during global motion), the central excitatory signal from bipolar cells will be cancelled out by the peripheral inhibitory signal from amacrine cells. But the central excitation does not coincide with the peripheral inhibition when the central motion is distinct from the peripheral one (i.e., during different motion), and finally results in firing activity in the ganglion cell. Similar with Y-cell, OMS cell only detects the difference in target- and background-motion, it is insensitive to motion direction and spatial property of the stimulation.

Apart from the above mentioned object motion across retina in horizontal or vertical direction, there is still another important motion along the "depth" – an approaching object will produce an expanding image on the retina, which can be detected by a certain subtype of Off-type ganglion cell. The underlying mechanism

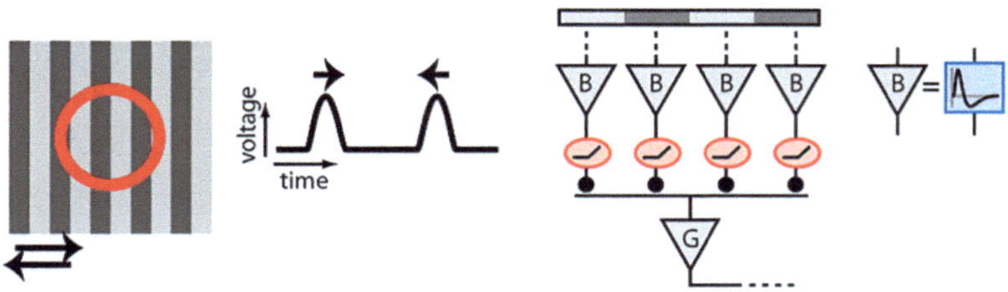

Fig. 5.3 The response property of Y-type ganglion cell. *Left*: a Y-type ganglion cell can be activated by moving gratings with high spatial frequency, without direction selectivity. *Right*: the underlying mechanism involves the presynaptic bipolar cells rectifying the input signals in parallel before sending their output to the ganglion cells (Adapted with permission from Gollisch and Meister 2010, Fig. 2)

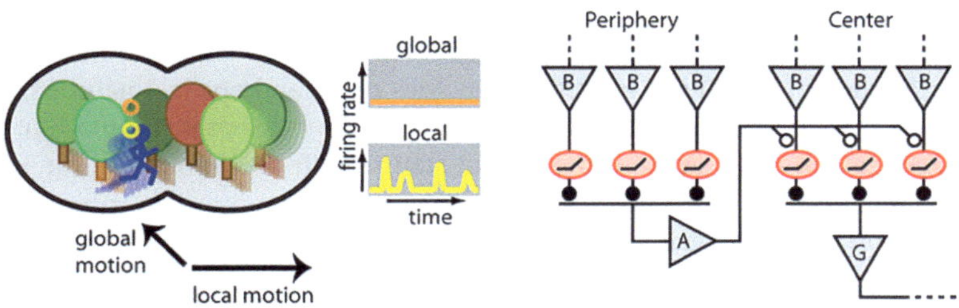

Fig. 5.4 The OMS GS is sensitive to local motion. *Left*: the OMS GS is sensitive to object motion which is distinct from the background. *Right*: the underlying mechanism is attributed to circuitry asymmetry for the central and peripheral inputs (Adapted with permission from Gollisch and Meister 2010, Fig. 2)

of such "depth" detection is related to the integration of excitatory input via the Off-pathway and the inhibitory input via the On-pathway (Fig. 5.5). When a dimming target is approaching, the cell receives intense excitatory input; while the target moves laterally, the excitation from one side will be attenuated by the inhibition at the other. What is different from the OMS cell is that the inhibition for this kind of cell occurs post-synaptically.

5.2 Reverse Correlation Estimation for Neuronal Response Properties

We discussed the receptive field models of retinal ganglion cells in the last section. The models describe the ganglion cells' firing activity changes in response to visual stimulations according to the receptive field properties of the cells. But in visual research, what is more frequently encountered is how to estimate the cell's receptive property according to the cell's activity elicited by certain stimulation. "Reverse correlation" is one of the effective methods. Such methods have been

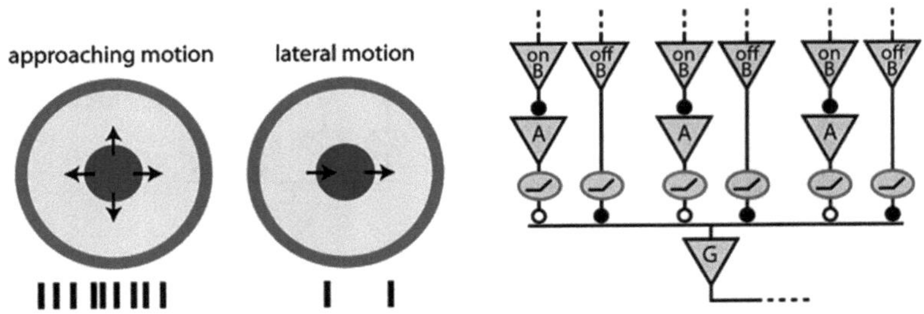

Fig. 5.5 Detection of approaching motion. *Left*: some off-type ganglion cell is sensitive to an expanding dark spot, which resembles an approaching object. *Right*: the underlying mechanism might be explained as the integration of excitatory/inhibitory signals from the off–on pathways (Adapted with permission from Gollisch and Meister 2010, Fig. 2)

successfully applied to study the signal process in various visual neurons, including retinal ganglion cells, lateral geniculate nucleus cells, and cortical neurons.

The firing activity of sensory neurons is affected by external stimulation. When no input is given, these neurons are in rest status, with background firing activity. Input stimulation alters the neurons' firing activity. From a system point of view, visual neuron can be treated as a cascade of a linear filter and a nonlinear filter (see Fig. 5.2). For such a system, we can calculate its system output according to its impulse response function, given a certain input stimulation.

For a linear system with its impulse response function being $h(t)$, when the input is $x_1(t)$, its output $y_1(t)$ is

$$y_1(t) = \int x_1\left(t'\right)h\left(t - t'\right)dt' \tag{5.6}$$

For input $x_2(t)$, the output $y_2(t)$ is

$$y_2(t) = \int x_2\left(t'\right)h\left(t - t'\right)dt' \tag{5.7}$$

When the input is $x_1(t) + x_2(t)$, the system output is

$$y(t) = y_1(t) + y_2(t) = \int \left(x_1\left(t'\right) + x_2\left(t'\right)\right)h\left(t - t'\right)dt' \tag{5.8}$$

For the visual stimulation, the input signal can be described as the spatial and temporal integration of a group of impulse inputs with different intensities; therefore, the system's response to the input can be presented as the integration of the system's response to these impulse signals.

There are various methods to measure the impulse response of a linear system. Intuitively, it can be directly measured by an impulse input. Another feasible way is to measure its frequency response by sine wave inputs. The impulse response function of a system can also be estimated using white noise. The "white noise" refers to a generalized stationary stochastic process, with zero mean and a constant spectral density A. The autocorrelation function of the white noise process is $A\delta(t)$. When the system is fed with a white noise as input, the mean level of its output $y(t)$ should be

$$E\{y(t)\} = \int E\left\{x\left(t'\right)\right\}h\left(t-t'\right)dt' = E\{x(t)\}\int h\left(t-t'\right)dt' = 0 \tag{5.9}$$

The cross-correlation function between the input and output is then

$$R_{XY}(\tau) = E\{x(t-\tau)y(t)\} = \int R_X(\tau)h\left(t-t'\right)dt' = Ah(t) \tag{5.10}$$

It is clear that when the system input is a white noise process, the system's impulse response property can be obtained by calculating the cross-correlation function between the output signal and the white noise input.

For the calculation of cross-correlation function, we can choose a time lag τ, multiply the stimulation at time t and system response at time $t+\tau$, and get $x(t)y(t+\tau)$; simply add the values with all time t; the averaged value is the cross-correlation value for time lag τ. Such algorithm is termed as "forward correlation." Similarly, we can do the same calculation by calculating the values of $x(t-\tau)y(t)$, which is termed as "reverse correlation." Intuitively, the results derived by the two calculations should be equivalent.

For a spiking neuron, its firing activity can be presented by a "0–1" sequence as presented in Fig. 5.6, where the system output equals to "1" when the neuron fires an action potential; otherwise, the system output equals to "0." Because the occurrence of "1" is far less than "0" in the firing sequence, therefore "reverse correlation" algorithm can greatly elevate the calculation efficiency in estimating the neuron's impulse response.

Because the neuronal response is composed of action potentials, therefore, according to the algorithm of the abovementioned "reverse correlation," there is a simple method for estimating the system's impulse response via the "white noise estimate" method which is by searching the stimulation signal at a time lag τ before

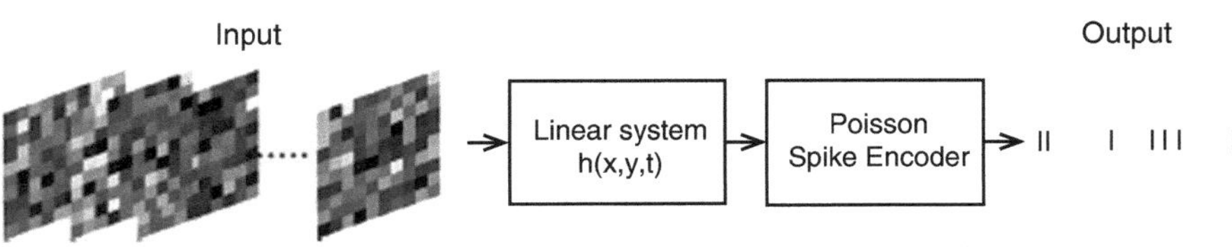

Fig. 5.6 Visual neuron described by a linear system. The impulse response $h(x,y,t)$ of the system can be estimated by the reverse correlation between a spatiotemporal white noise sequence and the cell's firing activities (Adapted with permission from Ringach and Shapley 2004, Fig. 3)

each action potential and summing up these stimulation signals, which provides the cross-correlation function of the input stimulation and the system's output $R_{XY}(\tau)$:

$$R_{XY}(\tau) = E\{x(t-\tau)y(t)\} \approx \frac{1}{N}\sum_{i=1}^{N} x(t-\tau) \tag{5.11}$$

This cross-correlation function is equivalent to the system's impulse response $h(t)$.

For a visual neuron, its properties in response to visual stimulation include both temporal properties and spatial properties. The visual neuron gets parallel inputs from the subregions of its receptive field and then gets the input signals integrated before forming its unified output. For describing such a system, an impulse response function which consists of both spatial and temporal components $h(x, y, t)$ is requested (Rodieck 1965). If the system is a linear one, its response to spatial integration of image input can be treated as the summation of its response to stimulation located in each subarea. Given the stimulation intensity of each spatial location (x, y) and the system's spatiotemporal impulse response $h(x, y, t)$, the system's firing activity can be estimated. Based on such properties, some practical algorithm of the reverse correlation technique was developed. The "checkerboard" method is one of the efficient algorithms. In this method, spatial patterns with "white noise" property are applied as input signals as presented in Fig. 5.6. In the input sequence, each spatial pattern is composed of equally divided sub-squares, with the light intensity in each subunit varying with time following a "white" sequence.

Given that, for a spiking neuron, the system's impulse response can be estimated using the reverse correlation algorithm, i.e., we can find all the spikes in the firing sequence and search for the stimulation images t ms before each spikes, and calculation for different time lags, we can get the spatiotemporal relationship between the visual stimulation and the neuron's response (Fig. 5.7), where the stimulation is denoted as I $(\mathbf{x}, t)$ with $\mathbf{x} = (x, y)$ being the spatial location of the stimulation and t being the time lag between the stimulation image and firing activity. Define the image in the sequence with the largest value as $I(\mathbf{x}_{\max}, t_{\max})$, the receptive field center $\mathbf{x}_{\text{center}}$ can be calculated as

$$\mathbf{x}_{\text{center}} = \frac{\sum_{\mathbf{x}} \mathbf{x} I(\mathbf{x}, t_{\max})}{\sum_{\mathbf{x}} I(\mathbf{x}, t_{\max})} \tag{5.12}$$

Although the property of retinal neuron system is with some linearity, the relationship between the central neuron's response and environmental stimulation signals is largely nonlinear. The structural property of the central visual part is that there are complex interconnections between the neurons. The LGN and cortical neurons not only receive feedforward inputs from their presynaptic neurons, but they also receive feedback signals from higher-order cortical areas, which form the sources of the nonlinearity of the central neurons. Given such nonlinearity, when

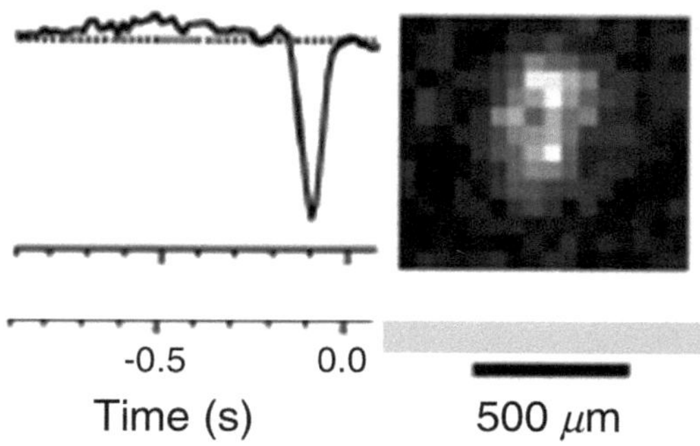

Fig. 5.7 Receptive field property of a ganglion cell from salamander retina (time delay is around 100 ms) (Adapted with permission from Schnitzer and Meister 2003, Fig. 6)

applying traditional light spot or slit to measure the neuron's spatial property in an experiment, the measurement error should be inevitable due to limited stimulation region. Therefore, while "spatiotemporal white noise" is applied as stimulation to estimate the response property of the neuron system via cross-correlation, it is with the advantage of activating the neuron being tested as well as other neurons, which allows for taking the interaction among neurons into account.

Figure 5.8 shows an early work applying such "reverse correlation" algorithm to measure the receptive field property of V1 neurons. In the process, small patches of dark or light stimulus were randomly determined and sequentially presented in randomly selected positions. Those patches which elicited one or more spikes from the neuron were added up (Fig. 5.8a), and the excitatory and inhibitory subareas of the neuron's receptive field which were respectively responsible to light and dark stimuli were estimated (Fig. 5.8b).

For the visual neurons, when the size, intensity, or color of the stimulation is altered, the cell's receptive field size, as well as the spatial distribution of the sensitivity, will change accordingly. Furthermore, in the visual signal pathway from the retina to the lateral geniculate nucleus to the visual cortex, the structure of the neuron's receptive field is getting more and more complicated. Therefore, when applying reverse correlation algorithm to measure the neuron's receptive field, it is necessary to select the stimulation pattern according to the cell's property under investigation. The cortical neurons are sensitive to some particular stimulation properties, such as orientation, motion direction, motion speed, or even some special spatial pattern. Therefore, simply applying randomized luminance stimulation may not be proper for studying these aspects of central neurons. But when the stimuli applied are well selected to test the neuron's response properties, better measurement results can be obtained. A "subspace reverse correlation" method was therefore developed, while the "subspace" reflexes the stimuli composed of a "subset" of particular stimulation patterns.

For instance, neurons in visual area V1 are sensitive to orientation of visual pattern. Therefore, to study these neurons' orientation selectivity, a set of light gratings with different orientations but with identical parameters for other aspects (i.e., mean luminance, contrast, spatial frequency, etc.) can be applied as proper stimulation (Fig. 5.9). In such a set of stimulation patterns, any two gratings with

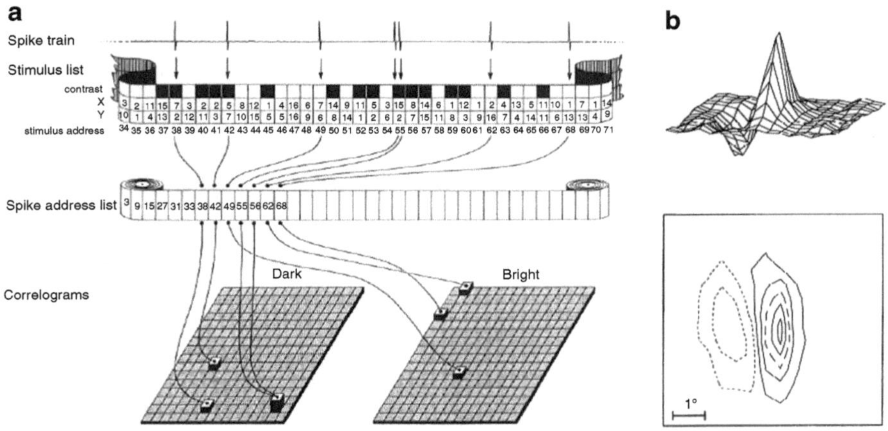

Fig. 5.8 Receptive field property of V1 cell measured using "reverse correlation" method. (**a**) Schematics of "reverse correlation" algorithm. (**b**) The excitatory (the positive part in the *upper panel* and *continuous curve* in the *lower panel*) and inhibitory (the negative part in the *upper panel* and *dashed curve* in the *lower panel*) subareas of a V1 cell's receptive field measured using the "reverse correlation" method (Adapted with permission from Jones and Palmer 1987, Figs. 2 and 7)

different orientations are spatially uncorrelated. Such randomized stimulation is pretty close to "white noise," with which the system's property can be calculated via the cross-correlation function between the input and output. Although such stimulation somehow differs from the original definition of "white noise," the orientation-selective spatiotemporal property of the central visual neuron can still be estimated via the reverse correlation method. For each time, the orientation distribution probability $p(\theta, \tau)$ can be calculated via reverse correlation: for N action potentials fired by a certain cell, one can get the orientation gratings (θ) for τ ms before each action potential and sum them up to obtain the optimal stimulation. Since the orientation selectivity of a neuron is time varying, the example given in Fig. 5.9 demonstrates the selectivity property distribution related to a certain time delay τ.

In the stimulation pattern given in Fig. 5.9, all the grating patterns are with a different orientation, and there is also one blank pattern without gratings but only with a full-field gray scale which is equivalent to the mean luminance intensity of the gratings. This allows for analyzing which orientation is responsible for the increment or decrement of the neuron's firing activity. By defining the neuron's response modulation function $R(\theta, \tau) = \log[p(\theta, \tau)/p(\text{blank}, \tau)]$, several parameters related to the tuning curve $R(\theta, \tau)$ can be investigated: (a) the preferred orientation θ_{pref} related to the maximum firing and the relevant firing rate change $R(\theta_{\text{pref}}, \tau)$; (b) the orientation $\theta_{\min}$ related to the minimum firing and the relevant firing rate changes $R(\theta_{\min}, \tau)$; (c) the orientation orthogonal to θ_{pref}, which is denoted as θ_{ortho}, and the related firing rate changes $R(\theta_{\text{ortho}}, \tau)$; (d) the "modulation

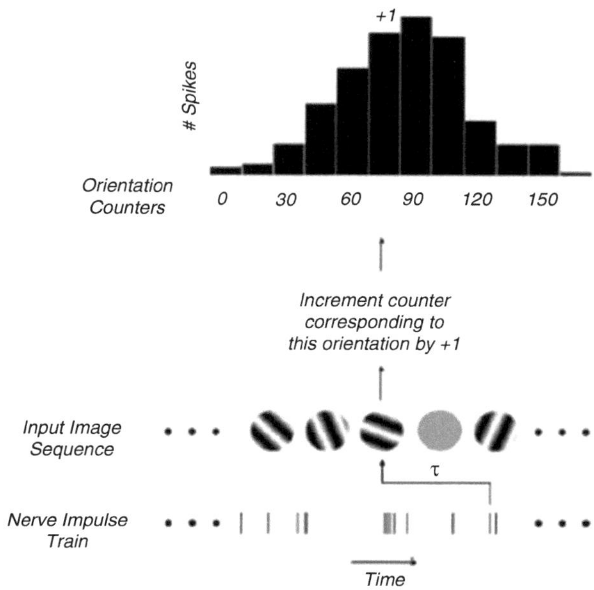

Fig. 5.9 "Subspace reverse correlation" method applied for central visual neuron's orientation selectivity measure (Adapted with permission from Ringach and Shapley 2004, Fig. 7)

depth" of the tuning curve as a function of time τ, $A(\tau) = R(\theta_{\text{pref}}, \tau) - R(\theta_{\min}, \tau)$; and (e) the half bandwidth of the tuning curve $1/2\left[R(\theta_{\text{pref}}, \tau) - R(\theta_{\text{ortho}}, \tau)\right]$, as illustrated in Fig. 5.10.

This subspace reverse correlation method can also be applied to study other aspects of cortical neurons (Bredfeldt and Ringach 2002), i.e., the stimulation patterns are constructed by a series of gratings with random spatial frequencies and phases (for fixed orientation) (Fig. 5.11).

When the time delay between the cell's firing activity and the stimulation imaging is τ ms, the modulation function for the neuron's response elicited by spatial frequency f can be expressed as $R(f, \tau) = \log[p(f, \tau)/p(\text{blank}, \tau)]$, which reflects the cell's firing frequency change in response to certain spatial frequency stimulation (as compared to the cell's background activity in response to full-field gray-scale stimulation). By choosing different τ values, the neuron's time-varying spatial frequency response properties can be investigated (with τ changing within a range of 0–150 ms).

The "receptive field" properties we have discussed include both spatial and temporal properties. Meanwhile, experimental evidence showed that the spatial and temporal characters are not necessarily separable in V1 neurons.

The examples given in Fig. 5.12 show the spatial frequency selectivity of V1 neurons, which is time varying, with the optimal spatial frequency shifting from low frequency to high frequency along with time (Fig. 5.12a, b) or with the selectivity changing along with time (Fig. 5.12a). Such time-dependent response property changes suggest that V1 neurons' behaviors do not follow the "label line" property; the information about the spatial frequency and temporal information share the same information channel and interact with each other.

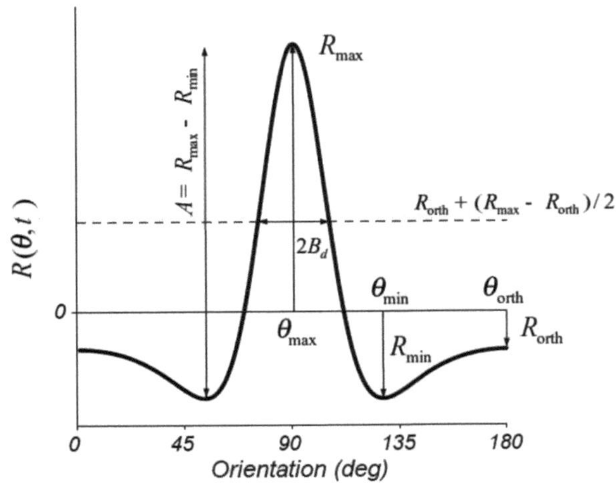

Fig. 5.10 The tuning curve of neuron's response $R(\theta, \tau)$ and relevant parameters (Adapted with permission from Ringach and Shapley 2004, Fig. 8)

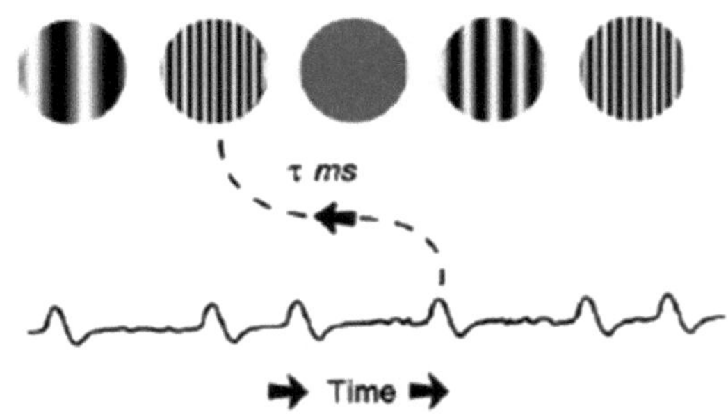

Fig. 5.11 Reverse correlation for measuring the spatial frequency preference—calculating the cross-correlation between the neuron's firing activity and the stimulation pattern presented τ ms before (Adapted with permission from Bredfeldt and Ringach 2002, Fig. 1)

In addition to orientation selectivity and spatial frequency selectivity, cortical neurons can also have motion direction selectivity. The "subregion reverse correlation" algorithm can also be applied to estimate the neuron's motion response by using a set of random stimulation images consisted of moving targets (Borghuis et al. 2003). By the reverse correlation function calculated for a cell's response and the random stimulation composed of quickly moving dots, the tuning curves of the cell's motion direction selectivity and motion speed selectivity can be measured.

Taken together, the cross-correlation method provides an efficient tool for measuring the neurons' activation properties. Such dynamic measure can be applied to both linear and nonlinear systems, which allows for measuring of the system's impulse response (or spatiotemporal impulse response) and for further investigation for the neuron and neural network properties.

5.3 Gain Control in Visual System

While the primary function of sensory systems is to encode/decode information about the external stimulation, signal transduction in sensory systems is characterized by adaptive adjustment of transfer function of the response process. Adaptation

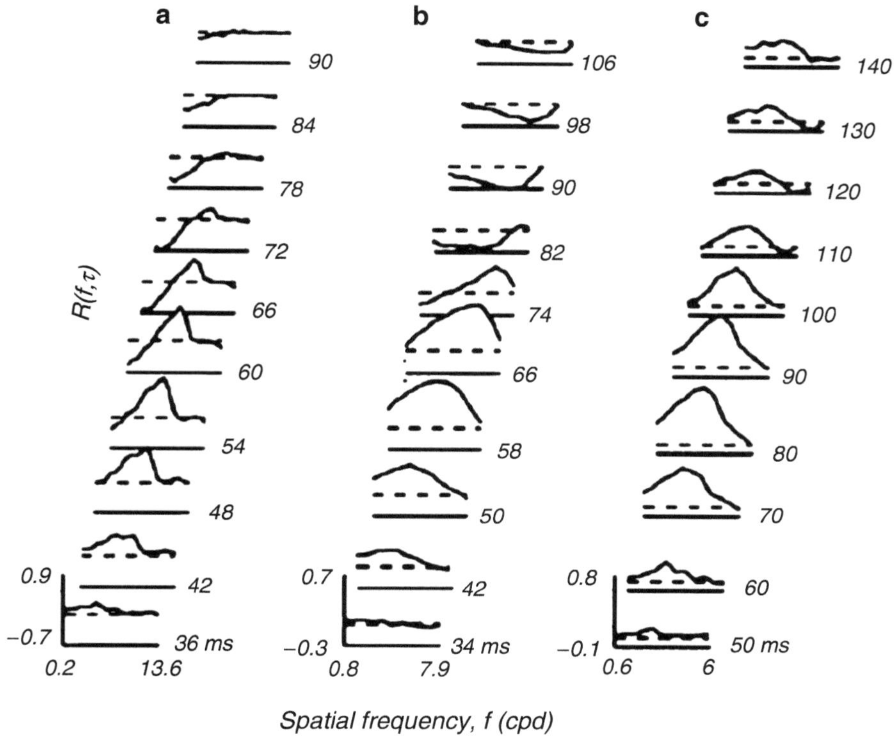

Fig. 5.12 Examples of dynamic changes in macaque monkey V1 neurons' spatial frequency tuning property. (**a**) Example of a neuron whose spatial frequency selectivity was increased over time; the optimal frequency was also changed, from 0.72 at the 36th ms to 2.63 at the 72nd ms. (**b**) Example of a neuron whose spatial frequency property was changed from low pass to band pass; the optimal frequency was changed from 1.51 at the 34th ms to 4.3 at the 74th ms. (**c**) In this example, the optimal frequency was not time varying, but the tuning curve was sharpened during a certain time period (70–110 ms) (Adapted with permission from Bredfeldt and Ringach 2002, Fig. 3)

to stimulation includes several aspects such as increasing responsiveness when input signal is weak, so as to improve signal-to-noise ratio, or decreasing responsiveness when input signal is strong, to prevent saturation and information loss.

In the visual environment, stimulation intensity shows great variety in both spatial and temporal domains. In natural visual environment, luminance intensity varies across several scales. In the meantime, stimulation contrast or movement speed of visual targets can also change. Adaptation to the environment enables the system to have a proper sensitivity and make an optimal response to the local stimulation (Fig. 5.13). Visual adaptation can thus occur in any stage of the visual neural system, including the retina, lateral geniculate nucleus, visual cortex, etc. Adaptation can also involve adaptation to various aspects of stimulation, including mean intensity of environmental luminance level, contrast which is the variance of intensity over time, or movement status of visual targets.

In vertebrates, retina is the first stage for visual neural information processing. It is highly specified and possesses very sensitive transducers—photoreceptors, together with organized neural circuitry constructed by layered neurons—and is equipped with multiple gain control mechanisms. Among these, the most commonly observed ones are gain controls in response to luminance and contrast stimulations.

The necessity for gain control comes from the fact that the dynamic range of visual neuron response is limited (up to $\times 10^2$), which does not match the dynamic range of stimulation intensity (e.g., up to $\times 10^9$ in luminance intensity). The existence of gain control mechanism enables retina to swiftly adjust its response dynamic range when stimulation condition is changed, so as to cope with the changing visual environment and to produce an efficient response (Fig. 5.13).

Luminance adaptation is characterized by dynamic adjustment of the gain of light responses. Such adjustment is dependent on the mean luminance level of background light and involves various mechanisms related to photoreceptors and retinal neuronal network. As for photoreceptors, vertebrate retinas possess rods and cones (see Fig. 5.14). Among these, rods have abundant membrane disks and with more condensed photopigments. This ensures that the rod system is more ready to be activated and with high sensitivity in response to dim light stimulation. But on the other hand, such structural property makes the rod system easier to be saturated in photopic condition. The function of the rod system is thus to mediate the visual process in mesopic condition. To the contrary, cones are with less membrane disks and with lower density of photopigments. Thus cones are with lower response sensitivity and not likely to be saturated. The main function of cones is thus to mediate the visual process in photopic condition. When the visual environment is switching between photopic and mesopic conditions, retina can adjust the relative activities of these two photoreceptor systems, so as to maintain normal visual activity.

In the meantime, while in exposure to sustained constant stimulation, the neuronal response often change along with time. Such stimulation-dependent gain change is important for sensory information processing and is particularly important for visual system. For example, on–off ganglion cells respond to both light-on and light-off transients. When environment illumination is suddenly changed, a remarkable increment in firing activity can be observed. But such activity increment is accompanied by a following gradual decrement. Prolonged light/dark stimulation cannot always induce sustained increase in the neuron's activity (Fig. 5.15); the ganglion cell's activity shows a gain adjustment process during its response to sustained stimulation.

Intuitively, such a process is remarked by a decrement in neuron's responsibility during sustained stimulation. In normal self-adaptation systems, a reduction in responsibility is often accompanied by a reduction in sensitivity. But the ganglion cell's adaptation process is equipped with more complex adjustment process. When the retina was exposed to a sustained light stimulation, transient responses and the following adaptation process similar to that illustrated in Fig. 5.15 could be observed from a retinal ganglion cell. However, if an additional impulse stimulation

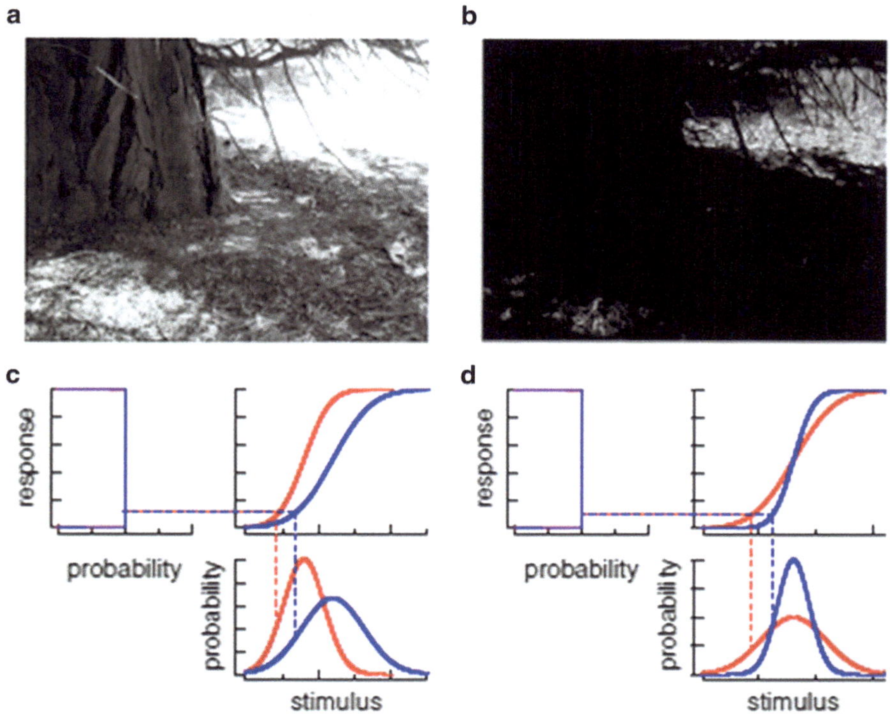

Fig. 5.13 Properties of natural visual environment and the request of visual adaptation. (**a** and **b**) Images of the same scene with high (**a**) and low (**b**) ambient luminance level. Visual adaptation occurs by compensating shift in either mean light intensity (**c**) or the contrast/width of the stimulation distribution (**d**). *Dashed lines* in (**c**) and (**d**) identify equivalent points on the stimulus distributions (Adapted with permission from Rieke and Rudd 2009, Fig. 1)

was given after the ganglion cell's adaptation to the sustained background lumination was completed, the cell's light response could still be elicited, with the response amplitude being positively correlated with the background light (Fig. 5.16). This phenomenon suggests that in the adaptation process of a ganglion cell's response, the changes of the neuron's activity and sensitivity may not be consistent with each other but instead involved more complicated processes. The adaptation to ambient light increment is characterized by a decrement in neuron's firing activity but at the same time accompanied by an increment in the neuron's sensitivity to additional stimulation. Adaptation to ambient light is to ensure the visual system to keep a reasonable dynamic range in response to additional changes in the environment. The decrease of firing rate does not mean information loss during adaptation. To the contrary, in this process, the neuronal information is likely "stored" in the retinal neuronal network in some manner and is presented by its response to additional stimulation.

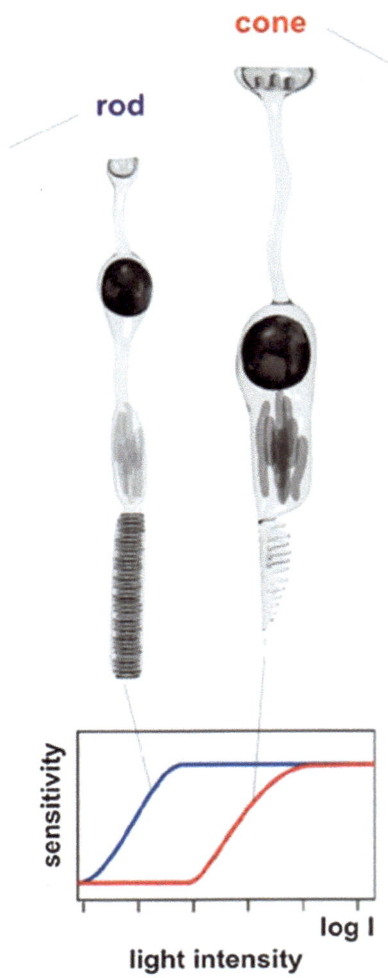

Fig. 5.14 Structure and light sensitivity of rod and cone (Adapted with permission from Pfaff 2013, Fig. 18.10)

Another type of gain control is related to response adaptation in exposure to contrast stimulation (defined as luminance changes around the mean level $C = L_{\max} - L_{\min}/L_{\max} + L_{\min}$, where $L_{\max}$ and $L_{\min}$ are the maximum and minimum levels of the stimulation, respectively). Although it is common that a stimulation signal with larger contrast can elicit a stronger response from sensory neurons, analytical results revealed that the seemingly increase in the neuron's firing rate in response to an increase in stimulation contrast was related to a reduced response sensitivity (Fig. 5.17).

Taking a retinal ganglion cell as a linear system, its transfer function can be constructed to describe the cell's response process under visual stimulation (Smirnakis et al. 1997). The analytical results show that when the stimulation

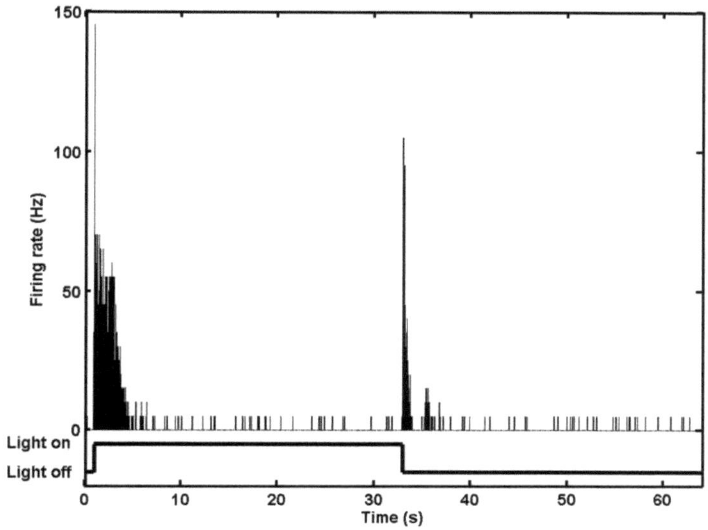

Fig. 5.15 An on–off ganglion cell's firing activity changes in response to sustained *light/dark* stimulation (From Chen et al. 2005, Fig. 1)

contrast is increased, the ganglion cell's response properties will also be changed. Define the first-order Wiener kernel of ganglion cell's response as

$$k(t) = \frac{1}{WC}\frac{1}{T}\int_0^T \left(I\left(t'\right) - M\right)R\left(t' + t\right)dt' \tag{5.13}$$

where $I(t)$ is the stimulation intensity at time t; M and W are the mean level and the standard deviation of the stimulation intensity, respectively; $C = W/M$ represents the stimulation contrast; and $R(t)$ denotes the firing rate at time t. The example presented in Fig. 5.17 shows that when the stimulation contrast was increased from $C = 0.09$ to $C = 0.35$, the first-order kernel of the system response was decreased accordingly. In addition to this reduction in sensitivity at the onset of high-contrast stimulation, the neuron's response was continuously decreased in exposure to sustained high-contrast stimulation, with time constant being more than 10 s, which is referred to as "slow adaptation"—we will get into more details about such "slow adaptation" in Sect. 5.4.

The visual neuron can also adapt to other aspects of stimulation signals. One example comes from a motion-sensitive neuron (H1 neuron) of the fly visual system. H1 neuron's response property is that its activity is enhanced while in exposure to visual target moving horizontally along with its preferred direction, and its activity is depressed by movement along with the opposite direction. While H1 neuron's firing sequence carries information about movement speed, it can adapt to changes in movement speed. Furthermore, when the distribution of stimulation speed is changed, H1 neuron's input–output relationship can be adaptively adjusted.

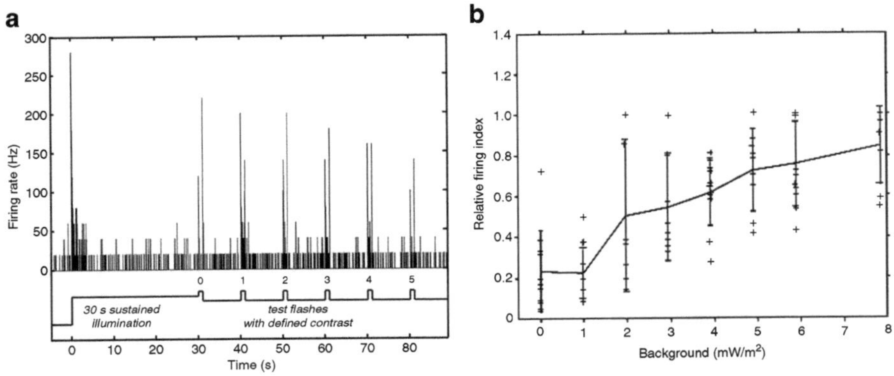

Fig. 5.16 Responsiveness during light adaptation and sensitivity adjustment in retinal ganglion cells. (**a**) A ganglion cell's adaptation in response to sustained light illumination and its response to additional light flash. (**b**) Relative firing activities to flashes with fixed intensity added on various levels of background illumination (Adapted with permission from Chen et al. 2005, Figs. 3 and 4)

Figure 5.18 illustrates such adaptive changes. In the experiments, the neuron was exposed to two varying stimuli with changing speeds which were normally distributed and with standard deviations $\sigma_1 = 2.3°/s$ and $\sigma_2 = 4.6°/s$. The relationship between the response firing rate and the movement speed was obtained for each case. The experimental results showed that the input–output relationship was not identical when the speed distribution range of the visual target was different (Fig. 5.18a). However, if the stimulus is normalized against its standard variation and the firing activity is normalized against its mean level, the input–output curves obtained during the two cases can be basically overlapped with each other (Fig. 5.18).

Further experiments compared the H1 neuron's input–output relationship when the neuron was in exposure to rapidly varying stimuli with Gaussian statistics whose standard deviation was taken between $\sigma = 18°/s$ and $\sigma = 180°/s$ (Fig. 5.19). According to dynamically changing speed signal, the dynamical acceleration signal can be deduced. The system's response elicited by movement speed and acceleration under these stimulation inputs can also be calculated. Analytical results show that, although the input–output relationship differed in response to stimuli with different parameters, the normalized curves for the relationship between firing rate and velocity/acceleration (with stimulation being normalized against its standard deviation and response being normalized against its mean value) are overlapped and matched well. This means that when the dynamic range of the stimulation is changed, the system can rescale its response dynamics to make an optimal match between the stimulation range and response dynamics, so as to ensure reliable and efficient signal transduction.

This example suggests that, via reasonable rescaling of the system's input–output relationship, the system's dynamic response range can be rescaled, so as to optimally match the input dynamic range. This might be one of the effective ways

Fig. 5.17 Gain control during contrast adaptation of salamander ganglion cell. (**a**) The firing activity adjustment of a ganglion cell in response to stimulation with low and high contrast. (**b**) Comparison of the first-order kernels calculated for different stage of the adaptation. (**c**) The peak-to-peak amplitude (*A value*) of the first-order kernel changes along with time (Adapted with permission from Smirnakis et al. 1997, Figs. 1 and 3)

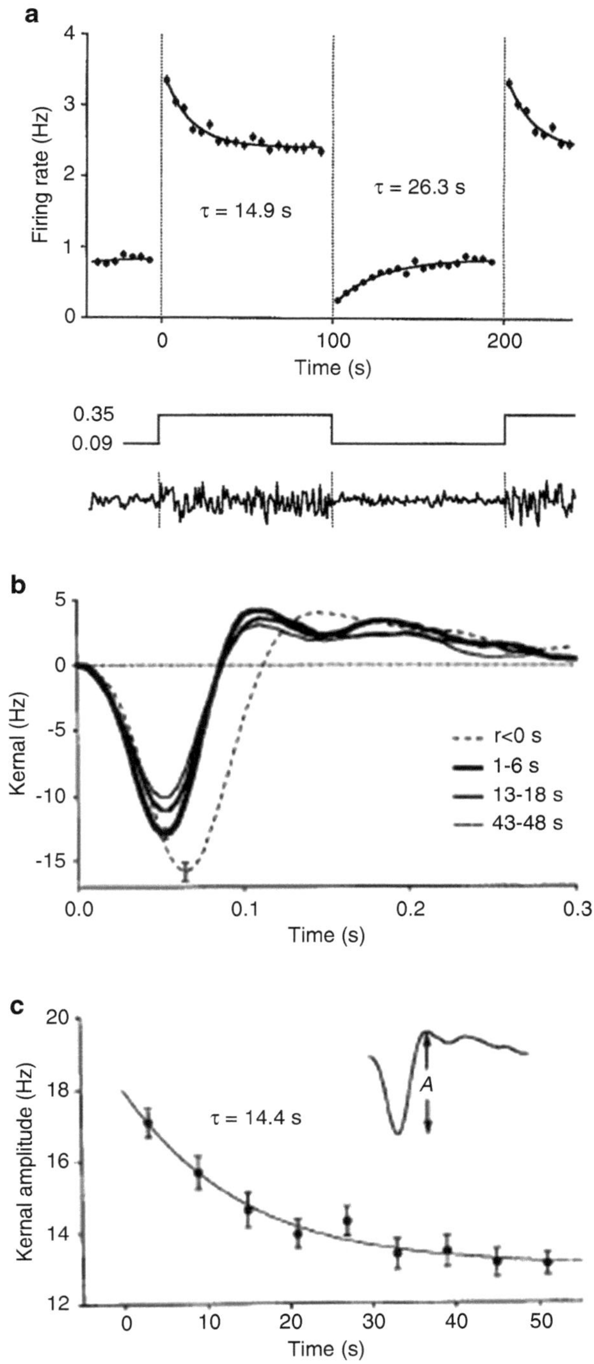

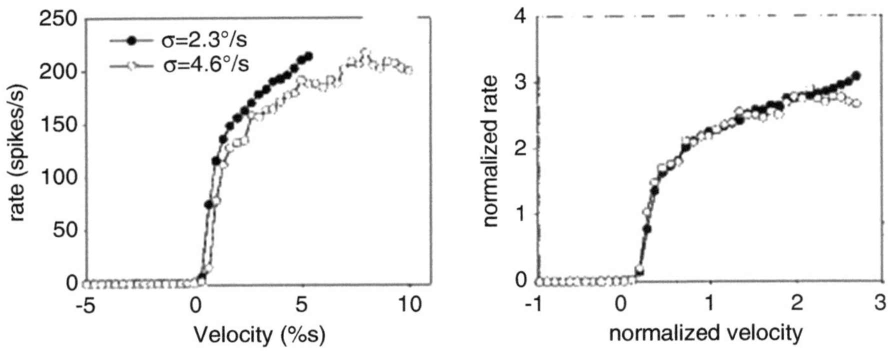

Fig. 5.18 H1 neuron's response function obtained under stimulation with different distributions $\sigma_1 = 2.3^\circ/s$ (*closed circles*) and $\sigma_2 = 4.6^\circ/s$ (*open circles*). (**a**) Relationship between stimulation speed (degree/s) and firing rate (times/s). (**b**) Normalized unit-less response curve (Modified from Brenner et al. 2000, Fig. 3)

to maximize the information transmission. This also suggests that neuronal system does not necessarily possess a single particular transfer function. When the stimulation condition is switched from one to another, the system can choose an optimal one among a cluster of transfer functions via the adaptive adjustment, so as to realize a reliable as well as effective transfer of the sensory information.

The statistical property of natural visual signal has a feature of being highly changeable in both temporal and spatial domains. From the abovementioned examples, we can see that to enhance coding efficiently, the nervous system should adjust its coding strategy according to the environmental condition, so that the system response can adapt to the local statistical property of the stimulation. Such adaptation process provides an adjustable coding manner to the environmental changes which was based on the "learning" process to the input parameters, to optimize the system's local behavior.

5.4 Information Processing During Sensory Adaptation

While changes in natural environment are almost unlimited, information capacity of the nervous system is definitely limited. This requests the nervous system to adjust its response property dynamically, so that the system's output property can ultimately satisfy effective signal transmission, and to realize efficient coding for the stimulation information.

The sensory neurons' response property can be dynamically adjusted according to the changes of stimulation properties. Such adaptation phenomenon exists in various sensory systems, which is to say that while performing cognition tasks, the sensory system does not simply and passively send the stimulation signals to the cortex but instead adjusts the system's coding strategy dynamically. Such

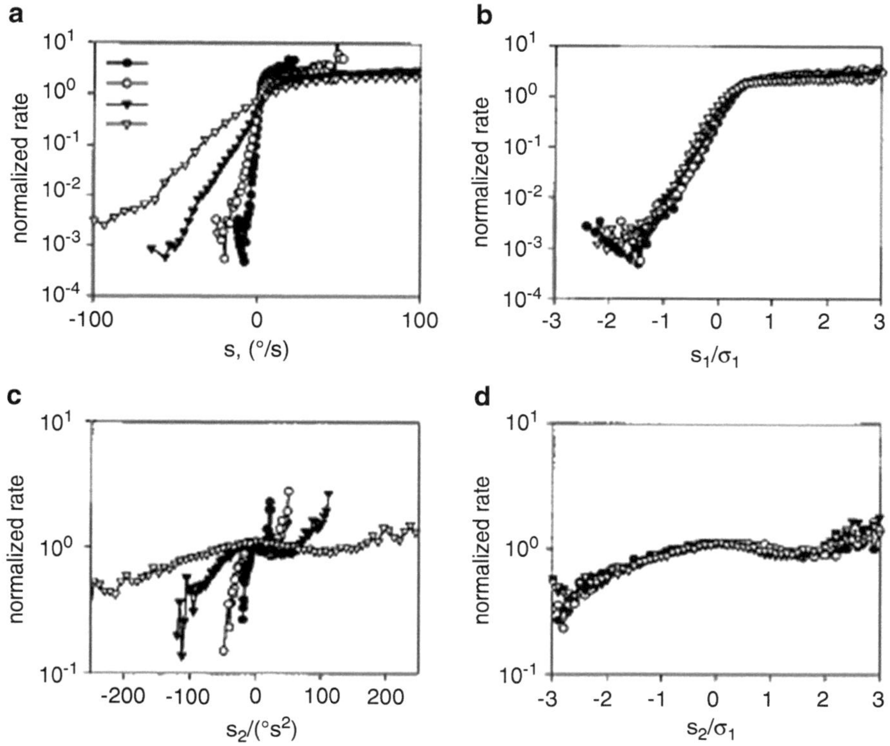

Fig. 5.19 H1 neuron's responses and changes under four sets of different stimulations. (**a**) Relationship between stimulation speed (degree/s) and firing rate (times/s). (**b**) Normalized results of (**a**). (**c**) Relationship between stimulation acceleration (degree/s^2) and firing rate (times/s). (**d**) Normalized result of (**c**) (Adapted with permission from Brenner et al. 2000, Fig. 6)

adaptation process allows the nervous system to make its limited dynamic range matching the environment and to enhance the efficiency of information transmission.

In the preceding section, we discussed the adjustment of visual neurons' responsiveness which is to match the statistical property of visual stimulation, including gain control during contrast changes as well as moving speed parameter changes. Actually, another important aspect of visual adaptation is that during sustained stimulation, the mean firing rate of the neuronal activity will be gradually decreased along with time, while the cell responds more actively to rarely appearing stimulation. Such adaptation helps to adjust the neuron's activation status and to avoid signal saturation and allows the cell to be responsive to newly emerging stimulation, which reflects that the system is more activated to stimulation that carries more information.

The example illustrated in Fig. 5.17 shows that, when the stimulation was switched between low-contrast and high-contrast light flickering, the retinal ganglion cell's sensitivity had a transient reduction at the onset of high-contrast stimulation and the cell's sensitivity was further decreased with the prolonged high-contrast stimulation. When described using a first-order system, the result was such that the amplitude of the first-order kernel was decreased during such adaptation process.

However, we have also introduced that the retinal ganglion cell's visual response can be more accurately described via a linear–nonlinear cascade (see Fig. 5.2). When the stimulation was switched from low contrast to high contrast, the neuron's firing activity was transiently increased, which was followed by a gradual reduction in the firing rate (Fig. 5.20). While using the linear–nonlinear model to analyze the adaptive process of the ganglion cell during a switch between different contrast levels, we can see that when the stimulation contrast was switched from low to high, the linear filter of the system was speeded up, which is presented by a left shift of the linear filter (Fig. 5.21a), but in the meantime, the threshold of the nonlinear component was also enhanced, with its sensitivity decreased, which is presented by a right shift of the nonlinear function (Fig. 5.21b). Furthermore, when the high-contrast stimulation was prolonged, the dynamic property of the system's linear filter was kept almost unaltered, while the sensitivity of the nonlinear component was further reduced, which was presented by a further right shift (Fig. 5.21). This shows that when the stimulation contrast is enhanced, the ganglion cell is subject to two separable adaptive changes. While the fast adaptation occurs within 0.1 s (which is also termed as contrast control), the slow adaptation takes more than 10 s. These two processes both contribute to the reduction of the cell's response sensitivity, but their functions in visual information processing might differ from each other.

In natural environment, light intensity of a certain spatial pixel rarely has dramatic changes in successive time; meanwhile, it is also true that adjacent pixels in space normally have similar light intensity. Therefore, for a certain pixel in the space at a certain time, the retinal neural circuit can "predict" its light intensity according to that of the adjacent points or of the preceding time and make comparison between this "predicted" intensity and its real value. This is to say that the ganglion cell detects the difference between local stimulation and the background, spatially and temporally. The benefit for such "predictive" coding is that the dynamic range of the local input signal is much smaller than that of the natural environment, so that the neuron's dynamic range can make good match with the input signal. It is therefore easy to understand that a ganglion cell does not only adapt to stimulation with temporal structure (such as adaptation to stimulation contrast as we just discussed) but can also adapt to signal with spatial structure—when the spatial structure of visual stimulation is changed, it will induce relevant changes in ganglion cell's spatiotemporal receptive field property, which ensures efficient and reliable information coding in the new visual environment.

In Fig. 5.22, the example shows the dynamic change of a ganglion cell's response property following the changes in the spatial correlation property of visual

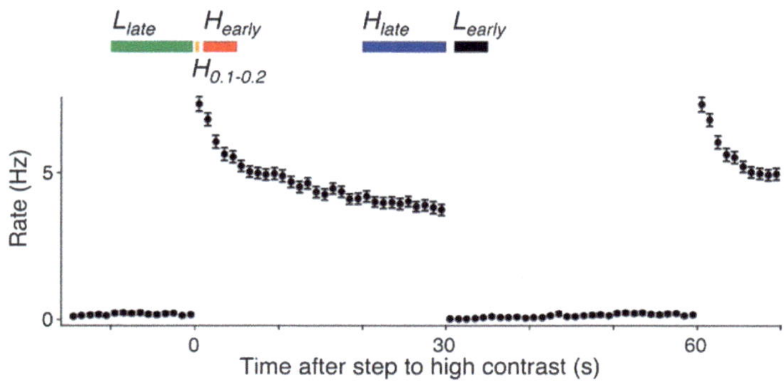

Fig. 5.20 Retinal ganglion cell's firing activity changes during switching between high contrast (0–30 s, contrast = 0.35) and low contrast (30–60 s, contrast = 0.05) flickering stimuli (Adapted with permission from Baccus and Meister 2002, Fig. 1)

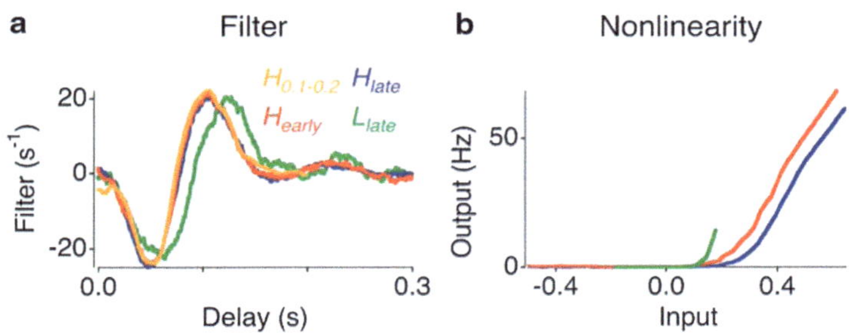

Fig. 5.21 The changes in model parameters of the ganglion cell's light response during contrast adaptation. (**a**) Linear filter property changes during the time period H_{early}, H_{late}, and L_{late} (as defined in Fig. 5.20); (**b**) nonlinear transfer function changes during H_{early}, H_{late}, and L_{late} (Baccus and Meister 2002, Fig. 3)

stimulation. In the experiment, two flickering stimulations with different spatial property were presented to elicit the cell's adaptation to spatial pattern. One stimulus is a spatially uniform flickering with adjacent pixels being positively correlated (stimulation A); the other is a checkerboard flickering with adjacent subareas being negatively correlated (stimulation B). When stimulation A (or B) was given and lasted for a certain period (13.5 s in the example), it induced the cell's adaptation to the stimulus. After the adaptation, a white noise stimulation (P) with spatial pixels uncorrelated was added on the adapting stimulation A (or B) to test the cell's receptive field property, and the cell's response to stimulations A and B can then be deduced, so as to investigate the influence of spatial pattern adaptation on the cell's responsiveness. The result showed that after adaptation to stimulation A, the cell's response was more sensitive to stimulus B; accordingly, the cell was more sensitive to stimulus A after being adapted to stimulus B. A possible explanation to this phenomenon is that the interneurons in the retinal

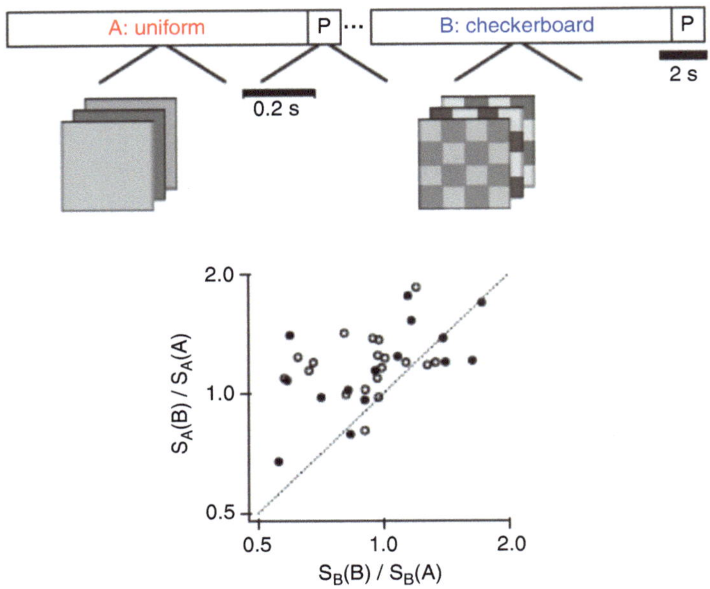

Fig. 5.22 Retinal ganglion cells' adaptation to stimulation with different spatial patterns. The *upper panel*: stimulation pattern with different spatial correlation property. The *lower panel*: most cells' response sensitivity to stimulus A (S_A) was enhanced by adaptation to B (indicated by the factor $S_A(B)/S_A(A) > 1$), while their response sensitivity to stimulus B (S_B) was enhanced by adaptation to A (indicated by the factor $S_B(B)/S_B(A) < 1$). Each data point represents one ganglion cell from experiments with tile size 400 μm (*filled*) or 200 μm (*open*). The axes are logarithmic (Adapted with permission from Hosoya et al. 2005, Fig. 1.)

neural network with different receptive field properties were activated to different levels and adapted under different stimulations and contributed to the network output in adaptation to different stimulations.

Similarly, another set of experiments measured the responsiveness changes of retinal ganglion cells in adaptation to horizontal or vertical gratings with identical parameters including mean luminance intensity, contrast and spatial frequency, etc. The results showed that the adaptation to horizontal gratings resulted in increment of the cell's response sensitivity to vertical gratings and vice versa (Fig. 5.23).

Retinal ganglion cells also adapt to motion pattern. In the natural environment, the visual system is actually often exposed to dynamic motion stimulation. When the visual scene is static, head movement or even eye movement will cause changes of the visual imaging projected on retina, which is equivalent to movement of the global visual scene. However, movement of particular object will result in differential movement pattern between the image on a small patch of retina and the background, which is termed as object motion. Part of the retinal ganglion cells are sensitive to such particular object motion and are called object-motion-sensitive (OMS) cells. OMS cells are insensitive to global motion and are only sensitive to local motion or differential motion between local and global areas and are adaptive to sustained differential motion. Experimental results showed that OMS cell's

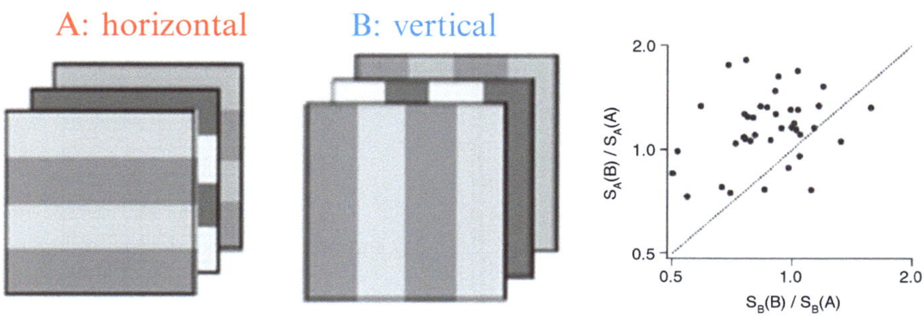

Fig. 5.23 Adaptation to horizontal gratings enhanced retinal ganglion cell's response sensitivity to vertical gratings and reduced the cell's response sensitivity to horizontal gratings and vice versa (Adapted with permission from Hosoya et al. 2005, Fig. 2)

response is most active at the stimulus onset and will be reduced gradually during the following several seconds. Such phenomenon is called objective motion adaptation, which has similar time constant as contrast adaptation.

In the example illustrated in Fig. 5.24, visual scene was divided into target area (which covers the central part of the tested ganglion cell's receptive field and part of its peripheral area) and background area (which is the area beyond the target area). The stimulation can be designed to mimic local motion by jitter in the target area and mimic global motion by jitter of the global scene, while the object motion in static background can be mimicked by differential jitter of the target area and the background area. Cell's firing activity was increased at the onset of stimulation switch between global motion and object motion and then adapted as the object motion was sustained. Similarly, when stimulation pattern was switched from global motion to local motion, cell's response was also increased at the onset, which was followed by an adaptation.

The functional significance of objective motion adaptation might be that the animal can be adapted to the ongoing motion, while keeping sensitive to novel movement that might occur.

In the visual system, the significance of neuronal firing activity is to encode visual information. From the above introduction, we can see that when a certain stimulus is sustained, the neuron's firing activity will be attenuated. In such adaptation process, the neuron's firing rate will be gradually decreased. But how is the information coding behavior adjusted during such a process?

In the example shown in Fig. 5.25, when a contrast stimulation was applied, the retinal ganglion cell was highly activated at the stimulation onset. But when the stimulation was sustained, the neuron's firing activity was gradually reduced. Following the information entropy calculation method introduced in §4.2, we can analyze the modulation of the neuronal information coding during the adaptation process (Fig. 5.26). The results show that during adaptation, due to the decrease in firing rate, the total entropy and noise entropy carried by the firing sequence were both decreased (a). Therefore, the information carried by the firing sequence was overall decreased along with the neuron's adaptation to stimulation (b). But during

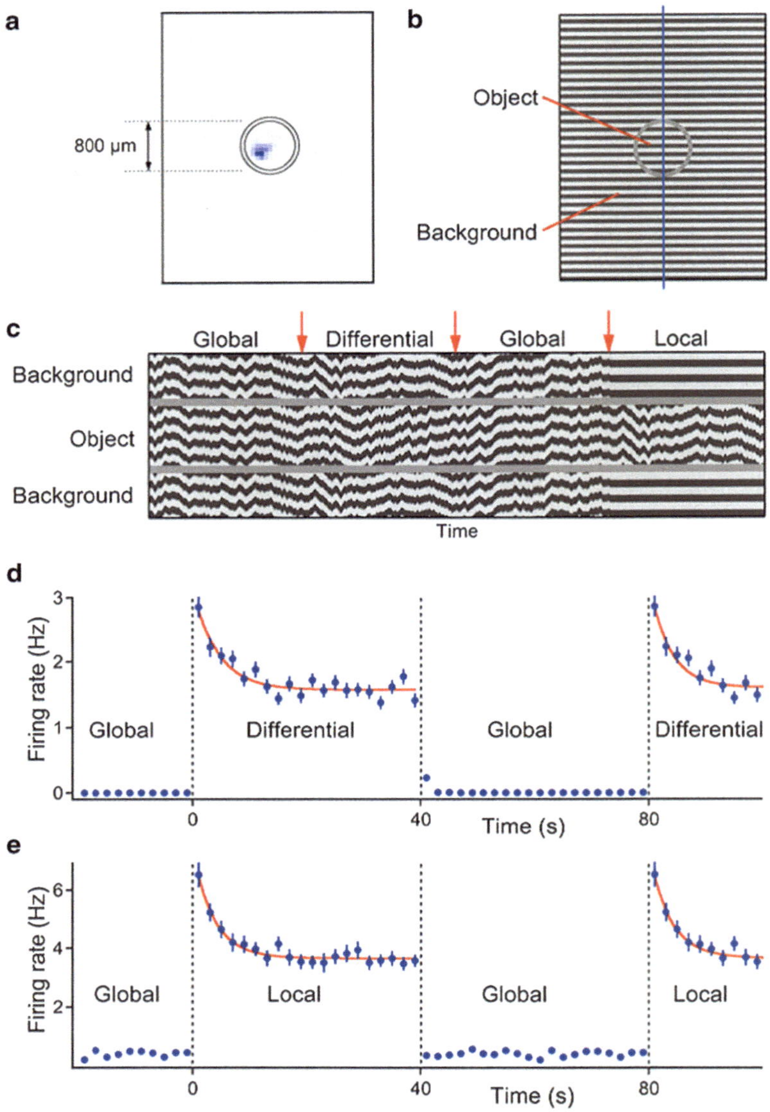

Fig. 5.24 Retinal ganglion cell's adaptation to object motion and local motion. (**a**) Receptive field profile of an OMS (salamander fast-off) ganglion cell. (**b**) An "object" grating projected in and around the cell's receptive field center and the background grating. (**c**) Motion pattern switched among "global motion," "differential motion," and "local motion"; the transitions are marked by *arrows*. (**d**) Average firing rate of the OMS cell in (**a**) to 50 successive trials of a stimulus switching between "global motion" and "differential motion" every 40 s. (**e**) Firing rate changes of another OMS cell in response to stimulus switching between "global motion" and "local motion" (Adapted with permission from Olveczky et al. 2007, Fig. 1)

this period, the decrease in firing amount was more severe than the decrease of the information entropy (c), so as the result, during the adaptation, each single spike firing carried more information (d). This seems to reflect the "energy-saving" principle of neuronal activity. The neuron's adaptation behavior does not only adjust the dynamic range of the response to stimulation but somehow increases the coding efficiency and decreases the energy consumption.

From these examples, one can tell that sustained stimulation can attenuate the activity of visual neurons, but such adaptive firing rate decrease is not necessarily accompanied by reduction in information transmission. In fact, such adaptation is not limited to the visual system but exists in other sensory systems as well.

In rat, whisker stimulation can activate particular cortical areas. While in exposure to periodic vibrissa deflections, the somatosensory cortical neuron's response also show gradual adaptive reduction. Before adaptation, the neuron's response to vibrissa deflection had high firing rate, but after being adapted to a 12-Hz sequence of punctuate vibrissa deflection (15 cycles, 800 deg/s), the neuron's response was obviously decreased. Such reduced response caused a reduction in the neurons' ability for detecting the vibrissa stimulation.

However, when the vibrissa deflection was with different parameters, it was found that adaptation enhanced the cortical neuron's ability in discriminating different stimulation. For any stimulation of a given deflection velocity, the neuron's response in the absence of preceding adapting stimulus was always higher as compared to that in the presence of the preceding adapting sequence, but on the other hand, the neuron's response before adaptation showed apparent tendency of saturation when stimulation intensity was increased, which caused the neuron's response to be similar to each other in response to stimulation with different velocities. To the contrary, the neuron's response after adaptation showed clear intensity dependency. Therefore, the stimulation intensity discrimination based on the neuron's activity was much better after adaptation as compared to that before adaptation (Fig. 5.27).

Taken together, these examples taken from different sensory systems during various sensory adaptations show that, although sensory adaptation is characterized by reduced responsibility, it is often accompanied by enhanced ability in information transmission. And such adaptive changes are ubiquitous for various sensory systems.

5.5 Population Information Coding During Sensory Adaptation

Animal's cognition about the environment is dependent on the adaptation ability of the neuronal network in response to stimulation. Sensory neurons have the property of changing their responsiveness and selectivity along with the environmental stimulation conditions, so as to match the environmental stimulation conditions.

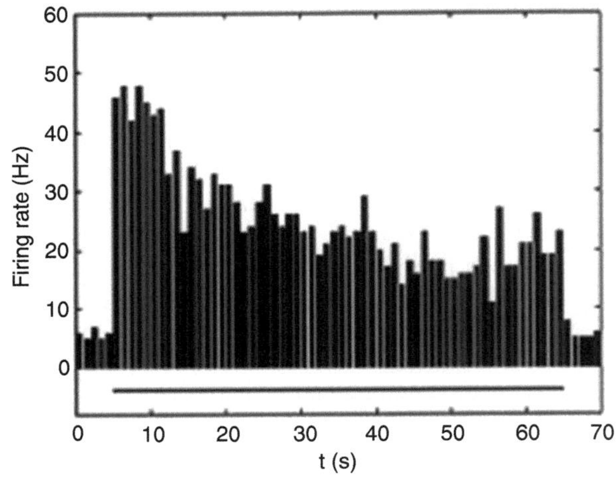

Fig. 5.25 A typical on-off ganglion cell's firing activity changes in response to sustained (5–65 s) contrast stimulation. At the On-transient of the stimulation, the neuron's firing rate was swiftly increased, reaching ~50 Hz. But as the stimulation was prolonged, the firing rate was gradually decreased, down to a level ~20 Hz (Adapted with permission from Jin et al. 2005, Fig. 3)

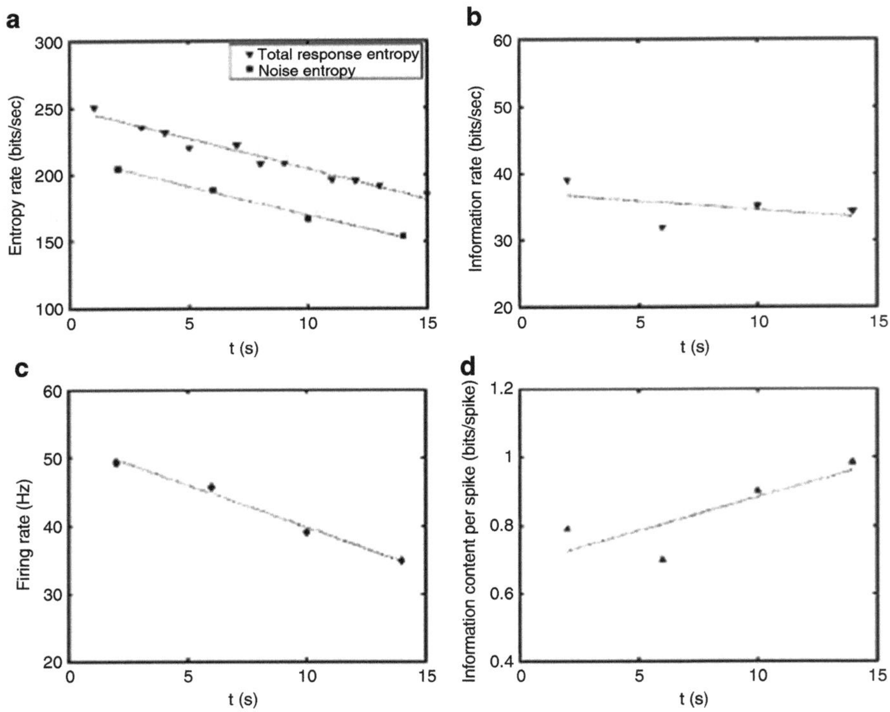

Fig. 5.26 Adjustment of retinal ganglion cell's coding behavior during visual adaptation. (**a**) The total entropy and noise entropy both showed decrease tendency during adaptation; (**b**) the information entropy was also decreased; (**c**) firing rate was dramatically decreased during adaptation; (**d**) information carried by single action potential spikes was increased during adaptation (Adapted with permission from Jin et al. 2005, Fig. 5)

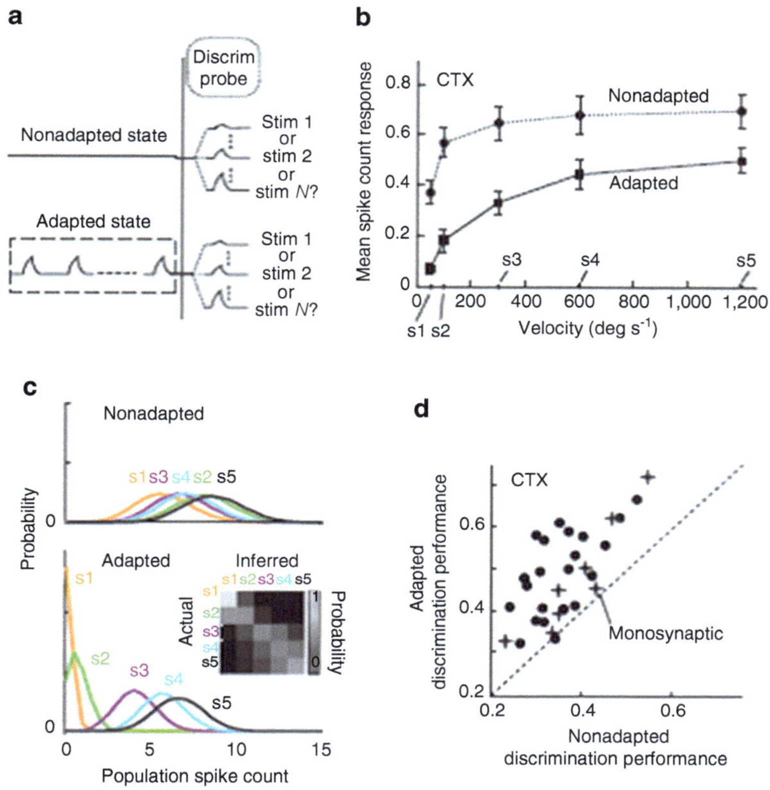

Fig. 5.27 Cortical neuron's ability of discriminating different stimulations before and after adaptation. (**a**) Cortical neuron's activity attributed to several possible stimuli (stim), in the presence (adapted) and absence (nonadapted) of a preceding adapting stimulus. (**b**) Mean spike count of the cortical (CTX) in response to the deflection velocity (s1–s5 as indicated on the abscissa) in nonadapted and adapted states ($n = 30$ cortical neurons; error bars are ±SE). (**c**) Spike count distribution of a cortical neuron during different stimuli (s1–s5) in nonadapted and adapted states. (**d**) The overall discriminability performance, quantified as the fraction of correct identifications (Adapted with permission from Wang et al. 2010, Fig. 3)

The functional significance of sensory adaptation might be (1) to adjust the sensitivity of the sensory system, to make it matching the stimulation range of the instant stimulation and (2) to promote the discrimination ability to novel stimulation or rare stimulation. However, we still know little about how the correlated activity changes during adaptation and their contribution to information coding.

In frog retina, there is a certain type of ganglion cells named dimming detector which is sensitive to dark stimulation and shows high firing activity in response to dark stimulation. Typically, these cells have adaptation property during sustained dark stimulation: the cell's firing activity is swiftly enhanced at light-off transient, which is followed by a gradual decreasing tendency. Such process normally lasts for several seconds (Fig. 5.28).

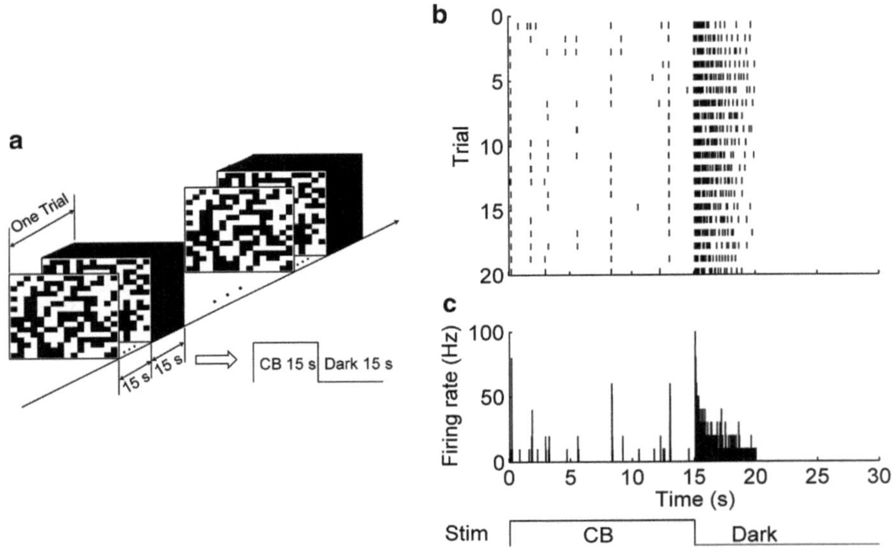

Fig. 5.28 Adaptation process of bullfrogs' dimming detector in response to sustained dark stimulation. (**a**) Visual stimulation pattern. CB is pseudorandom checkerboard stimulation. (**b**) One example dimming detector's firing sequences in response to 20 repetitive stimulation trials. (**c**) Firing frequency changes of the example ganglion cell in B (Adapted with permission from Xiao et al. 2013, Fig. 1)

We have introduced that concerted activities are widely existed among retinal ganglion cells. In dimming detectors, the firing activities of nearby cells also show concerted activities. In Fig. 5.29, it shows the concerted activities of two groups of dimming detectors whose activities were obtained by multielectrode recordings in response to sustained dark stimulation. In this example, 10 neurons recorded by adjacent electrodes of planar multielectrode arrays (64 electrodes arranged in an 8×8 matrix, with distance between adjacent electrodes being 150 μm in horizontal and vertical directions) can be grouped into two groups (cells #1–#5 and cells #6–#10). The concerted activity between pairwise neurons show two different patterns: synchronized activity (with sharp peak at 0 lag in cross-correlation function which was due to a gap-junction connection between the cells) was found in neuron pairs within each group, while correlated activity (with distributed time lags in cross-correlation function which was due to a common presynaptic input) was found between neurons from different groups.

On looking into the changes of single cell's firing activity and the two kinds of concerted activities during the sustained dark stimulation, the result shows that along with time, the single cell's firing activity had a typical adaptive reduction (Fig. 5.28b, c), and the correlated activity induced by common presynaptic input also showed a reduction with the decrease in firing activity, but the synchronized activity mediated by electric synapse showed a gradual increase during this process (Fig. 5.30).

For information coding and signal transmission, what we are concerned is how much change in the neurons' activity reflects the stimulation change or, in other words, if we can discriminate the stimulation property according to neurons' activity changes. We modify the experimental protocol given in Fig. 5.28, to present 10-s dark screen (light intensity = 0.0015 nW/cm^2) or gray screen (light intensity = 19.5 nW/cm^2) after the checkerboard stimulation (Fig. 5.31) and to look into the neuronal activity property changes during the 5-s adaptation changes afterward. The observation was such that both sustained dark screen and gray screen caused adaptive reduction in the cell's response with a tendency as illustrated in Fig. 5.28b, c.

Comparing the difference between the cell's responses to the two different stimulations, it was found that the cell's responses to the two stimuli showed some difference at the stimulation onset (the 1st second), with the firing rate being higher for the response to the dark screen. However, as the stimulation was prolonged, the difference between the cell's responses was gradually decreased, and the firing rates were similar with each other after the 2nd second (Fig. 5.32a). On the other hand, the synchronized activity showed the opposite tendency. During the 1st second of stimulations, the cells concerted activity during response to the two different stimuli were quite similar to each other and were both weak. But during the adaptation process, the synchronizing strength was gradually increased, with the response to gray screen changing more significantly. Therefore, the difference of the pairwise cells' response pattern during different stimulation was increased along with the adaptation process (Fig. 5.32b). This result suggests that it might be possible to discriminate these two kinds of stimulations according to the synchronized activities among neurons.

Practically, the neural ability of discriminating stimulation can be measured using a method called receiver operating characteristic (ROC), which is a coordinate-based classification model. When a neuron's response elicited by certain stimulation has a distribution which differs from that of its background activity, it can be judged whether or not the neuronal activity is a stimulation-related response or just a background activity, according to a preset threshold value. For each given threshold value, four possible classification results can be obtained: true positive (TP), false positive (FP), true negative (TN), and false negative (FN) (Fig. 5.33). Thus for each classification thresholds, one can obtain a value of a FP ratio (FP/(FP + TN), false alarm rate) and a TP ratio (TP/(TP + FN), hit rate), the relationship between which forms an ROC curve (with FP ratio for the abscissa and TP ratio for the ordinate) (Fig. 5.33). The area under the curve of ROC (AUROC, normally between 0.5 and 1) can be used to measure the performance of the classification model.

The analysis using the ROC method shows that during the 1st second of the response, the discrimination performance based on cell's firing count was better than that based on the synchronization strength, but the performance was reversed during the 2nd second, and during the 3rd second, the discrimination ability based

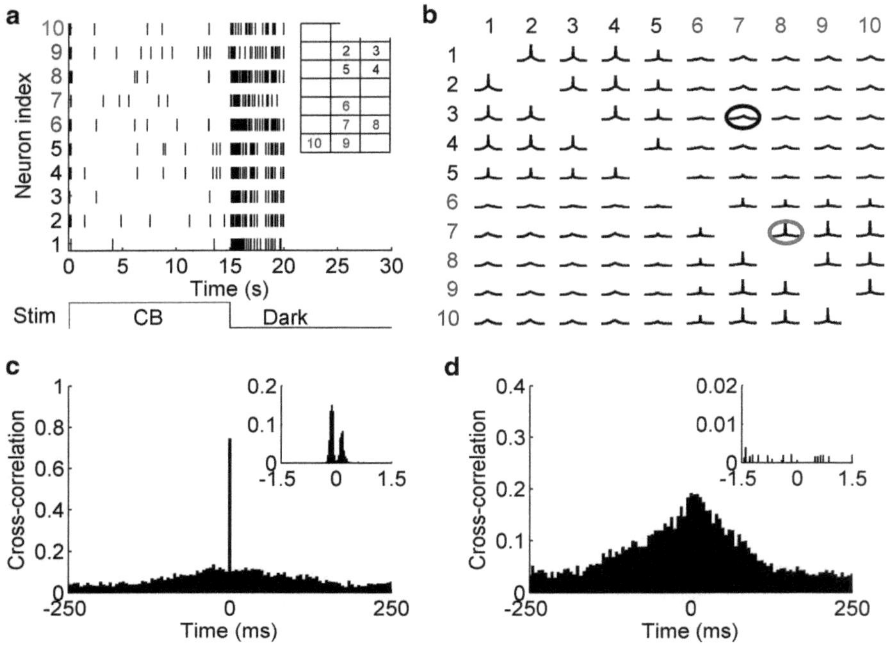

Fig. 5.29 Correlation patterns of dimming detectors in response to dark stimulation. (**a**) Firing sequences of 10 dimming detectors in response to repetitive stimulation. The *inset* in the *upper-right* corner shows the relative position of the electrodes by which these 10 neurons' activities were recorded. (**b**) Cross-correlation function between pairwise neurons' activities of the 10 neurons given in (**a**). (**c**) An example for synchronized activity (the neuron pair indicated by *gray circles* in (**b**); bin size = 5 ms). The *inset* in the *upper-right* corner shows the cross-correlation function calculated using bin size of 0.025 ms. (**d**) An example for correlated activity (the neuron pair indicated by *black circles* in (**b**), bin size = 5 ms). The inset in the *upper-right* corner shows the cross-correlation function calculated using bin size of 0.025 ms (Adapted with permission from Xiao et al. 2013, Fig. 2)

on synchronizing strength was significantly better than that based on single cell's firing rate (Fig. 5.32c). Such an analytical result was compatible with the direct impression provided by Fig. 5.32a, b.

This example shows that although a single neuron has an adaptation property, the synchronized activity of pairwise neurons will be enhanced during adaptation, which is helpful for increasing the coding accuracy. Such phenomenon is also observed during visual adaptation to other kinds of stimulation. We have introduced in Sect. 5.4 that some retinal ganglion cells respond to objective motion, and their response also has the time-dependent adaptation property—the cell's firing rate will be gradually reduced when the objective motion stimulation is sustained. But for the pairwise cell's correlated activities, it shows that for those cells sharing correlated activities, the correlation strength was increased after adaptation (Fig. 5.34).

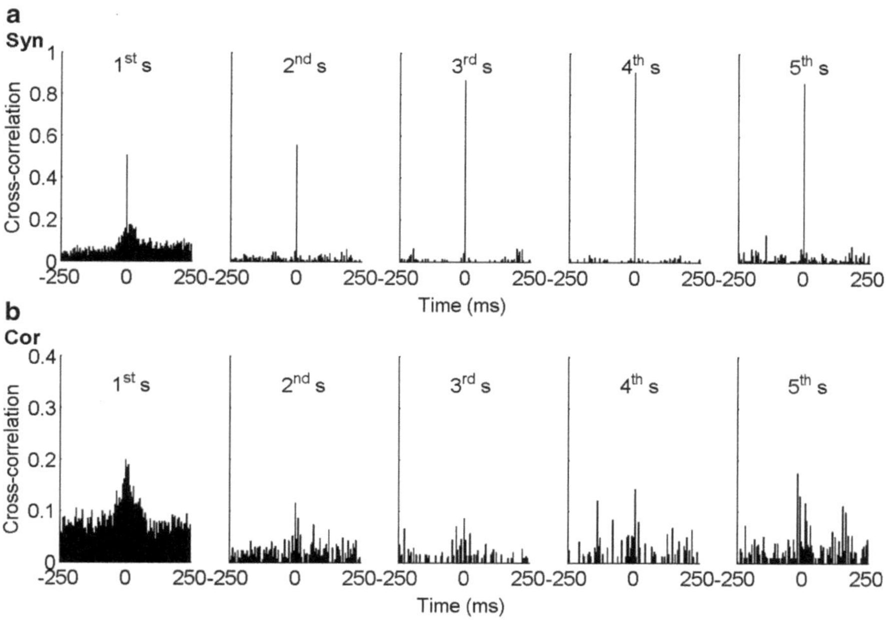

Fig. 5.30 Changes in correlation strengths between neurons over time during the luminance adaptation. (**a**) Synchronized activity was increased during the 5-s adaptation. (**b**) Correlated activity was attenuated during the 5-s adaptation (Xiao and Liang, unpublished data)

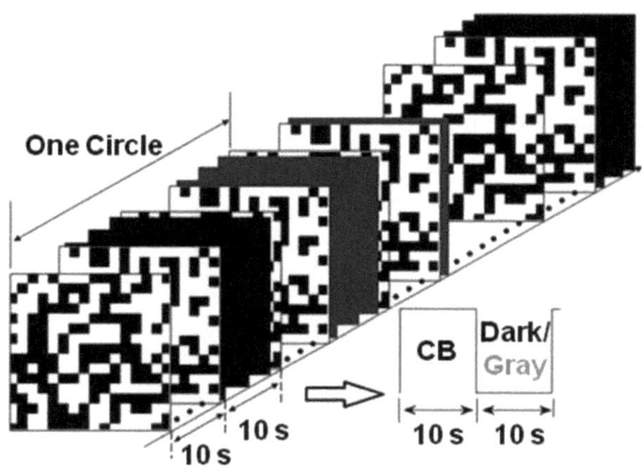

Fig. 5.31 Coding strategy of RGCs during the luminance adaptation process. The stimulation pattern in which 10-s pseudorandom checkboard stimulation was followed by 10-s *dark screen* or *gray screen*. Each experimental trial includes two checkboard flickering, one *dark screen*, and one *gray screen*; 20 repeats were given

As introduced in preceding sections, adaptation to sustained stimulation is ubiquitous in sensory systems. In Sect. 5.4, we have introduced that rat somatosensory cortical neuron's response to vibrissal deflection was reduced after adaptation, but in the meantime, the adaptation enhanced the cortical neuron's ability in discriminating different stimulations. Because in rat somatosensory system, the

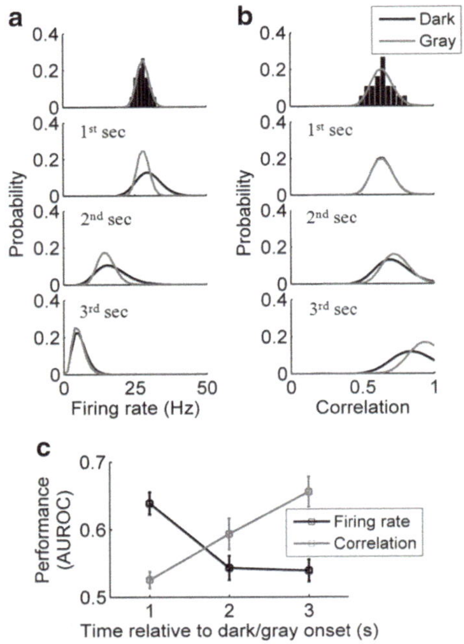

Fig. 5.32 Discrimination performances based on firing count and neural correlation quantified by receiver operating characteristic (ROC) analysis. (**a**) Firing count distribution of a ganglion cell's activity during different stages (1st–3rd seconds) in adaptive response to *dark screen* (*black*) and *gray screen* (*gray*). (**b**) Synchronization strength distribution of the ganglion cells' activities during different stages (1st ~ 3rd seconds) in adaptive response to *dark screen* (*black*) and *gray screen* (*gray*). The *top panels* in (**a**) and (**b**) provide examples for fitting the spike count histogram and correlation strength histogram with gamma distributions. (**c**) The time-dependent AUROC during light adaptation calculated based on firing count and synchronization strength (Adapted with permission from Xiao et al. 2013, Fig. 6)

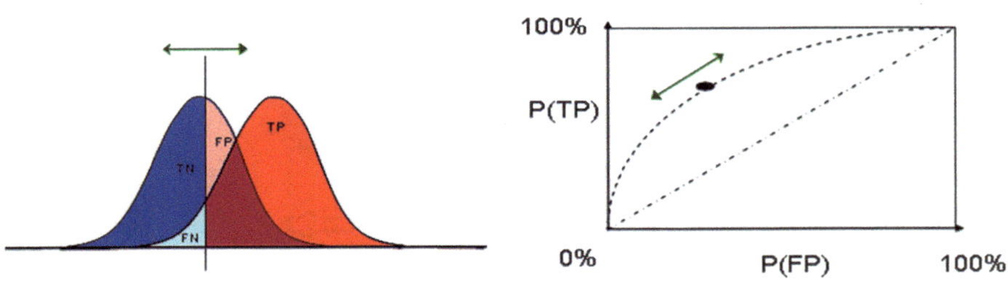

Fig. 5.33 ROC method

whisker stimulation information is transferred from the periphery to the thalamus, being encoded by the ventral posteromedial (VPm) nucleus in the thalamus and further transmitted to the somatosensory cortex; therefore, changes in the cortical cell's response property might be due to changes in the thalamic neuron's response

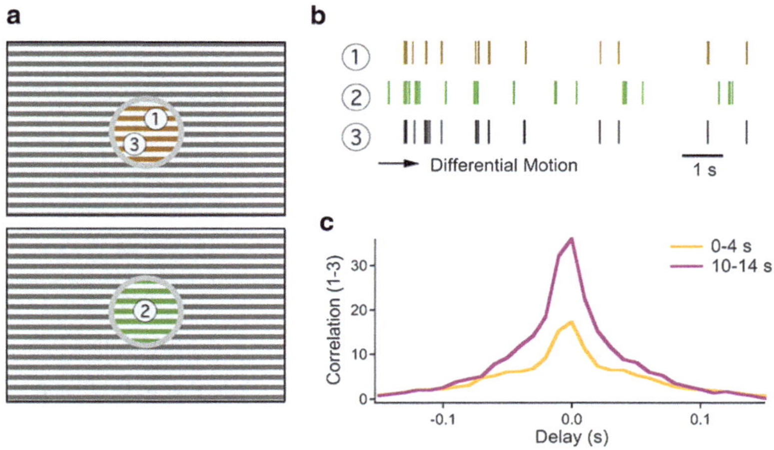

Fig. 5.34 The changes in correlated activity between OMS cells during adaptation. (**a**) Two moving objects with different trajectories (indicated by *different gray scales*) presented on jittered background. (**b**) The three OMS cells' (shown in (**a**)) firing activities in response to moving objects, among which cell 1 and cell 3 were exposed to the same motion trajectory, while cell 2 was exposed to another moving object. During adaptation to differential motion, these cells' firing activities all showed a reduction tendency. (**c**) Cross-correlation function between the spike trains recorded from cell 1 and cell 3, which shows that the correlation strength of these two cells' activities was significantly increased after adaptation (Adapted with permission from Olveczky et al. 2007, Fig. 9)

property. By examining the VPm neuron's activity, it was found that the single neuron's responses to different stimulation intensity were all reduced due to adaptation, and such reductions were with similar degree (Fig. 5.35a). Therefore, the adaptation did not alter the discrimination ability based on single neuron's activity.

However, in the adaptation process, the synchronizing strength between pairwise VPm neurons activity was also reduced; in the meantime, as compared to the situation that the synchronization strengths in response to different stimulation intensities were similar with each other before adaptation, the synchronization strengths after adaptation show clear intensity dependence (Fig. 5.35b), so that thalamic output had better also intensity dependence after adaptation, which allowed cortical neurons to enhance their discrimination ability along with the adaptation process (see Fig. 5.27).

5.6 Efficient Coding in Visual System

In the preceding sections, we have seen that sensory neurons adapt to sustained or repeated stimulation with reduced firing activity, and in the meantime, different coding strategies are adopted, so as to ensure efficient coding during adaptation.

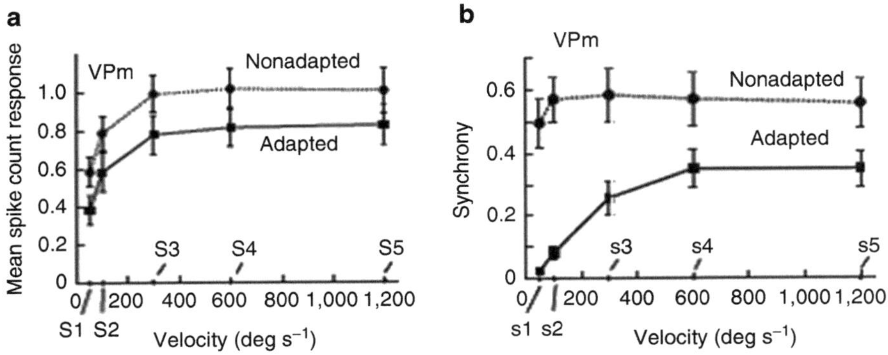

Fig. 5.35 Thalamus neuron's activity in response to vibrissal deflection with different velocities before and after adaptation. (**a**) VPm neurons adapted to persistent ongoing periodic vibrissal deflection with different velocities. In contrast to the sensitivity curve in cortex (see Fig. 5.27), the sensitivity curve in VPm retained its shape after adaptation, resulting in little or no change in overall sensitivity and thus no change in performance. Error bars are ±SE, $n = 32$ cells. (**b**) Synchronous firing across VPm pairs as a function of deflection velocity, before and after adaptation. Error bars are ±SE; $n = 19$ pairs (Adapted with permission from Wang et al. 2010. Figs. 4 and 6)

On the other hand, sensory adaptation may involve different aspects. For instance, visual neurons adapt to sustained stimulation including luminance (mean light intensity) and contrast (standard deviation of intensity relative to the mean), and visual gain control also involves luminance gain control, contrast gain control, etc. Although luminance adaptation and contrast adaptation processes occur in parallel, luminance gain control and contrast gain control are independent to each other.

In natural images, the distributions of luminance and contrast are both with a wide range, and local luminance or contrast in an image typically varies by a factor over tenfold. When selecting a group of small patches from a natural image (Fig. 5.36a) and measuring the mean light intensity and contrast of these patches, the results are such that the mean luminance varies from several hundred to several thousand cd/m^2 and contrast changes over a range from 0 to ~0.35 (Fig. 5.36b).

Statistical results obtained from many natural images show that distributions in mean luminance and contrast in natural images are independent to each other: different stimulation contrasts are with a similar distribution pattern for all the given luminance levels, and different luminances are with a similar distribution pattern for all the given contrasts (Fig. 5.37).

In the visual system, the lateral geniculate nucleus (LGN) receives input signals from retinal ganglion cells and sends output signals to visual cortex. As expected, LGN neurons have gain control mechanisms for both varying luminance and contrast levels. Given that the distributions of luminance and contrast in natural image are independent to each other, it is intriguing that whether or not the luminance gain control and contrast gain control in LGN neurons are independent to each other.

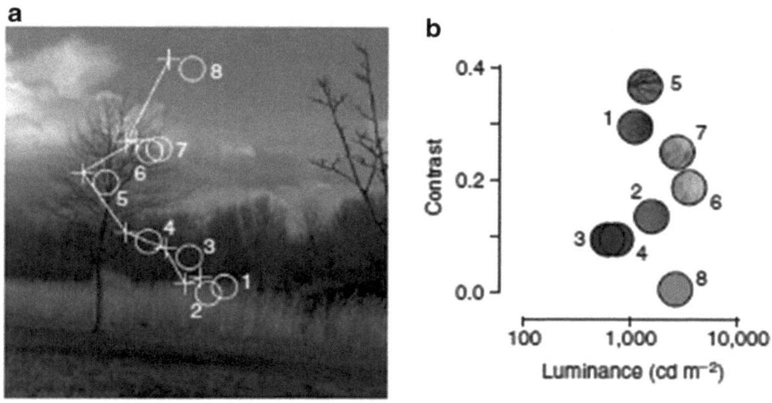

Fig. 5.36 Luminance and contrast distribution in a natural image. (**a**) A sequence of fixations (*crosses*) and the corresponding locations of the arbitrary receptive field (*circles*, diameter = 1°). (**b**) Enlarged image patches as indicated by *circles* in (**a**) and their contrast level (*ordinate*) and average luminance (*abscissa*) (Adapted with permission from Mante et al. 2005, Fig. 1)

To answer this question, LGN neuron's response activity can be elicited by different combinations of luminance level and contrast (drifting gratings with mean luminance level covering a range from 6 to 64 cd/m^2, contrast changing from 3 % to 100 %, Fig. 5.38a, background), and the neuron's impulse response can be analyzed via a descriptive model. The results were such that the neuron's sensitivity was not constant but was gradually reduced when luminance or contrast was increased (Fig. 5.38a, curves). The neuron's response character can also be presented by its spectral response property via Fourier transform which converts the systems' impulse response in the time domain to the transfer function in the frequency domain (Fig. 5.38b).

As similar to retinal ganglion cells, LGN's response can be described by a descriptive model which contains a spatial receptive field and a nonlinear rectifier. If the luminance and contrast are processed separately in the LGN neuron, then the linear receptive field component in the traditional descriptive model can be separated into three cascade components: (i) a linear filter with fixed gain, (ii) a luminance gain filter, and (iii) a contrast gain filter. By singular value decomposition (SVD) performed on the cell's transfer function presented in Fig. 5.38b, three separable components (a fixed transfer function (lower-left panel in Fig. 5.38c), a luminance-dependent transfer function (lower-right panel in Fig. 5.38c), and a contrast-dependent transfer function (upper-left panel in Fig. 5.38c)) can be obtained (Fig. 5.38c). Finally, the impulse response of each component in time domain can be obtained via inverse Fourier transform which converts the transfer function back to the impulse response (Fig. 5.38d). As expected, when luminance or contrast is increased, the amplitude of the relevant filter is getting smaller.

While distributions of luminance levels and contrasts are basically independent in natural images, the luminance gain control and contrast gain control in the LGN

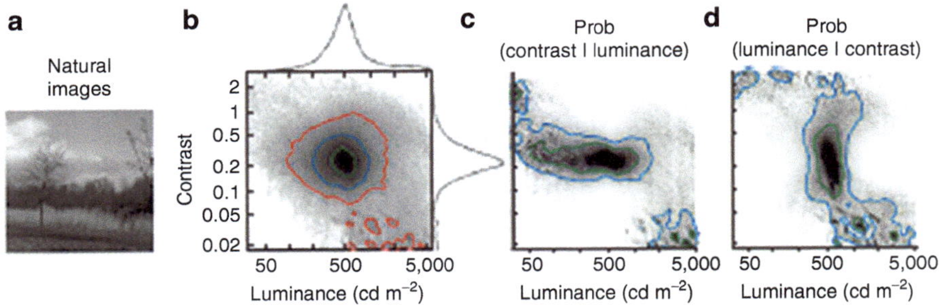

Fig. 5.37 Statistics of local luminance and contrast in natural images. (**a**) A natural image (same as in Fig. 5.36a). (**b**) Joint distribution of luminance and contrast as sampled from 300 natural images. The *contours* indicate the regions containing 90 % (*red*), 65 % (*blue*), and 40 % (*green*) of the observations. The *curves* on the sides of the joint distribution indicate the marginal distributions of luminance and contrast. (**c**) Conditional probability of observing a certain contrast given a specified luminance. (**d**) Conditional probability of observing a certain luminance given a specified contrast (From Mante et al. 2005, Fig. 2)

neuron are also independent to each other, which reflects the match of neural process and the natural environment. The system's resources for information transmission are remarkably saved, and its capacity for information transmission is greatly enhanced given the separability of the adaptation processes. For a set of combinations of m luminance levels and n contrasts, the traditional predictive model needs $m \times n$ filters to predict the responses, while a model with separable components only requires $m + n + 1$ filters, which greatly reduces the degree of freedom and ensures efficient coding for the natural visual stimulation.

In natural scene, apart from the temporally redundant signals (including sustained or repeated stimulation), the images can also contain uninformative correlations between adjacent pixels. Therefore, spatial filtering is required to remove spatial correlation and to optimize information transmission. The center–surround antagonistic receptive field structure in a visual neuron is considered as a spatial filter to remove spatial correlations in a sense that the firing activity of visual neurons (including retinal ganglion cell, LGN neuron, and cortical neuron) is normally attenuated when the center and surround subareas of the receptive field are equally stimulated.

In addition to a single neuron's receptive field property, the spatial correlation in visual images can also be reduced by "de-correlation" between pairwise neurons' activities. Although correlated activities have been frequently observed in visual neurons and are confirmed to be contributive to visual information processing, when the correlation in the neurons' firing activity is compared to that contained by visual stimulation, it shows that the neurons' correlation is much weaker, which is to say that neurons' firing activities are "de-correlated."

To test such "de-correlation" effect, two different kinds of stimulation can be applied to elicit the neurons' activities: one is naturalistic stimulation which consists of a pseudorandom Gaussian flicker with long-range spatiotemporal

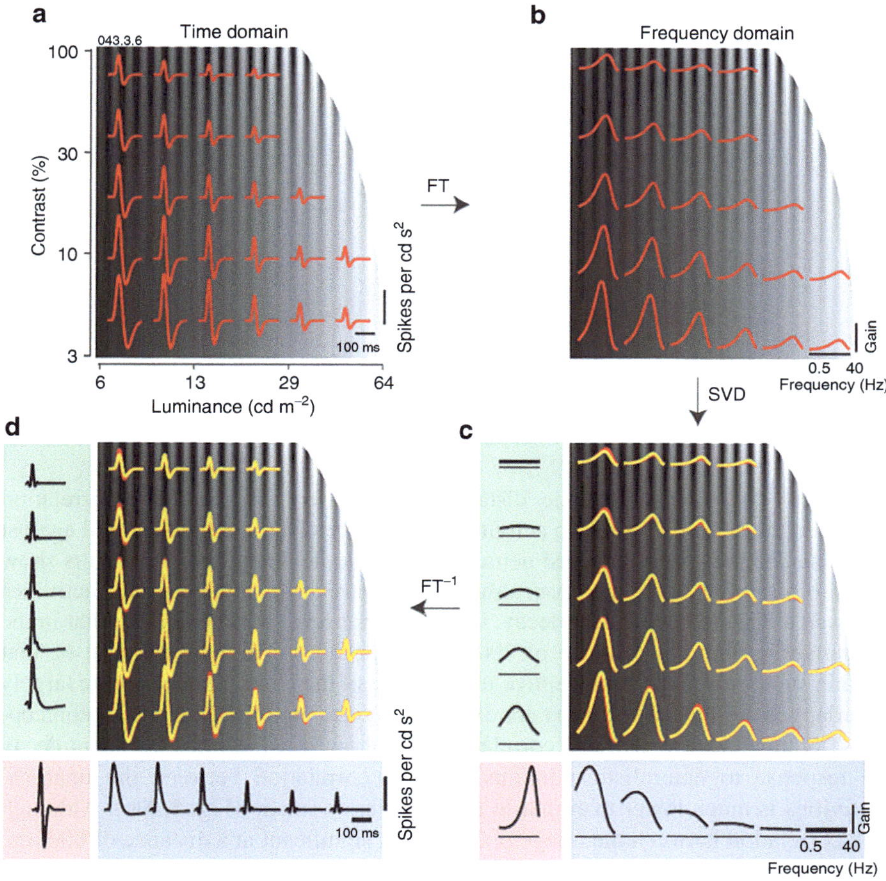

Fig. 5.38 The separability of luminance gain control and contrast gain control. (**a**) Impulse responses of an LGN cell, measured for various combinations of mean luminance and contrast. (**b**) The transfer functions related to the impulse responses presented in (**a**), which are obtained by Fourier transform. (**c**) Singular value decomposition (*SVD*) performed on the neuron's transfer function (*upper-right panel*) results in separation of three components: a fixed transfer function (*lower-left panel*), a luminance-dependent transfer function in the appropriate column (*lower-right panel*), and a contrast-dependent transfer function in the appropriate row (*upper-left panel*). (**d**) Impulse responses predicted by the separable model (*upper-right panel*), compared to those predicted by the descriptive model (as derived from (**a**) for comparison). Each impulse response (*upper-right panel*) is the convolution of the three components obtained by reversed Fourier transform performed on the relevant components presented in (**c**) (Adapted with permission from Mante et al. 2005, Fig. 6)

correlations (Fig. 5.39a) and another white noise which consists of uncorrelated flicker (Fig. 5.39b).

Spatial correlation between pixels with different distance can be calculated for both stimuli. The results show that, for the white noise, pixels with long distance are uncorrelated (Fig. 5.40a), while for the naturalistic stimulus, the correlation is

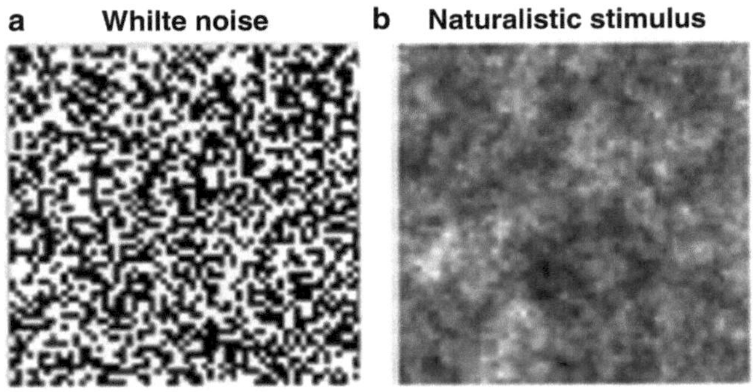

Fig. 5.39 Sample frames of (**a**) white noise and (**b**) naturalistic stimuli used to measure the "de-correlation" of retinal neurons' activity (Adapted with permission from Pitkow and Meister 2012, Fig. 1)

gradually decreased when the distance is increased (Fig. 5.40b). Correlation between the neurons' response activities can also be calculated and plotted against the distance between the paired neurons' receptive field centers. The results show that, for the white noise response, although the correlation of the neurons' activities showed a distance-dependent decay, it has a higher level as compared to that in the stimulation, within a distance of 300 μm (Fig. 5.40a). The reason might be that within such a distance, the receptive field centers of the pairwise neurons are largely overlapped, so that the neurons are exposed to correlated inputs from photoreceptors. Meanwhile, the neurons' correlation also showed a distance-dependent decay in response to naturalistic stimulus, but the correlation between the neurons' activities is much lower than that in the stimulation image (Fig. 5.40b). Although the correlation between the image pixels is still significant at a distance of 600 μm, the neurons' correlation is very weak when two cells are separated by a distance of 400 μm.

Given that, for a given neuron, its firing activity is basically determined by its spatiotemporal receptive field and the correlated activity of pairwise neurons is dependent on the similarity between the spike sequences, therefore the "de-correlation" of neurons' firing activities might be attributed to several factors: event timing, sparseness of firing, and noise. While event timing is determined by the neuron's spatiotemporal receptive field properties, sparseness might be related to nonlinear rectification of the input–output relationship. Again, the mechanism underlying the neuronal "de-correlation" can be tested by quantitative analysis.

To test whether the spatiotemporal receptive field structure contributes to the de-correlation, the neurons' responses can be calculated by convolving each spatiotemporal receptive file with the naturalistic image, and the remaining correlations can be analyzed accordingly. The results show that the correlation was even extended when only the receptive field center was considered, but the addition of the antagonistic surround effectively reduced the correlation, although not completely. But when nonlinear rectification of the input–output relationship and

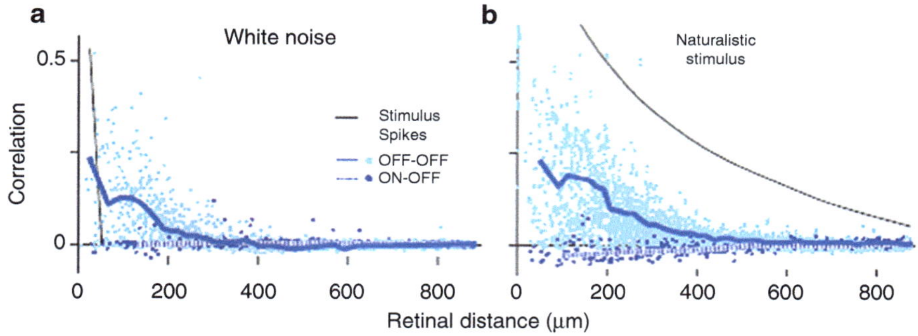

Fig. 5.40 Distance-dependent correlation coefficient between stimulus pixels and correlation between pairwise ganglion cells' responses. (**a**) White noise. (**b**) Naturalistic stimulus. Each point represents correlation of a cell pair. Comparisons are made within a cell type (*solid lines*) or across cell types (*dashed lines*). *Thin lines* are the correlation between stimulus pixels (Adapted with permission from Pitkow and Meister 2012, Fig. 1)

trial-to-trial variation was introduced in, it resulted in further de-correlation (Fig. 5.41a). Comparison among all these factors' contributions to the de-correlation showed that, at a distance of 300 μm, the de-correlation obtained by model output was up to 92 % of the measured level, among which the antagonistic receptive field structure contributed ~25 %, the nonlinearity contributed ~60 %, and the noise contributed ~15 % (Fig. 5.41b). Therefore, both receptive field property and the nonlinearity in ganglion cell's input–output relationship contribute to remove the spatial redundancy to enhance coding efficiency, with the nonlinearity being the dominant effect.

5.7 Olfactory Information Processing

The key function in olfactory information processing is to recognize and discriminate odors. Olfactory systems have evolved over millions of years to solve this "object-identification" problem, which is essential for mammals and insects to survive in natural environments, such as for feeding and reproduction (Fig. 5.42). To elucidate how animals generate olfaction-related behaviors, it is necessary to understand how odorant information is transmitted, processed, memorized, and retrieved in the olfactory system. This section introduces the studies related to the following questions: (1) How is odor information represented and transformed in different brain regions? (2) What is the underlying neural circuitry responsible for these information representations and transformation? (3) What are the properties of these circuitry organizations?

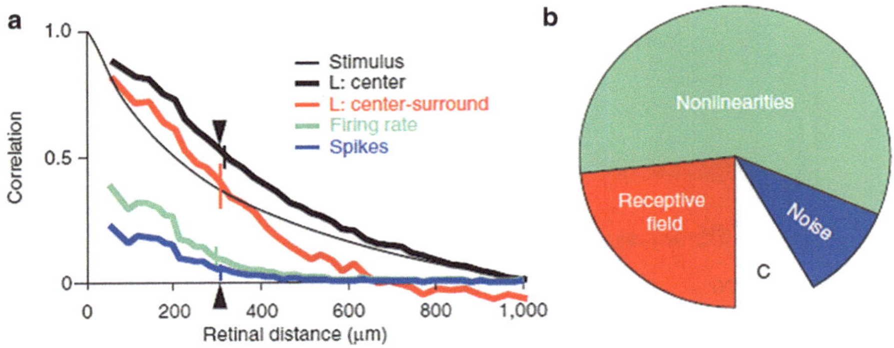

Fig. 5.41 Nonlinearity makes the most significant contribution to de-correlation. (**a**) Correlation strength as function of distance calculated for neurons and models during naturalistic stimulation. (**b**) The contribution of each response components to the de-correlation (Adapted with permission from Pitkow and Meister 2012, Fig. 2)

5.7.1 Neural Circuitry Underlying Olfactory Information Processing

Compared to other sensory systems, the olfactory system is extremely flat. It does not relay information through the thalamus but instead passes signal directly from receptor neurons, via the olfactory bulb (OB, the analogy of insect antennal lobe (AL)), to the olfactory cortex. In the below, we briefly review the neural anatomy of olfactory systems for both vertebrate and invertebrate.

Olfactory Receptor Neurons

The olfactory information is first represented by olfactory receptor neurons (ORNs). Volatile odor molecules bind olfactory receptors (ORs) at the surface of the dendrites of olfactory receptor neurons, inducing currents flowing into the cell membranes of ORNs. Therefore, the odorant stimulus is transduced into electrical signals by ORNs. It is likely that most ORNs express a single OR out of the entire repertoire, although it has not been proved in experiments. In *Drosophila*, there is good evidence that most ORNs express only one OR; some ORNs express two or three ORs. However, the same OR is never expressed by more than one ORN type.

ORs constitute a large and diverse gene family, and therefore ORNs could be classified into different types according to expressed OR type. There are about 1000 types of ORNs for moth, 830 types of ORNs for locust, and about 50 types of ORNs for *Drosophila*.

The same type ORNs converge at the same subregion of the OB (invertebrate analogy AL), called a glomerulus (Fig. 5.43). In each glomerulus, ORNs make excitatory synapses onto mitral and tufted cells (M/T, insect projection neurons

Fig. 5.42 Olfactory information processing. Olfactory information processing is essentially an "object-identification" problem. The olfactory system transforms chemical signals into electrical signals, which are then processed step by step before transmitted to the brain's advanced cortex, where the final perception of the odor is formed (Adapted from http://www.nytimes.com/2008/08/05science/05angier.html)

(PNs)). M/T cells are the only projection neurons in OB. Generally, each M/T cell innervates its dendrite into a single glomerulus. Thus, ORNs and M/T cells constitute the direct information pathway for olfactory information transmission.

Generally, there are several projection neurons innervating their dendrites into a single glomerulus. These cells receive the same feedforward synaptic inputs and are electrically coupled with each other. As a result, their spiking activities are highly correlated with each other. We usually call these neurons "sister cells" for simplicity.

A prominent feature of this ORN–PN projection is its high convergence ratio. The number of ORNs is much larger than that of M/T cells. It is reported that the convergence could make M/T cells (invertebrate analogy PN) to pool olfactory information from different spatial points and also increase the signal-to-noise ratio of olfactory information.

Lateral Interactions in AL (OB)

As we have noted, the ORN–M/T (PN) pathway constitutes the direct olfactory information transmission channel. In fact, there are abundant direct and indirect glomerulus-to-glomerulus interactions for both vertebrate and invertebrate. Below we will discuss the circuitry underlying lateral interactions in the first relay station of the olfactory system.

Lateral interactions in OB occur in two distinct layers, the glomerular layer and the external plexiform layer (Fig. 5.44). These two layers may be involved in two stages of information processing. In the glomerular layer, the interneurons are collectively termed juxtaglomerular cells, in which the largest class is the periglomerular (PG) cells. ORNs form direct excitatory synapses onto PG cells, and each PG cell usually extends its dendrite into a single glomerulus. A PG cell

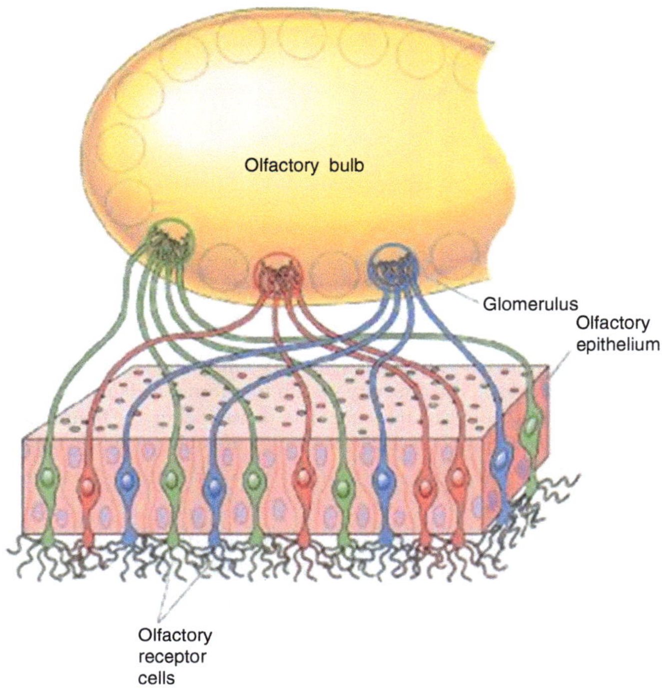

Fig. 5.43 Projection of olfactory receptor neurons (*ORNs*) to the olfactory bulb (*OB*). Each olfactory bulb only receives inputs from receptor cells expressing specific protein genes. Color marked for different receptors expressing specific genes (Adapted with permission from M. F. Bear, et al. *Neuroscience: Exploring the brain*, Chap. 8, Fig. 8.18)

may inhibit ORN–M/T synapses through presynaptic inhibition at the ORN axon terminal. Within a glomerulus, PG cells also form reciprocal dendrodendritic synapses with M/T cells. There are other forms of lateral interactions in the glomerular layer that needs to be further investigated.

In the external plexiform layer, M/T cells send several secondary dendrites through this layer to contact with GABAergic granule cells. Granule cells are restricted in this layer. There are reciprocal interactions between M/T cells and granule cells through dendrodendritic synapses. It is unknown about the functional role of this form of lateral interactions between M/T cells and granule cells. Apart from inhibitory interactions in this layer, there are also excitatory inter-glomerulus interactions between M/T cells. M/T cells do not synapse directly with each other but may excite each other via diffusion of glutamate.

Lateral interactions in invertebrate AL are relatively simple (Fig. 5.45). Contrast with vertebrate OB, the AL does not receive feedback from higher brain regions, suggesting that olfactory information processing in AL is largely feedforward. There are mainly two types of neurons in AL, PNs and LNs. PNs are the only neurons conveying olfactory information to higher centers, the lateral horn and

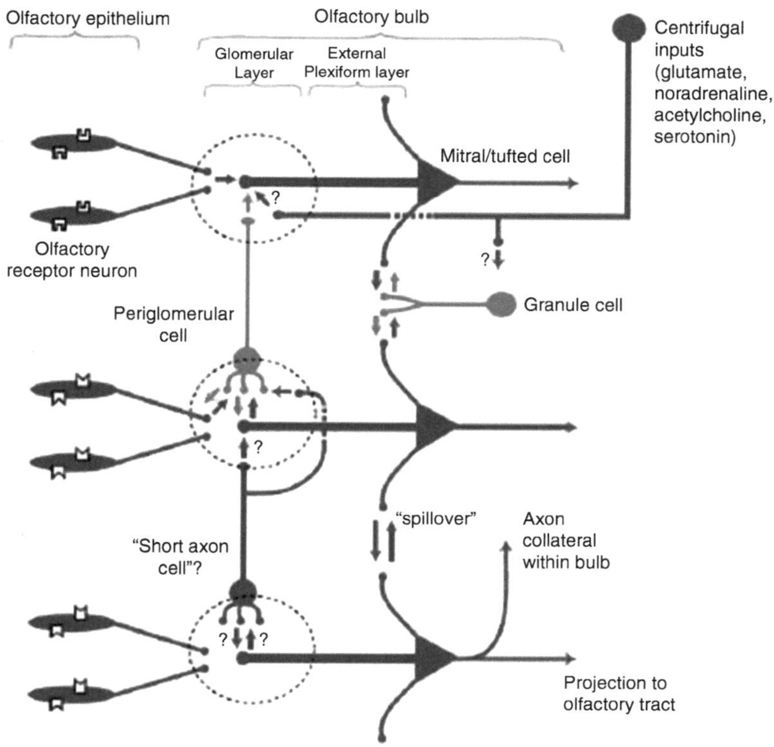

Fig. 5.44 Lateral interactions in AL (OB). Lateral interactions in OB occur in two distinct layers, the glomerular layer and the external plexiform layer (Adapted with permission from R. I. Wilson, et al. *Annu. Rev. Neurosci.*, Fig. 2)

mushroom body. Each PN sends a dendrite into a single glomerulus and receives direct input from ORNs expressing the same ORs. The synapses from ORN to PN are strong and exhibit strong short-term-depression. LNs, which receive synaptic inputs from both ORNs and PNs in AL, can be inhibitory or excitatory, releasing GABA or acetylcholine, respectively. Excitatory inter-glomerulus interactions are mainly mediated by eLNs. Notably, eLNs make specialized synapses with iLNs and PNs, and the synapses from eLNs to iLNs are mixed chemical electrical synapses. By contrast, synapses from eLN to PNs are purely electrical. Inhibitory inter-glomerulus interactions are mediated through iLNs. iLNs lack axons and communicate with PNs through dendrodendritic interactions. The major form of lateral inhibition occurs at the ORN axon terminal, which is presynaptic inhibition. Interestingly, both GABA-A and GABA-B presynaptic inhibition occurred at this synapse (Olsen and Wilson 2008).

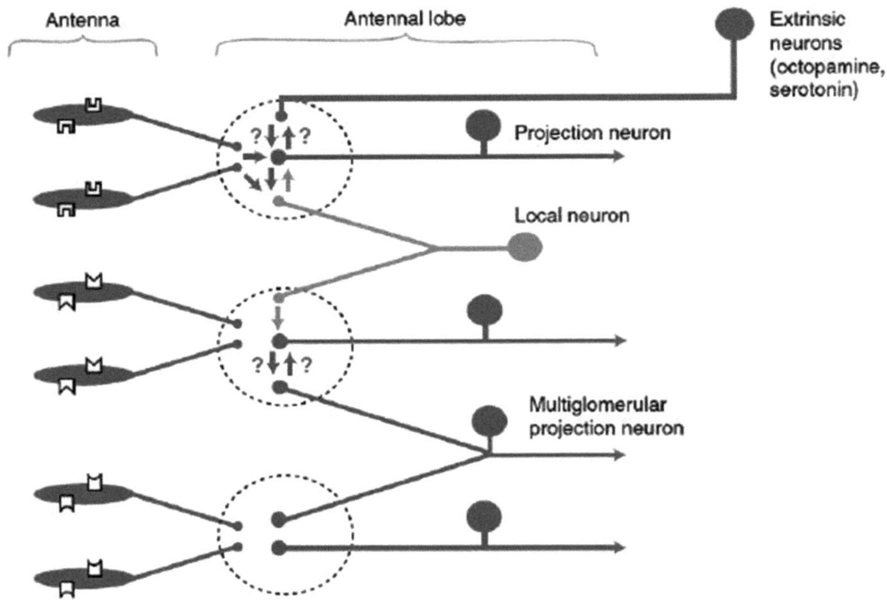

Fig. 5.45 Antennal lobe circuitry. The interactions in antennal lobe include excitatory interactions and inhibitory lateral interactions (Adapted with permission from R. I. Wilson, et al. *Annu. Rev. Neurosci.*, Fig. 3)

5.7.2 Information Representation in Olfactory Systems

A chemical odor contains identity and intensity information, which is represented by projection neurons with different codes layer by layer. Here, we introduce how odor information is represented at both the single neuron level and the neural population level.

Information Representation by ORNs

Odor information is first represented by ORNs in the olfactory system. A study carried in the Carlson lab has systematically investigated ORN responses to odor stimuli in *Drosophila* antenna (Hallem and Carlson 2006). Twenty-four out of about 50 types of ORN responses are measured under varying odor identity and intensity. Their results could be mainly summarized as follows: (1) an odorant stimulus could evoke both excitatory and inhibitory response by ORNs. As the spontaneous firing rate of ORNs varies greatly among different types of ORNs, some neurons will show elevated firing to odorant stimulation while others with high spontaneous firing rates will display degraded firing to odor stimulation. (2) ORN tuning width varies continuously from broadly to narrowly tuned ORN.

Some receptors respond to many tested odors while others respond strongly to only one odor, such as for CO2- and pheromone-sensitive ORNs. (3) Intensity coding would depend on both the strength of activation of individual receptors as well as the total number of receptors activated. The firing rate change of an ORN will be larger, and meanwhile, some ORNs are activated at higher concentrations. Experimental and modeling work has shown that the dynamics of ORNs could be separated as a transduction and spike generation process.

Information Representation by OB (AL)

Olfactory information is converged into the first relay station of the olfactory system, the OB (insect analogy AL). In this region, the olfactory information is recoded to be easily decoded by downstream neurons. How is odor information encoded in this region? Here, we mainly introduce odorant information representation at the locust antenna lobe, with a special emphasis to the following questions: (1) How is odorant information represented in the level a single neuron? (2) How is odor information represented in the level of neuronal population? Does odor identity and intensity information interfere with each other? (3) How is a repeatedly presented odor represented in AL? (4) How are temporally mixed odors represented?

Studies from the Laurent lab have investigated single-neuron responses to odor stimulus in the locust AL (Laurent et al. 1996) (Fig. 5.46). Odor stimulus is applied through a tube to a locust, while intracellular and extracellular signals in locust AL are recorded at the same time. The results show that the odor could evoke temporal patterns in PNs modulated with a slow and a fast time scale. On a slow time scale, the PN responds with characterized epochs of increased and decreased firing on the time scale of 200–300 ms. This slow temporal pattern is odor and neuron specific, implying that the odorant information could be contained in the temporal pattern of PN activity. On the fast time scale, the PN spiking activity is synchronized to other PNs and to the LFP at around 20 Hz, and PN to PN synchronization occurs at a few oscillation cycles. This suggests that odors can evoke specific sequence activities in neuronal ensemble, and this sequence develops at every oscillatory cycle.

O. Mazor and G. Laurent explored how an odor stimulus is represented in the locust AL at the level of neuronal population (Laurent 1999) (Fig. 5.47). In the experiment, PN population activities are recorded extracellularly in vivo with tetrodes. It is found that odor responses to stimuli were characterized as trajectories in the PN state space with three main phases: an on transient, lasting for 1–2 s; a fixed point, stable for at least 8 s; and an off transient, lasting a few seconds as activity returns to baseline. All of the three phases are odor specific, suggesting that the neural responses at all these three phases are odor informative. Moreover, by measuring KC responses to odor stimulus, which is downstream to PN, it shows that the KC is most responsive when PN population activity is in its transient dynamics. This means that the population transient information is decoded by downstream neurons. Therefore, PN transient activity is more relevant for olfactory behavior.

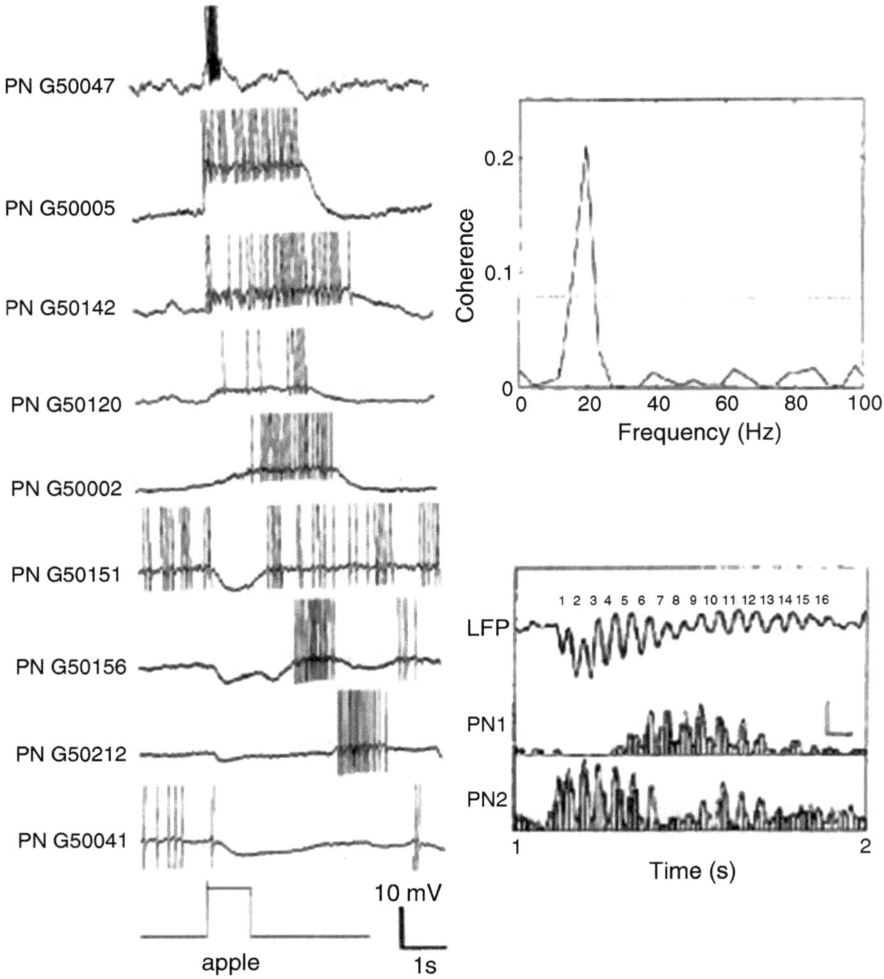

Fig. 5.46 Range of temporal patterns of response to single odor across neurons. *Left*: temporal response patterns of nine different antennal lobe PNs in response to the odor apple. *Right upper*: the LFP shows 20-Hz oscillation. *Right bottom*: comparison between odor-evoked PNs with LFP (Adapted with permission from G. Laurent, et al. *The Journal of Neuroscience*, Fig. 1)

Stopfer et al. investigated how odor intensity information is encoded and how odor identity and intensity information interfered with each other (Stopfer et al. 2003) (Fig. 5.48). This paper examined this issue in the locust olfactory system by recording the activities of the PN ensemble and Kenyon cells (KCs). They found that the oscillatory power of LFP around 20 Hz increases significantly with concentration. The firing phase of PN does not change with odor concentration. The summed population response varies little with odor identity and intensity, indicating that the increase in LFP power is due to a tighter synchronization of PNs.

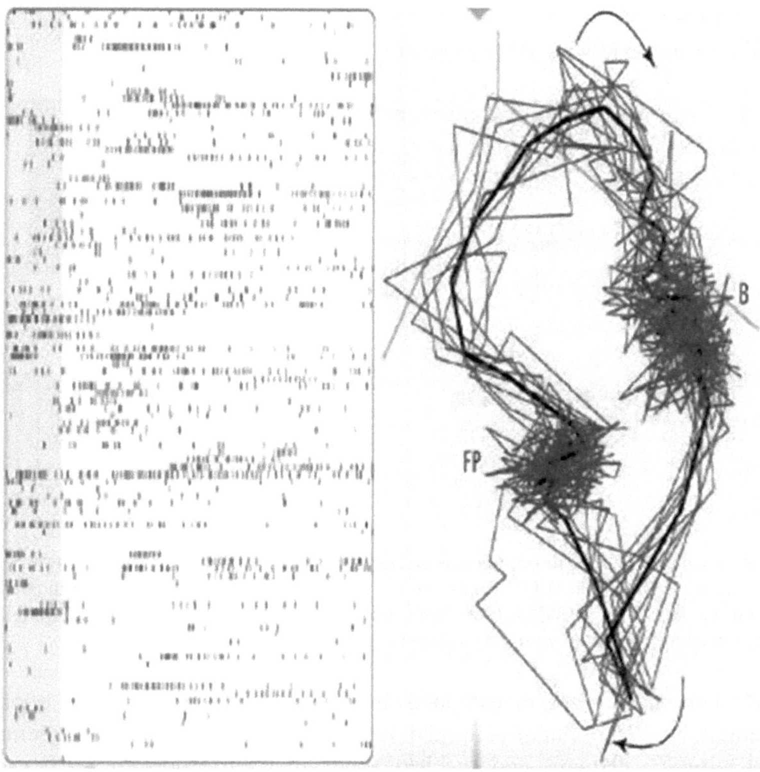

Fig. 5.47 PN population response to the odor information. *Right*: single-trial response of 99 locus antennal lobe principal neurons to one odor can be recorded. *Left*: projections of principal-neuron trajectories, representing the succession of states visited by this neural network in response to one odor. *Gray lines*, individual trials; *black line*, average of 10 trials. *B* baseline state; *FP* fixed point, reached after 1.5 s (Adapted with permission from M. Rabinovich, et al. *Science*, Fig. 1)

The slow temporal pattern of PNs changes substantially and, sometimes, abruptly with concentration. In the population level, odor representations can be described as sequences of odor- and concentration-specific trajectories. Trajectories representing different concentrations of the same odor stay next to each other. For the downstream neuron of PNs, KCs' response to odor stimuli shows some degree of concentration invariance.

PN responses to repeatedly odor stimulation have also been investigated by Stopfer et al. (1999) (Fig. 5.49). It is reported that the firing rate of PN rapidly decreases in intensity over time. However, PN shows an increase in spike time precision and interneuronal oscillatory coherence around 20 Hz over repeated presentation of a same odor. Once this enhanced correlation is established, this precise response could last for several minutes. The experience-related response change is assumed to be a form of short-term memory.

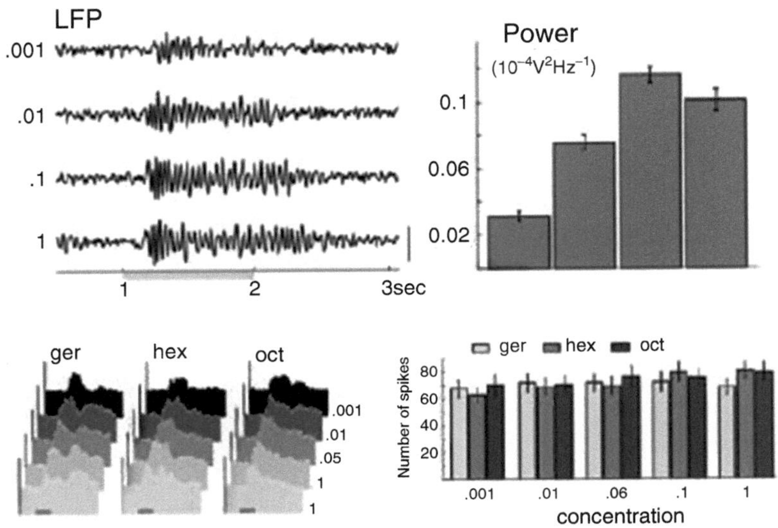

Fig. 5.48 Response of PNs to odor concentrations. *Left upper*: higher odor concentrations elicit greater oscillatory power in LFP. *Right upper*: oscillatory power in 20 Hz significantly with concentration. *Left bottom* and *right bottom*: mean response intensity is constant across odors and concentrations (Adapted with permission from M. Stopfer, et al. *Neuron*, Fig. 2)

B.M. Broome et al. investigated how temporally intervening odors are represented by PN population in the locust AL. PN ensemble representations tracked stimulus changes and reached states that corresponded neither to the representation of either odor alone nor to the static mixture of the two.

Information Representation in MB

Classical lesion experiments and genetic blockade of Kenyon cell synaptic output show that mushroom bodies are essential for memory formation and retrieval (Heisenberg 2003). Studies on locust and *Drosophila* MB show that KC responses in MB are sparse, which means that a KC can only respond to much a few odor stimuli and an odor can only activate a few KCs in MB at a trial. KCs rarely evoke a spike at its spontaneous state and generate several spikes when responding to odors. Thus, each spike evoked by KC is highly informative.

The underlying mechanism for generation of sparse code has been investigated for both locust and *Drosophila* (Jortner et al. 2007; Turner et al. 2008). Several mechanisms are recognized to be relevant to the generation of a sparse code, such as neuronal intrinsic properties, neuronal projections from AL to MB, and network oscillations. Among these species, a theoretical paper has proposed that the divergence projection from AL to MB is essential to the formation of sparse code. As the number of KCs in MB is much larger than that of its presynaptic neurons, an odor

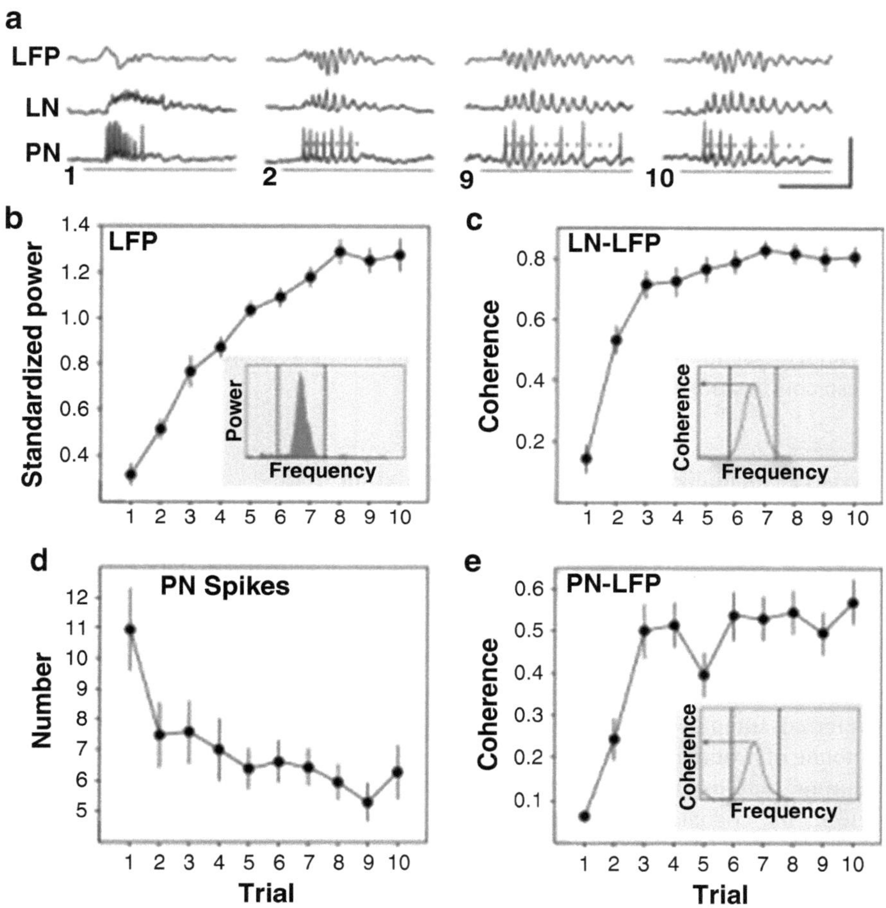

Fig. 5.49 Response intensity decreases, while coherence and spike time precision increase, over repeated odor presentations. (**a**) Simultaneous local field potential (LFP) and intracellular recordings from a local neuron (LN) and projection neuron (PN). (**b**) LFP power spectrum increased during the first 6 or 7 trials before reaching asymptote. (**c**) Coherence between LNs and LFP increased rapidly over stimulus trials. (**d**) The number of odor-elicited action potentials in PNs decreased markedly by the second trial. (**e**) Coherence between PN spike time and LFP increased over trials (Adapted with permission from M. Stopfer, et al. *Nature*, Fig. 1)

stimulus will activate complex spatial pattern in PNs and excite KCs in MB. A very few KCs who receive enough amount of EPSCs within a time window will respond to odor stimulus. Thus, a sparse code is formed in MB. The sparse code is assumed to be beneficial to memory. To load a new odor memory in local circuits, the synaptic connection weights will be modified according to synaptic learning rule. A sparse code, as each memory is represented by only a few active neurons, and few synapses need to be modified to restore a new memory. Thus, the newly stored memory will not interfere much with other memories, which enables a large network memory capacity.

5.7.3 From Neural Circuits to Neural Computation

Before understanding how neural circuits lead to behaviors, it is useful to make sure what computations the brain implements and how neural circuits are organized to realize these computations. This is an intermediate step of understanding neural mechanisms of behaviors. In the below, we introduce a few computations the olfactory system may employ and discuss the biophysical mechanisms underlying these computations.

Divisive Normalization in Olfactory Population Code

In the sensory system, it is frequently observed that a neuron's response to its preferred stimulus could be suppressed by superimposing a non-preferred stimulus to these neurons (Olsen et al. 2010). This form of suppression has been successfully fitted by an operation of divisive normalization, in which a neuron's response to external drive could be divisively normalized by a summed activity of a pool of neurons. In the divisive normalization model, the neuronal response curve to its preferred stimulus is input modulated by other neurons' activity, in which the amount of external input that needed to drive saturation is modulated by other neurons' activity. In a study of Olsen et al., it has been reported that the PN firing rate to the cognate ORN input could be well fitted in an operation of divisive normalization, where the denominator is determined by total activities of PNs in AL (Olsen et al. 2010).

As this form of operation has been widely observed in many cortical regions, it has been believed to be a canonical neural computation that the neural systems recruited. A lot of literatures have explored the computational roles of this form of operation in terms of neural information processing. It is reported that divisive normalization could effectively reduce redundancy of stimulus inputs.

The underlying mechanisms that give rise to divisive normalization may be different for different brain regions and sensory modalities. In the olfactory system, it has been reported that the operation of divisive normalization is induced by GABAergic lateral presynaptic inhibition by iLN in AL (Olsen and Wilson 2008).

Gain Control

Gain control is widely existed in neural system; it could be changed depending on the input stimulus or the behavioral state of an animal. It is useful for a neural system to use a limited channel resource to transmit as much information as possible in terms of neural information transmission (Martin et al. 2011).

In the olfactory system, gain control is also widely observed in the level of a single neuron or of a population of neurons (Fig. 5.50). In a single ORN, it responds weakly to a wide range of odor stimulus and yet responds strongly to a few high concentration of odor stimulus. To best transmit the odor stimulus information, the projection neuron in AL should amplify weak inputs and suppress strong inputs. This nonlinear transformation emphasizes the differences of weak inputs and suppresses strong inputs. The nonlinear transformation maybe is largely mediated by short-term depression of ORN to PN synapse, which enables an intra-glomerular gain control of inputs. In the level of a neuronal population, the mean firing rate of PNs varies little with odorant concentration while total afferent into PN increases with odor concentration (Stopfer et al. 2003). This implies that an adaptive control mechanism exits in the neuronal population. Inhibition and excitatory synaptic connections in LNs that may integrate many or all glomeruli may adaptively regulate the dynamical range of PN relative to the total input into the AL. In *Drosophila*, both presynaptic and postsynaptic inhibitions from inhibition LNs scale with the total input from ORNs. Excitatory LNs also integrate inputs from many glomeruli and may increase PN output that can be detected by downstream neurons. These two forms of excitation and inhibition may supply a neural substrate for the adaptive control of neural population responses.

Trajectory

How input information is represented by a population of neurons is a fundamental problem in theoretical neuroscience. Experimental data in the olfactory system has revealed that odor-generated distributed spatiotemporal patterns are odor and concentration specific (Mazor and Laurent 2005; Rabinovich et al. 2008). Such successive states of excitation could be discarded as trajectories in a high-dimensional state space. Experimental data has also shown that trajectory information is decoded by downstream neurons. The above experimental finding illustrates that transient dynamics may serve as a basic form of brain computations, with its trajectory controlled by incoming inputs.

Signal and Noise Correlations

There are several PNs (2–5) in each glomerulus. These PNs receive almost the same synaptic inputs from ORNs and are electrically coupled with each other. Therefore, the spiking activities are highly correlated both in the spontaneous state and in response to odorant stimuli. In this study, we will call PNs within the same glomerulus as sister cells. A great interest is to investigate the strong correlations of sister cells' activities in neural information processing. K. Padmanabuhan and N.N. Urban have investigated the role of a sister cell to transmit neural information (Padmanabhan and Urban 2010). They reported that different sister cells in a glomeruli transfer common but also different parts' information of inputs, although

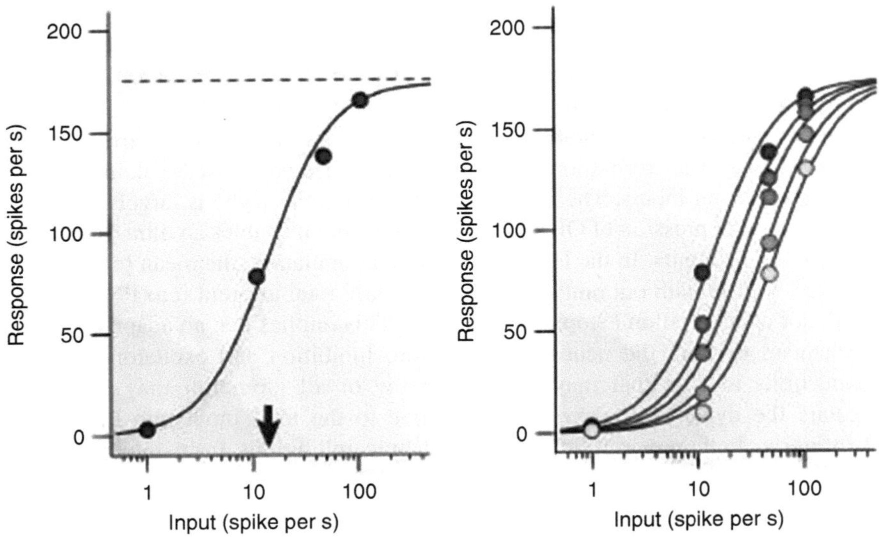

Fig. 5.50 Normalization in the olfactory system of the fruitfly. *Left*: responses of olfactory neurons in the antennal lobe to single test odorant, as a function of activity in the presynaptic receptor neuron. *Right*: response of olfactory neurons in the antennal lobe to test odorant in the presence of mask odorants of increasing concentration (lower concentrations are shown by *darker color*) (Adapted with permission from M. Carandini and D. Heeger, *Nature Reviews Neuroscience*, Fig. 1)

these cells receive identical stimuli. This is largely due to the intrinsic biophysical diversity, which enables PNs to maximize information transmission. To uncover the computational roles of signal and noise correlation in sister cells, it is helpful to see in what forms PN activities are decoded by downstream neurons. The odorant information is represented as sparse code KCs, with each odor represented by a few proportion of KCs' spikes. It is assumed that sister cells may promote the robustness of the KCs' sparse code (Fig. 5.51).

Oscillation

Oscillation in the band of 20–30 Hz has been observed in odor-evoked LFP of the locust AL. Intracellular recordings in vivo from locust antennal lobe projection neurons (PNs) revealed that individual PNs phase-lock with population activities at times that depend on the stimulus, and this fine structure of population activity could be eliminated by application of GABAA antagonist picrotoxin (MacLeod and Laurent 1996). This oscillatory synchronization has been declared to be related to fine odor discrimination. There has been a lot of modeling works to stimulate the effect of GABAA inhibition to network oscillatory dynamics. It shows that slow inhibitory connections between inhibitory LNs and PNs are responsible for the generation of temporal patterning of PNs, whereas fast inhibitory dynamics enables oscillatory synchronization of PNs.

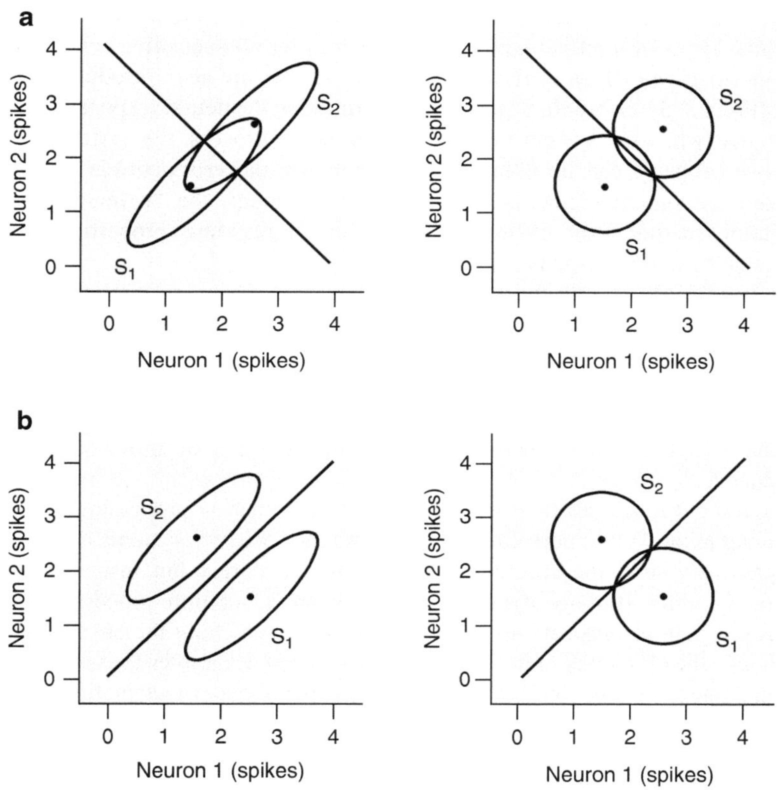

Fig. 5.51 Effects of correlations on information encoding. In all cases, we show the response distributions for two neurons that respond to two different stimuli. (**a**) A larger fraction of the ellipses lie on the "wrong" side of the decision boundary for the true, correlated responses than for the independent responses. (**b**) A smaller fraction of the ellipses lie on the wrong side of the decision boundary for the correlated responses (Adapted with permission from B. B. Averbeck, et al. *Nature Reviews Neuroscience*, Fig. 1)

5.8 Summary and Advanced Readings

Information processing in different sensory systems might have different characteristics; e.g., "receptive field" is specified for visual neurons and somatosensory neurons, while auditory neurons are with "tonotopic" character. On the other hand, different systems might share common rules for information coding and transmission which can be analyzed using similar computational tools.

This chapter is devoted to introduce some aspects of information coding in different sensory systems.

Section 5.1 is focused on models of retinal ganglion cell's receptive field, which quantitatively describe the relationship between the visual stimulation and firing activity of then cells. Models describing the response and spatiotemporal structure of stimulation are introduced. More complicated models describing some particular types of a retinal ganglion cell such as motion-sensitive cell and object-motion-sensitive cell, as well as depth-detection cell, are also introduced.

Section 5.2 describes the algorithm for estimating the neural response properties. Given that when the system input is a white noise process, the system's impulse response property can be obtained by calculating the cross-correlation function between the output signal and the white noise input, the "reverse correlation estimation method" for estimating the neuronal response properties based on white noise-elicited response is introduced.

Section 5.3 deals with the adaptive adjustment of transfer function of neuronal response process. Because the dynamic range of the visual neuron response does not match the dynamic range of stimulation intensity, adaptation to the environment enables the system to have a proper sensitivity and make proper response to local stimulation. "Gain control" in visual neurons in adaptation to stimulation parameters such as luminance intensity, stimulation contrast, or movement speed is introduced.

Section 5.4 discusses neuronal strategy for information processing during sensory adaptation. While performing cognition tasks, sensory systems do not simply and passively send the stimulation signals to the cortex but instead adjust the systems' coding strategy dynamically. Such an adaptation process allows the nervous system to make its limited dynamic range matching the environment and to enhance the efficiency of information transmission. Examples for visual neurons' coding strategy during contrast adaptation and spatial pattern adaptation, as well as motion pattern, are given. Adaptive change in somatosensory neurons' coding strategy is also introduced.

Section 5.5 goes further for population information coding during sensory adaptation. Examples from visual and somatosensory systems are given to show that during sensory adaptation, coding ability based on multiple neurons' concerted activity is enhanced, although the single neuron's activity has an adaptive decrease, which is often accompanied by reduced coding ability.

Section 5.6 particularly discusses efficient coding strategies in visual neurons. Visual process involves response to luminance and contrast changes, while in natural images, distributions in luminance intensity and contrast are independent to each other. Accordingly, luminance gain control and contrast gain control in visual neurons are also independent to each other, which reflects the match in temporal property of neural process and the natural stimulation. On the other hand, reduced correlation in adjacent neurons' responses to spatially correlated visual imaging contributes to remove the spatial redundancy and enhance coding efficiency.

Section 5.7 is focused on information processing in the olfactory system. Compared to the visual system, the function of the olfactory system is relatively simple, which is mainly about recognizing and discriminating odors, a pattern recognition problem. This section introduces the anatomic structure of the olfactory system and some recent progresses on understanding its computational principles.

5.8.1 Advanced Readings

Dayan P and Abbott LF (2001) Theoretical Neuroscience – Computational and Mathematical Modeling of Neural Systems, The MIT Press

Galizia CG & Lledo P-M, eds (2013) Neurosciences – From Molecule to Behavior: A University Textbook, Springer

Mallot HA (2013) Computational Neuroscience – A First Course, Springer

References

Averbeck BB, et al. Neural correlations, population coding and computation. Nat Rev Neurosci. 2006;7(5):358–66.

Baccus SA, Meister M. Fast and slow contrast adaptation in retinal circuitry. Neuron. 2002;36:909–19.

Bear MF, Connors BW, Paradiso MA. Neuroscience: exploring the brain. Philadelphia: Lippincott Williams & Wilkins; 2007.

Borghuis BG, Perge JA, Vajda I, van Wezel RJA, van de Grind WA, Lankheet MJM. The motion reverse correlation (MRC) method: a linear systems approach in the motion domain. J Neurosci. Methods. 2003;123:153–66.

Bredfeldt CE, Ringach DL. Dynamics of spatial frequency tuning in macaque V1. J Neurosci. 2002;22:1976–84.

Brenner N, et al. Adaptive rescaling maximizes information transmission. Neuron. 2000;26:695–702.

Chen AH, et al. Luminance adaptation increased the contrast sensitivity of retinal ganglion cells. NeuroReport. 2005;16:371–5.

Gollisch T, Meister M. Eye smarter than scientists believed: neural computations in circuits of the retina. Neuron. 2010;65:150–64.

Hallem EA, Carlson JR. Coding of odors by a receptor repertoire. Cell. 2006;125(1):143–60.

Heisenberg M. Mushroom body memoir: from maps to models. Nat Rev Neurosci. 2003;4 (4):266–75.

Hosoya T, et al. Dynamic predictive coding by the retina. Nature. 2005;436:71–7.

Jin X, et al. Information transmission rate changes of retinal ganglion cells during contrast adaptation. Brain Res. 2005;1055:156–64.

Jones JP, Palmer LA. The two-dimensional spatial structure of simple receptive fields in cat striate cortex. J Neurophysiol. 1987;58:1187–211.

Jortner RA, Farivar SS, Laurent G. A simple connectivity scheme for sparse coding in an olfactory system. J Neurosci. 2007;27(7):1659–69.

Laurent G. A systems perspective on early olfactory coding. Science. 1999;286(5440):723–8.

Laurent G, Wehr M, Davidowitz H. Temporal representations of odors in an olfactory network. J Neurosci. 1996;16(12):3837–47.

MacLeod K, Laurent G. Distinct mechanisms for synchronization and temporal patterning of odor-encoding neural assemblies. Science. 1996;274(5289):976–9.

Mante V, et al. Independence of luminance and contrast in natural scenes and in the early visual system. Nat Neurosci. 2005;8:1690–7.

Martin JP, Beyerlein A, Dacks AM, Reisenman CE, Riffell JA, Lei H, Hildebrand JG. The neurobiology of insect olfaction: sensory processing in a comparative context. Prog Neurobiol. 2011;95:427–47.

Mazor O, Laurent G. Transient dynamics versus fixed points in odor representations by locust antennal lobe projection neurons. Neuron. 2005;48(4):661–73.

Meister M, Berry MJ. The neural code of the retina. Neuron. 1999;22:435–50.
Olsen SR, Wilson RI. Lateral presynaptic inhibition mediates gain control in an olfactory circuit. Nature. 2008;452(7190):956–60.
Olsen SR, Bhandawat V, Wilson RI. Divisive normalization in olfactory population codes. Neuron. 2010;66(2):287–99.
Olveczky BP, et al. Retinal adaptation to object motion. Neuron. 2007;56:689–700.
Padmanabhan K, Urban NN. Intrinsic biophysical diversity decorrelates neuronal firing while increasing information content. Nat Neurosci. 2010;13(10):1276–82.
Pfaff DW, editor. Neuroscience in the 21st century: from basic to clinical. New York: Springer; 2013.
Pitkow X, Meister M. Decorrelation and efficient coding by retinal ganglion cells. Nat Neurosci. 2012;15:628–38.
Rabinovich M, Huerta R, Laurent G. Transient dynamics for neural processing. Science. 2008;321 (5885):48–50.
Rieke F, Rudd ME. The challenges natural images pose for visual adaptation. Neuron. 2009;64:605–16.
Ringach D, Shapley R. Reverse correlation in neurophysiology. Cognit Sci. 2004;28:147–66.
Rodieck RW. Quantitative analysis of cat retinal ganglion cell response to visual stimuli. Vision Res. 1965;5:583–601.
Schnitzer MJ, Meister M. Multineuronal firing patterns in the signal from eye to brain. Neuron. 2003;37:499–511.
Smirnakis SM, et al. Adaptation of retinal processing to image contrast and spatial scale. Nature. 1997;386:69–73.
Stopfer M, Laurent G, et al. Short-term memory in olfactory network dynamics. Nature. 1999;402 (6762):664–8.
Stopfer M, Jayaraman V, Laurent G. Intensity versus identity coding in an olfactory system. Neuron. 2003;39(6):991–1004.
Turner GC, Bazhenov M, Laurent G. Olfactory representations by drosophila mushroom body neurons. J Neurophysiol. 2008;99(2):734–46.
Wang Q, et al. Thalamic synchrony and the adaptive gating of information flow to cortex. Nat Neurosci. 2010;13:1534–43.
Wilson RI, Mainen ZF. Early events in olfactory processing. Annu Rev Neurosci. 2006;29:163–201.
Xiao L, et al. Shifted encoding strategy in retinal luminance adaptation: from firing rate to neural correlation. J Neurophysiol. 2013;110:1793–803.

Chapter 6
Network Models of Neural Information Processing

Abstract Neurons and synapses are the basic building units of the brain. They form neural circuits of various structures and hence implement different functions. Understanding how neural networks achieve brain functions is at the core of modeling studies. In this chapter, we will introduce some network models, including classical Hopfield model, continuous attractor neural network, and reservoir network. We will also discuss the studies on how short-term plasticity of neuronal synapses affects the dynamics and computations of a neural network.

Keywords Network models • Hopfield model • Continuous attractor network • Reservoir networks • Dynamical synapses

Neurons and synapses are the basic building units of the brain. They form neural circuits of various structures and hence implement different functions. To some extent, the computation of a neuron is rather "simple," whose essential role is to integrate synaptic inputs and fire an action potential if necessary (although the dendritic computation of a neuron can be quite complicated). It is the network that carries out the complicated computations involved in higher cognitive functions of the brain. Thus, understanding how neural networks achieve brain functions is at the core of modeling studies.

Over the years, a large number of network models have been proposed in the literature to elucidate brain functions. In this chapter, we introduce some of them. We start to introduce the classical Hopfield model for associative memory. This model is rather abstract, but captures some fundamental features of how a network system may utilize its dynamics mediated by structured neuronal interactions to realize a brain functions. It is one of the early models that proposed the important concept of "emergent computation," i.e., a neural network performs computation by updating its state in response to external inputs in a parallel distributed way. We then introduce the model of continuous attractor neural networks (CANNs). Compared with the Hopfield model, a CANN contains more biological elements, but it is still an idealized mathematical model. The key structure of a CANN is the translational invariance of neuronal interactions, a property which allows a CANN to hold a continuous family of stationary states and hence endows the network with the capacity of processing continuous stimuli. A common feature of Hopfield model and CANNs is that they assume information is encoded in the attractors (stationary

P. Liang et al., *An Introduction to Neural Information Processing*,
DOI 10.1007/978-94-017-7393-5_6

states) of the network. The third network model called reservoir networks, which we will introduce in this chapter, abandons the attractor idea. By taking into account the facts that the number of neurons involved in a brain function is huge and that the dynamics of individual neurons and synapses are stochastic and diverse, a reservoir network assumes that there is no well-designed "neural code"; instead, the stimulus information is encoded causally in the large state space of the network. The downstream system learns to read out this stimulus information, which is a much easier problem in practice. It is hard to say which network model, attractor or reservoir, is biologically more plausible. Both of them capture a few fundamental characters of neural circuitry. The forth network model we introduce is the networks aiming to mimic evidence accumulation in the decision-making process. Finally, we introduce the studies on how short-term plasticity of neuronal synapses affects the dynamics and computations of a neural network.

6.1 Hopfield Model for Associative Memory

Memory is an old yet one of the most active research topics in the field of neuroscience. To some extent, it is the collection of memories embedded in the brain that determine whom we are in the society. Conceptually, memory refers to the process of knowledge representation, storage, and retrieval, while the process of acquiring memory is learning. Although knowledge is accumulated about the brain regions related to different forms of memory, it is yet far from clear on how a memory item is actually stored in the brain. Neurons are the basic building units of the brain, which generate action potentials to pass on information to thousands of other neurons via synapses. The manifest of a memory is a function of the dynamics of a network formed by a large number of neurons. Thus, how neurons are organized to represent and retrieve memory is the key to understanding memory. In the mid of twentieth century, Donald Hebb postulated that memory could be stored if the synaptic connections that repeatedly drive activity in a cell are potentiated. As a result, cells in a specific group are unified and fire together in an assembly to represent a memory item. This strengthening of synaptic connections between coactive neurons is known as Hebbian plasticity and was later validated by experiments. In this section, we will introduce a specific form of memory, associative memory, which has been extensively studied in computational models (Hertz et al. 1991). Associative memory refers to that given partial information of data, the entire knowledge of data can be retrieved through dynamics of a network system. One of the most influential models for associative memory is the Hopfield model in name of John J. Hopfield (1982). Despite its simplicity, the model gives us insightful understanding of how memory may be formed in biological systems.

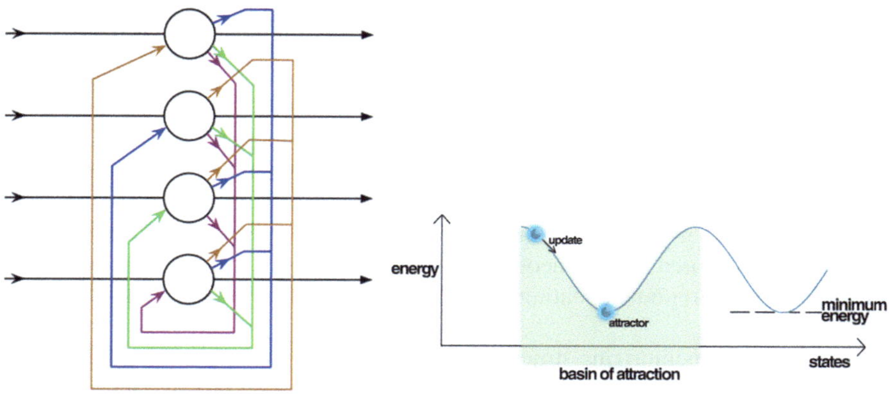

Fig. 6.1 The Hopfield model. (**a**) The structure of Hopfield model with four nodes. Each neuron receives synaptic inputs from other neurons, with connection strength depending on Hebbian learning rule. (**b**) Energy landscape of Hopfield network. Note that each state in the model has an energy defined by energy function, and the memorized item is a local minimum in the energy landscape and thus can have basin of attraction (Adapted from Wikipedia: Hopfield network, Figs. 1,2)

6.1.1 The Hopfield Model

Each neuron is modeled as a binary state variable S_i, with an active state (+1) and an inactive state (−1), respectively. The dynamics of the neuron in the network is given by (Fig. 6.1)

$$S_i = \text{sgn}\Big(\sum_j w_{ij} S_j - \theta\Big) \tag{6.1}$$

where the neuronal activation function is chosen as the sign function,

$$\text{sgn}(x) = \begin{cases} +1, x \geq 0; \\ -1, x < 0. \end{cases} \tag{6.2}$$

Without loss of generality, we usually drop the threshold term with $\theta = 0$.

The synaptic connection strength from neuron j to neuron I, w_{ij}, is set to be

$$w_{ij} = \frac{1}{N} \sum_{\mu=1}^{\mu=P} \xi_i^{\mu} \xi_j^{\mu} \tag{6.3}$$

where N is the number of neurons, P the total number of stored patterns labeled by μ, and ξ_i^{μ} the i-th element of stored pattern μ. With this setting of connection weights, it could be seen that connection weights are symmetric, i.e., $w_{ij} = w_{ji}$. This setting of connection weights is usually called Hebbian learning rule, because of its similarity to the Hebb's postulate that cells fire together, wire together. The difference is that when both two neurons are inactive, there is also a positive effect on the connection weights, which is not biologically plausible. One restriction in the connection weights in the model is that the self-connection weights w_{ii} is set to be 0, which is beneficial for the network dynamics converging into its stored pattern.

There are two ways to update the neuronal state, namely, asynchronous and synchronous. For asynchronous updating, each time one neuron is selected randomly to update its state according to Eq. (6.1). For synchronous updating, all neurons are updated simultaneously. Comparing these two forms of updating methods, asynchronous updating is more natural and realistic for biological systems.

In contrast to memorizing structured patterns, the stored patterns in the Hopfield model are normally assumed to be random with each neuron drown from +1 or −1 with equal probability.

The Hopfield model recalls its stored pattern through the network dynamics with an initial noisy version of the stored pattern. Each stored pattern corresponds to an attractor in the dynamical state space; therefore, all initial states that are near to the stored pattern will converge into that stored pattern by the network dynamics. Such a retrieval process is illustrated in Fig. 6.2. Memory retrieval begins from a clue, which is a noisy version of the stored pattern. The third panel in the figure shows intermediate states during the retrieval process, and the final stable state is shown in the last panel. Note that the final state is the stored pattern that is embedded in the network through Hebbian learning.

Sometimes, the retrieved pattern does not match to the stored one. One can check the stability of a memorized pattern. The condition for a memorized pattern to be stable is

$$\operatorname{sgn}\left(\sum w_{ij}\xi_j^{\nu}\right) = \xi_i^{\nu}, \quad \text{for}\, i = 1, \ldots, N. \tag{6.4}$$

The total input received by a neuron is given by

$$h_i^{\nu} = \sum_j w_{ij}\xi_j^{\nu} = \frac{1}{N}\sum_j\sum_{\mu}\xi_i^{\mu}\xi_j^{\mu}\xi_j^{\nu} \tag{6.5}$$

which can be further separated into two terms, i.e.,

$$h_i^{\nu} = \xi_i^{\nu} + \frac{1}{N}\sum_j\sum_{\mu\neq\nu}\xi_i^{\mu}\xi_j^{\mu}\xi_j^{\nu} \tag{6.6}$$

If the second term, called the cross-talk term, is small enough (less than 1) compared to the first term, all noisy patterns close to the stored pattern will relax to the stored one. Thus, the stored pattern is stable and associative memory is achieved.

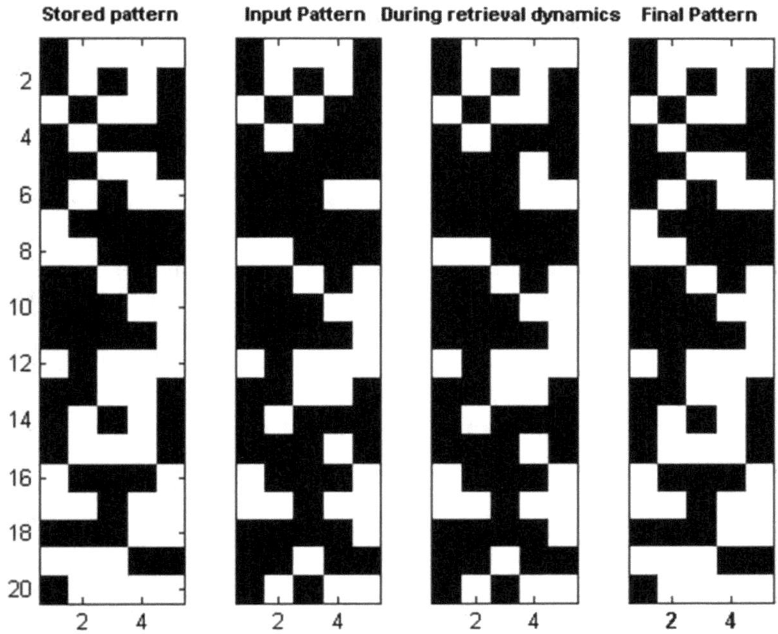

Fig. 6.2 Retrieval process of one stored pattern from its noisy version. Each memory item is embedded according to Hebbian learning rule so that each memory pattern is a local minimum of energy landscape. Starting with a noisy pattern of a memory item, the dynamics would enter into a memory pattern which is nearest to this starting noisy pattern, which could be seen in the *right panel* compared with the *left panel*

Hopfield model is famous for introducing the idea of an energy function into the neural network theory. It is defined as

$$E = -\frac{1}{2}\sum_{ij} w_{ij} S_i S_j \tag{6.7}$$

The energy function assigns a value to each network state and ensures that the value of energy function will decrease or stay the same when updating. Thus, computation in the neural system can be regarded as that starting from an initial state with high energy, the network evolves to low energy states, until it reaches a local minimal in its energy landscape. This local minimal usually corresponds to one attractor in the dynamical state space (Fig. 6.2).

Considering the network state is updated from S_i to $S_i^{'}$, we have

$$S_i^{'} = \text{sgn}\left(\sum_j w_{ij} S_j\right) \tag{6.8}$$

The change of the energy function is given by

$$E^{'} - E = -\sum_{j} w_{ij} S_{i}^{'} S_{j} + \sum_{j} w_{ij} S_{i} S_{j} = \left(S_{i} - S_{i}^{'}\right) \sum_{j} w_{ij} S_{j} \tag{6.9}$$

Note that if $S_i = S_i^{'}$, the energy function stays the same; otherwise, $E' - E < 0$ always holds.

In general, an energy function can only exist for a network having symmetric connection weights and a monotonic neuronal activation function. That's why in modeling studies people prefer to use symmetric connection weights to simplify theoretical analysis, although this is not biologically plausible.

From Eq. (6.6), we see that the cross-talk term critically affects the retrieval process. To determine in which condition the Hopfield model will retrieve to the right pattern, we define the following quantity:

$$C_{i}^{\nu} = -\xi_{i}^{\nu} \frac{1}{N} \sum_{j} \sum_{\mu \neq \nu} \xi_{i}^{\mu} \xi_{j}^{\mu} \xi_{j}^{\nu} \tag{6.10}$$

which is a product of $-\xi_i^\nu$ with the cross-talk term. If C_i^ν is negative, the cross-talk term has the same sign with the stored item ξ_i^ν, and then successful retrieval occurs. If C_i^ν is positive and larger than 1, then the state of neuron I is flipped, leading to wrong recall.

With purely random patterns, we can estimate the probability of wrong recall, which is given by

$$P_{\text{error}} = \text{Prob}\left(C_{i}^{\nu} > 1\right) \tag{6.11}$$

Note that C_i^ν is a random variable which is $1/N$ times with the sum of NP independent random binary numbers. In a large network with relatively small stored pattern, C_i^ν can be approximated with a Gaussian random variable with the zero mean and variance $\sigma^2 = P/N$. Thus, we have

$$P_{\text{error}} = \text{Prob}\left(C_{i}^{\nu} > 1\right) = \frac{1}{\sqrt{2\pi}\sigma} \int_{1}^{\infty} e^{-x^2/2\sigma^2} dx = \frac{1}{2}[1 - erf(N/2P)] \tag{6.12}$$

where *erf*(.) is the error function. With the above equation, we can find the relationship between P_{error} and the maximal number of stored pattern $P_{\max}$, i.e., the storage capacity. For example, if we set the criterion $P_{\text{error}} < 0.01$, we get $P_{\max} = 0.15N$.

To conclude, the storage capacity $P_{\max}$ is proportional to the network size N, if a small percentage of error is permitted for randomly stored patterns. Note that for nonrandom patterns the storage capacity will be changed.

The above analysis only gives a rough estimation of the memory capacity; a detailed estimation of the memory capacity is reported to be $P_{\max} \approx 0.136N$ if a small recall error is allowed (Amit et al. 1985, 1987).

In the Hopfield model, apart from the attractors that correspond to the stored patterns, there are other attractors existed in the model that are not useful for memory, which are called spurious states. For example, one type of spurious states is the reversed states with respect to the stored patterns, and they have the same energy as that of the original states (Kanter and Sompolinsky 1987).

6.2 Continuous Attractor Networks

Continuous attractor neural networks (CANNs) are an idealized computational model for neural information representation. The key structure of a CANN is the translational invariance of neuronal recurrent interactions. With the proper parameters, a CANN can hold a continuous family of localized stationary states (often called bumps), each of which corresponding to one feature value encoded by the neural system. These bump states form a manifold in the network state space, on which the network is neutrally stable, and this property endows a CANN with the capacity of tracking the changes of external inputs smoothly. CANNs have been successfully applied to describe the encoding of continuous stimuli, e.g., orientation (Ben-Yishai et al. 1995), moving direction, head direction (Zhang 1996), and spatial location of objects (Samsonovich and McNaughton 1997), in neural systems. They are also widely used in other modeling studies, such as for working memory and decision-making, although sometimes people may not emphasize that their models have the structure of CANNs. Recent experimental data has indicated that CANNs capture some fundamental features of neural information representation. In this section, we introduce the fundamental dynamics and computational properties of CANNs.

6.2.1 The Model

There are several mathematical formulations of CANNs. Here, we present the one which allows us to pursue theoretical analysis and hence elucidates the key features of a CANN clearly. The results are, however, applicable to other CANN models.

Consider that a one-dimensional continuous stimulus x, e.g., head direction or orientation, is encoded by an ensemble of neurons. The value of x is in the range of $(-\pi, \pi]$ with the periodic condition imposed. Neurons are aligned in the network according to their preferred stimulus values. Denote $U(x,t)$ as the synaptic input at time t of the neurons whose preferred stimulus is x and $r(x,t)$ the corresponding firing rate. The dynamics of $U(x,t)$ is determined by the recurrent input from other neurons, its own relaxation, and external input $I^{\text{ext}}(x,t)$, which is written as

$$\tau\frac{\partial U(x,t)}{\partial t} = -U(x,t) + \rho\int_{x'} J\left(x,x'\right) r\left(x',t\right) dx' + I^{\text{ext}}(x,t) \tag{6.13}$$

where τ is the synaptic time constant, typically of the order 1–2 ms, and ρ is the neuron density. $J(x,x')$ is the interaction strength from neurons at x' to neurons at x and is chosen to be

$$J\left(x,x'\right) = \frac{J_0}{\sqrt{2\pi}a}\exp\left(-\frac{\left(x-x'\right)^2}{2a^2}\right) \tag{6.14}$$

where the parameter a controls the neuronal interaction range. Note that $J(x,x')$ is a function of $(x\text{-}x')$, i.e., the neuronal interactions are translation invariant in the space of natural preferred stimuli. This property is crucial for a CANN to track moving input smoothly. Since the neuronal interaction range we consider is much smaller than the range of stimulus values, i.e., $a \ll 2\pi$, we can effectively regard $-\infty < x < +\infty$ in the analysis below.

The neuronal firing rate $r(x,t)$ is determined by the synaptic input according to

$$r(x,t) = \frac{U_+(x,t)^2}{1 + k\rho\int_{x'} U_+\left(x',t\right)^2 dx'}, \tag{6.15}$$

where $U_+ = U$ for $U > 0$ and otherwise $U_+ = 0$. The neuronal firing rate first increases with the input and then saturates gradually due to the divisive normalization by the overall network activity.

The dynamics of a 2D CANN are described similarly, except that the 1D variable x is replaced by a 2D vector $\mathbf{x} = (x,y)$ with $x, y \in (-\infty, +\infty)$, and the neuronal interaction is expressed as $J(\mathbf{x},\mathbf{x}') = J/(2\pi a^2)\exp\left[-(x-x')^2/(2a^2)\right]$ $\exp\left[-(y-y')^2/(2a^2)\right]$.

6.2.2 *Intrinsic Dynamics of a CANN*

We first introduce the intrinsic dynamics of a CANN in the absence of external inputs. It is straightforward to check that for $0 < k < k_c \equiv \rho J_0^2/\left(8\sqrt{2\pi}a\right)$, the network holds a continuous family of stationary states (Fig. 6.3), which are

$$\overline{U}(x|z) = U_0\exp\left[-\frac{(x-z)^2}{4a^2}\right], \quad \overline{r}(x|z) = r_0\exp\left[-\frac{(x-z)^2}{2a^2}\right] \tag{6.16}$$

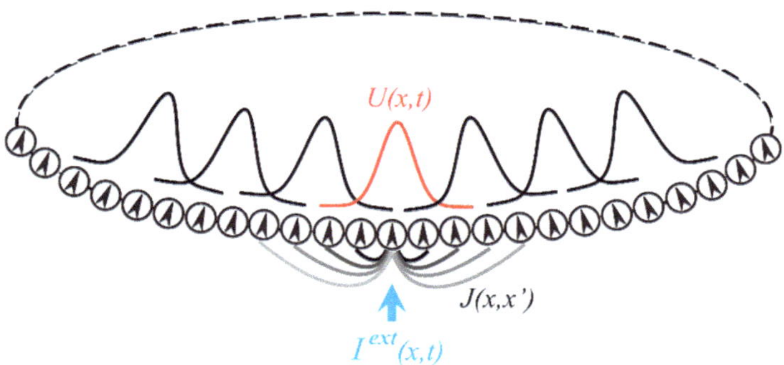

Fig. 6.3 (**a**) CANN encodes a continuous stimulus, e.g., head direction. Neurons are aligned in the network according to their preferred stimuli. The neuronal interaction $J(x, x')$ is translational variant in the space of stimulus values. The network holds a continuous family of stationary states (bumps)

where $U_0 = \left[1 + (1 - k/k_c)^{1/2}\right] J_0/(4\sqrt{\pi}ak)$ and $r_0 = \left[1 + (1 - k/k_c)^{1/2}\right] J_0/\left(2\sqrt{2\pi}ak\rho\right)$.

These stationary states are translationally invariant among themselves and have a Gaussian shape with free parameter z indicating their positions.

The stability of the Gaussian bumps can be studied by considering the dynamics of fluctuations (Fung et al. 2010). Consider the network state $U(x,t) = \overline{U}(x|z) + \delta U(x,t)$. Using the linearly stability analysis, we have

$$\tau \frac{\partial}{\partial t} \delta U(x,t) = \int_{-\infty}^{+\infty} dx' F\left(x, x'|z\right) \delta U\left(x', t\right) - \delta U(x,t), \tag{6.17}$$

where the interaction kernel $F\left(x, x'|z\right)$ is given by

$$\begin{aligned} F\left(x, x'|z\right) &= \frac{2\sqrt{2}U\left(x'\right)}{J_0 U_0} \left[J(x,x') - k\rho \int_{-\infty}^{+\infty} dx'' J\left(x, x''\right) r\left(x''\right)\right] \\ &= \frac{2}{a\sqrt{\pi}} \exp\left[-\frac{\left(x - x'\right)^2}{2a^2}\right] \exp\left[-\frac{\left(x' - z\right)^2}{4a^2}\right] - \frac{1 + \sqrt{1 - k/k_c}}{\sqrt{2\pi}a} \exp\left[-\frac{(x-z)^2}{4a^2}\right] \exp\left[-\frac{\left(x' - z\right)^2}{4a^2}\right] \end{aligned} \tag{6.18}$$

We are interested in the eigenfunctions and eigenvalues of the kernel $F(x,x'|z)$. By using the basic wave functions of quantum harmonic oscillators,

$$v_n\left(x|z\right) = \frac{(-1)^n \left(\sqrt{2}a\right)^{n-1/2}}{\sqrt{\pi^{1/2} n! 2^n}} \exp\left[\frac{(x-z)^2}{4a^2}\right] \left(\frac{d}{dx}\right)^n \exp\left[-\frac{(x-z)^2}{2a^2}\right] \tag{6.19}$$

We can calculate the eigenvalues and the first four right eigenfunctions of the kernel $F(x,x'|z)$, which are

$$\lambda_0 = 1 - \sqrt{1 - k/k_c}, \quad \lambda_n = 2^{1-n}, \quad \text{for } n \geq 1, \ldots,$$
$$u_0^R(x|z) = v_0(x|z),$$
$$u_1^R(x|z) = v_1(x|z),$$
$$u_2^R(x|z) = \frac{\sqrt{1/2}}{\left[\left(1 - 2\sqrt{1 - k/k_c}\right)^2 + 1/2\right]^{1/2}} v_0(x|z)$$
$$+ \frac{1 - 2\sqrt{1 - k/k_c}}{\left[\left(1 - 2\sqrt{1 - k/k_c}\right)^2 + 1/2\right]^{1/2}} v_2(x|z), u_3^R(x|z)$$
$$= \sqrt{\frac{1}{7}} v_1(x|z) + \sqrt{\frac{6}{7}} v_3(x|z).$$

Figure 6.4 illustrates the first four right eigenfunctions of F. The right eigenfunctions of F correspond to the various distortion modes of the bump. For example, $u_0^R(x|z)$ corresponds to the amplitude distortion of the bump, $u_1^R(x|z)$ to the positional shift, and $u_2^R(x|z)$ to the width. Starting from $n = 3, u_n^R(x|z)$ are linear combinations of $v_k^R(x|z)$, where $k = n, n-2, \ldots$. Some of them may be dominated by a specific type of distortion, such as $u_3^R(x|z)$ being dominated by skewness.

We note that $u_1^R(x|z) = v_1(x|z)$ and $\lambda_1 = 1$, corresponding to the positional shift of the bump, is sustained under the network dynamics. This is the mathematical expression of the neutral stability of the stationary states of the network. The eigenvalues for all other eigenfunctions are all less than one. Since these other modes of bump distortions decay exponentially with the time constant $\tau/(1 - \lambda_n)$, their contributions on the network dynamics are small when compared with that due to the positional shift.

The fact that the dynamics of a CANN is dominated by only a few motion modes suggests that we may use this property to simplify the description of the network dynamics; that is, we can project the network dynamics onto its dominating modes. By projecting functions $f(x)$ onto a mode $\phi(x)$, we mean computing $\int f(x)\phi(x)dx / \int \phi(x)^2 dx$. Typically, by including one or two leading motion modes, the simplified dynamics is adequate to capture the main features of a CANN. By this, we can analytically quantify the tracking behaviors of a CANN.

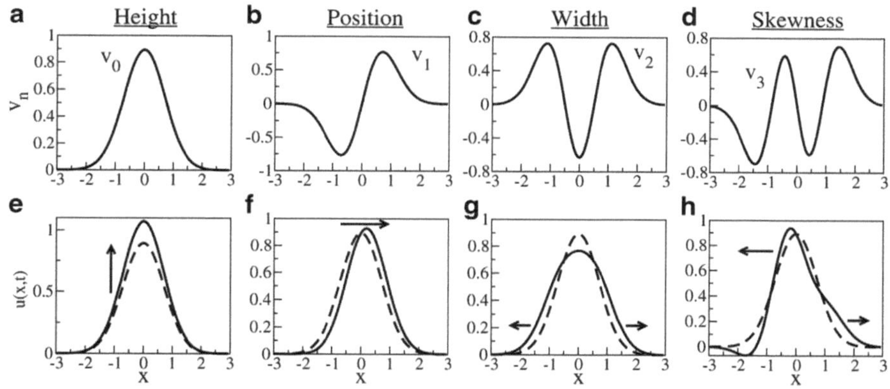

Fig. 6.4 (**a–d**) The first distortion modes of the bump state. (**e–h**) Their effects of producing distortions in the height, position, width, and skewness of the Gaussian bump. *Solid* and *dashed lines* represent distorted and undistorted bumps, respectively (Adapted with permission from Fung et al. (2012), Fig. 1)

6.2.3 Tracking Property of a CANN

The bump states of a CANN form a sub-manifold in the state space of the network, on which the network is neutrally stable. This property enables a CANN to track a moving stimulus smoothly, provided that the stimulus speed is not too large. To investigate the tracking performance of CANNs, we consider the external input to be of the form

$$I^{\text{ext}}(x,t) = \alpha \exp\left[-\frac{(x-z_0)^2}{4a^2}\right], \tag{6.20}$$

where the stimulus position $z_0(t)$ is time dependent.

Denoting the lag of the bump with respect to the stimulus as $s = z_0 - z$, whose value is calculated to be

$$\frac{ds}{dt} = \frac{dz_0}{dt} - \frac{dz}{dt} = v_{\text{ext}} - \frac{\alpha s}{\tau}\exp\left(-\frac{s^2}{8a^2}\right) = v_{\text{ext}} - g(s), \tag{6.21}$$

The size of lag s is determined by two competing factors: the first term represents the movement of the stimulus, which tends to enlarge the lag, and the second term represents the collective effects of the neuronal recurrent interactions, which tends to shorten the lag. The tracking is achieved when these two forces match each other: $v = g(s)$; otherwise, s diverges. The function g(s) is concave and has the maximum value of $g_{\max} = 2\alpha a/(\tau\sqrt{e})$ at $s = 2a$, which defines the maximum trackable speed of a stimulus to the network.

6.2.4 Anticipative Tracking of a CANN

A CANN is able to track moving inputs smoothly; however, the network bump is always lagging behind the instant position of the moving stimulus due to the delay of neuronal responses. This delayed tracking cannot explain the anticipative behavior observed in the experiments. Time delays are pervasive and significant in neural information processing. For instance, visual signal transmitting from the retina to the primary visual cortex takes about 50–80 ms, and the time constant for single neurons responding to synaptic input is of the order 10–20 ms. If these delays are not compensated properly, our perception of a fast moving object will lag behind its true position in the external world significantly, impairing our vision and motor control. A straightforward way to compensate for time delays is to anticipate the future position of a moving object, covering the distance the object will travel through during the delay period (Berry et al. 1999). A large volume of experimental data has revealed that the internal representation of head direction of a rat, corresponding to a bump state in a CANN, is always leading the instant direction of the rat head by approximately constant time (Blair and Sharp 1995; Goodridge and Touretzky 2000).

It turns out that by incorporating slow negative feedback modulation in the dynamics of neurons, a CANN is able to achieve anticipative tracking. Here, we consider one such feedback modulation, spike-frequency adaptation (SFA). SFA is a dynamical feature commonly observed in the activities of neurons when they have experienced prolonged firing. It may be generated by a number of mechanisms (Benda and Herz 2003; Gutkin and Zeldenrust 2014). In one mechanism, neural firing elevates the intracellular calcium level of a neuron, which induces an inward potassium current and subsequently hyperpolarizes the neuronal membrane potential. In other words, strong neuronal response induces a negative feedback to counterbalance itself. Incorporating SFA, the dynamics of a CANN is written as

$$\tau \frac{\partial U(x,t)}{\partial t} = -U(x,t) + \rho \int_{x'} J\left(x,x'\right) r\left(x',t\right) dx' - V(x,t) + I^{\text{ext}}(x,t) \tag{6.21}$$

where the synaptic current $V(x,t)$ represents the effect of SFA, whose dynamics is given by

$$\tau_v \frac{\partial V(x,t)}{\partial t} = -U(x,t) + mU(x,t) \tag{6.23}$$

where τ_v is the time constant of SFA, typically of the order 40–120 ms. The parameter m controls the SFA amplitude. This equation gives rise to $V(x,t) = m\int_{-\infty}^{t} \exp\left[-\left(t-t'\right)/\tau_v\right] U\left(x,t'\right) dt'/\tau_v$, that is, $V(x,t)$ is the integration of the neural synaptic input (and hence the neural activity) over an effective period of τ_v. The negative value of $V(x,t)$ is subsequently fed back to the neuron to suppress its response. The higher the neuronal activity level is, the larger the negative

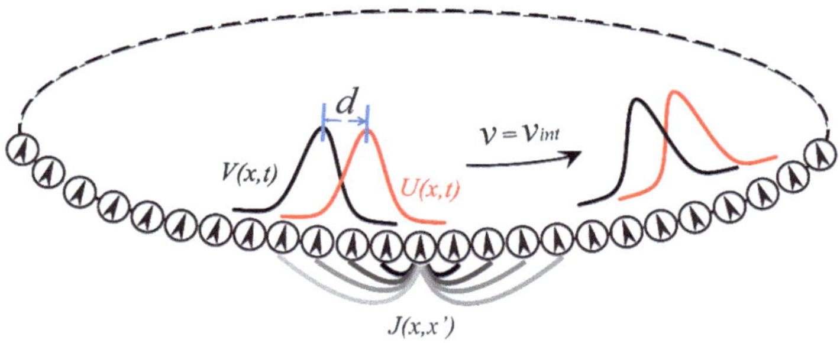

Fig. 6.5 An example of traveling wave in a 1D CANN with SFA. The neuronal activity level is represented by the bump of synaptic inputs $U(x, t)$, which moves with a speed v_{int} spontaneously without relying on an external moving input. Since SFA is a slow process, the feedback current $V(x, t)$ lags behind $U(x, t)$ with a constant distance d.

feedback will be. The time constant $\tau_v \gg \tau$ indicates that SFA is slow compared to neural firing.

A CANN with SFA can hold a traveling wave state (Fig. 6.5). The underlying mechanism is intuitively understandable. Suppose that a bump emerges at an arbitrary position in the network. Due to SFA, those neurons which are most active receive the strongest negative feedback, and their activities will be suppressed accordingly. Under the competition (mediated by recurrent connections and divisive normalization) from the neighboring neurons which are less affected by SFA, the bump tends to shift to the neighborhood; and at the new location, SFA starts to destabilize neuronal responses again. Consequently, the bump will keep moving in the network like a traveling wave.

The condition for the network to support a traveling wave state can be theoretically analyzed. Consider the following Gaussian ansatz for the traveling wave state:

$$\begin{aligned}
\overline{U}(x,t) &= A_u \exp\left\{-\frac{[x-z(t)]^2}{4a^2}\right\}, \\
\overline{r}(x|z) &= A_r \exp\left\{-\frac{[x-z(t)]^2}{2a^2}\right\} \\
\overline{V}(x|z) &= A_v \exp\left\{-\frac{[x-z(t)+d]^2}{4a^2}\right\}
\end{aligned} \tag{6.24}$$

where $dz(t)/dt$ is the speed of the traveling wave and d is the separation between $\bar{U}(x,t)$ and $\overline{V}(x,t)$. Without loss of generality, we assume that the bumps move from left to right, i.e., $dz(t)/dt > 0$. Since $V(x,t)$ lags behind $U(x,t)$ due to slow SFA, $d > 0$ normally holds.

To solve the network dynamics, we utilize an important property of CANNs, that is, the dynamics of a CANN are dominated by a few motion modes corresponding to different distortions in the shape of a bump. We can project the network dynamics onto these dominating modes and simplify the network dynamics significantly (Fung et al. 2012). Applying the projection method, we solve the network dynamics and obtain the traveling wave state. The speed of the traveling wave and the bumps' separation are calculated to be

$$v_{\text{int}} \equiv \frac{dz(t)}{dt} = \frac{2a}{\tau_v}\sqrt{\ln\frac{m\tau_v}{\tau}}, \quad d = 2a\sqrt{\ln\frac{m\tau_v}{\tau}} \tag{6.25}$$

The speed of the traveling wave reflects the intrinsic mobility of the network, and its value is fully determined by the network parameters. Hereafter, we call it the intrinsic speed of the network, referred to as v_{int}, and v_{int} increases with the SFA amplitude m. The larger the value of v_{int}, the higher the mobility of the network.

From the above equations, we see that the condition for the network to support a traveling wave state is $m > \tau/\tau_v$. We note that SFA effects can reduce the firing rate of neurons significantly. Since the ratio τ/τ_v is small, it is expected that this condition can be realistically fulfilled.

SFA induces intrinsic mobility of the bump states of a CANN, manifested by the ability of the network to support self-sustained traveling waves (Sato et al. 2012). When the network receives an external input from a moving stimulus, the tracking behavior of the network will be determined by two competing factors: the intrinsic speed of the network v_{int} and the speed of the external drive ($v_{\text{ext}} = dz_0(t)/dt$). Define $s = z(t) - z_0(t)$ to be the displacement of the network bump relative to the external drive. We consider that the network is able to track the moving stimulus, i.e., the network dynamics will reach a stationary state with $dz(t)/dt = dz_0(t)/dt$ and s a constant. Since we consider that the stimulus moves from left to right, $s > 0$ means that the network tracking is leading the moving input; whereas $s < 0$ means the network tracking is lagging behind.

Using the Gaussian ansatz for the network state as given above and applying the projection method, we solve the network dynamics and obtain

$$\begin{aligned} &d = v_{\text{ext}}\tau_v, \\ &\frac{s}{a}\exp\left(-\frac{s^2}{8a^2}\right) = \frac{A_u}{\alpha}\frac{v_{\text{ext}}\tau_v}{a}\left[m\exp\left(-\frac{{v_{\text{ext}}}^2{\tau_v}^2}{4a^2}\right) - \frac{\tau}{\tau_v}\right] \end{aligned} \tag{6.26}$$

From the above two equations, we see that the condition for $s > 0$, i.e., the network state leads the external input, is given by $v_{\text{int}} > v_{\text{ext}}$. Similarly, when $v_{\text{int}} < v_{\text{ext}}$, $s < 0$ holds (Fig. 6.6a, b). That is, when $v_{\text{int}} > v_{\text{ext}}$, the network bump leads the instant position of the moving stimulus for sufficiently weak stimuli, achieving anticipative tracking. Figure 6.6c presents the simulation results.

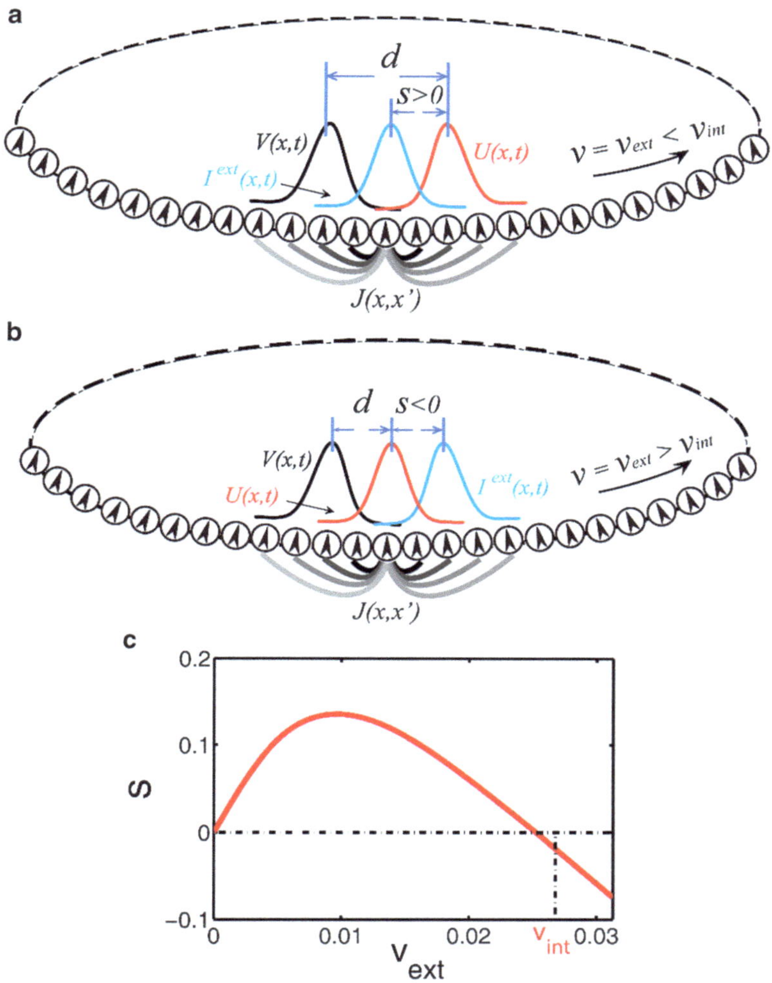

Fig. 6.6 (**a**) An example of anticipative tracking in the reference frame of the external drive. (**b**) An example of delayed tracking. In both cases, the profile of $V(x, t)$ is lagging behind the bump $U(x, t)$ due to slow SFA. (**c**) The separation s vs. the speed of the external input v_{ext}. Anticipative tracking $s > 0$ occurs when $v_{\text{int}} > v_{\text{ext}}$ (Adapted with permission from Mi et al. (2013), Fig. 3)

6.3 Reservoir Networks

Reservoir network computation is a computational framework for real-time computation (Maass et al. 2002; Buonomano and Maass 2009). In contrast with traditional offline computation, such as Tuning machines, that maps an input to an output with specified steps of calculations, reservoir computing maps input streams to output streams. This type of computation is more close to biological systems that the external input is continuously applied to our nervous system, and at the same time our brain continuously responds to the surrounding environments by sending motor signals to muscles. Of notable interest to the field of neural computation, reservoir computing is famous in its adaptive power for implementing a large volume of desired tasks by only tuning its readout part, while the other part is fixed.

Typically, in the framework of reservoir computing, the input stream is faded into the reservoir. Due to the nonlinear dynamics of the reservoir, multiple aspects of information about the input, including the past and the current information of the input, are embedded in the extremely high-dimensional space of the reservoir state. Consequently, a simple memoryless readout, i.e., a perceptron, is adequate to extract the necessary information for a computational task. This division of computation into reservoir and readout is efficient, in that the reservoir generally provides almost all desirable information about the input stream and many aspect of the input information can be easily accessed and decoded by a simple readout. For different tasks, the needed information from the input stream may be different; therefore, a different readout is required without adjusting the other part of the reservoir.

There are mainly two widely used reservoirs, namely, liquid state machines (LSMs) and echo state networks (ESN), and both of them are recurrent networks with random connection weights. The main difference is that the neuron in LSM is a spiking neuron with only two states, while in ESN each neuron is an analog one. In this section, we use LSM as an example to elucidate the computational properties of reservoir networks.

6.3.1 Liquid State Machines

LSMs are one type of recurrent spiking neural networks in which neurons are randomly and sparsely connected (Fig. 6.7). Also, each neuron receives time-varying inputs from external sources. The recurrent nature of the connections turns the time-varying inputs into a spatiotemporal pattern of activations in the network. Finally, linear readout neurons extract the spatiotemporal activation patterns of the LSM. LSM is a generic recurrent circuit of spiking neurons to model a cortical microcircuit. The general point can be intuitively understood by making an analogy between neural networks and a liquid. A stone is thrown into a still body of water and creates a spatiotemporal pattern of ripples. Thus, the pattern in the liquid is the result of interactions between subsequent falling stones and internal state of the liquid. Similarly, the response of a population of neurons in a network is determined by the interplay of the external stimulus and the internal state of the network.

An LSM consists of liquid and readout. The liquid L^M in biological models is typically a generic recurrently connected local network of neurons; the readout is to extract information from the liquid and report the results. The input function $u(t)$ to the liquid can be a continuous sequence of disturbances, and the target output can be

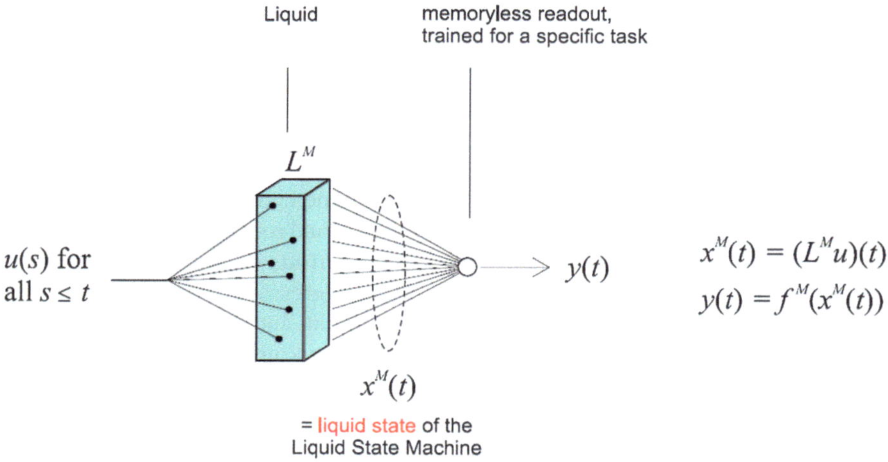

Fig. 6.7 Structure of an LSM, which transforms input steams $u(.)$ into output streams $y(.)$. L^M denotes a liquid (e.g., some dynamical system) and the "liquid state." $x^M(t)$ is the input to the readout at time t. More generally, $x^M(t)$ is that part of the current internal state of the liquid that is "visible" for the readout. Only one input and output channel is shown for simplicity (Adapted with permission from Maass, W et al. Computability in Context: Computation and Logic in the Real World, Fig. 4)

some chosen function $y(t)$ of time that provides a real-time analysis of this sequence. LSM maps input functions of time $u(t)$ to output functions $y(t)$ of time. Here, $x^M(t)$ is an internal "liquid state" at time t, which constitutes its current response to preceding perturbations, that is, to preceding inputs $u(s)$ for $s < t$, the liquid state consists of analog values that may change continuously over time. Here, in a physical implementation, this liquid state $x^M(t)$ consists of all information about the current internal state of a dynamical system which is accessible to readout modules.

In a mathematical view, L^M is like a filter or some operator which can map input function $u(.)$ onto functions $x^M(t)$. The second part of LSM is a memoryless readout map f^M and, at every time t, transforms $x^M(t)$ into output $y(t)$. In contrast to the liquid filter L^M, the readout map f^M is in general chosen in a task-specific manner.

In contrast to other computational models such as attractor-based models, Turing machines, LSM is a real-time computing system on continuous streams of data, such as spike trains which provide external input to the microcircuit. This property is important because it helps model process continuous stream of inputs with stereotypical recurrent circuits of spiking neurons in real time. In LSM, the input $u(.)$ and output $y(.)$ are both data streams and could be multidimensional. The operator or filter L^M maps input data stream $u(.)$ to output data stream $y(.)$ rather than numbers to numbers. The characteristic feature of such computational processing is that the target output $y(t)$ depends not only on the values of the input $u(s)$ at a time point s but also the past input at many preceding time points.

Comparing to other computational models, another difference is that LSM is a model for an adaptive computing system. This is apparent when putting the LSM in the context of a learning framework. And the LSM model is motivated by the hypothesis that the learning capability of an information processing device is its most delicate aspect. However, training a RNN and controlling its complex dynamics are difficult. What is more, in biological models, the recurrently connected neural circuits always have noise and make many methods break down. Here, the architecture of LSM makes the learning very simple and robust. It treats the RNN as a reservoir network and concentrates the learning efforts on the connections from the liquid to readout, which is a simple linear regression. The readout neuron typically represents a periphery network of the liquid, and it can be modeled by a linear gate, a perceptron, by a sigmoid gate, or by a spiking neuron. The liquid here is like a preprocessor, which amplifies the range of possible functions of the input streams $u(t)$ that can learn. A liquid L^M can serve a large number of different readout neurons that each learns to extract a different summary of information from the same liquid. And it promises that several real-time computations be carried out in parallel within the same circuitry, which appears to be a generic feature of neural microcircuits.

LSM can have wide applications in information processing. David Sussillo uses a reservoir network to realize a running and walking task as shown in Fig. 6.8. They implemented this example using all the architectures in both sparse and fully connected configurations with similar results. Using a force learning rule, they successfully trained networks that produced both motions demonstrating that a single chaotic recurrent network can generate multiple, high-dimensional, nonperiodic patterns that resemble complex human motions.

6.3.2 Scale-Free Network

What type of reservoir network is optimal for a given range of tasks is a common research question in the field. The conventional studies often consider that reservoir networks are random networks, and not much is known about the computational power of a reservoir network having specific large-size topology. Mi et al. recently studied a reservoir network with the scale-free topology and found that such a network can generate a broad range of long-period oscillations.

The model has two crucial structures. The first is the scale-free topology, meaning that the network consists of a few neurons having many connections, called hub neurons, and a large number of neurons with few connections, called low-degree neurons. The second is that hub neurons are hard to be activated to avoid epileptic firing of the network, and this can be realized in two ways, one is to consider electrical coupling between neurons and the other is to consider the efficacy of a chemical synapse decreases with the connectivity of the postsynaptic neuron.

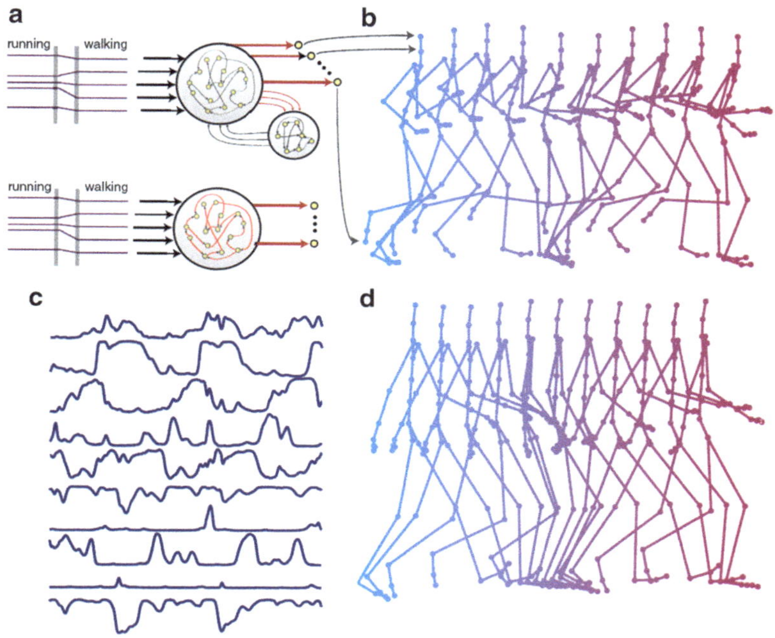

Fig. 6.8 Networks that generate both running and walking human motions. (**a**) Either of these two network architectures can be used to generate the running and walking motions (modifiable readout weights shown in *red*), but the upper network is shown. Constant inputs differentiate between running and walking (*purple*). Each of 95 joint angles is generated through time by one of the 95 readout units (*curved arrows*). (**b**) The running motion generated after training. Cyan frames show early and magenta frames late movement phases. (**c**) Ten sample network neuron activities during the walking motion. (**d**) The walking motion, with colors as in (**b**) (Adapted with permission from Sussillo D et al. Neuron, Fig. 8)

With two properties above, the network has the capacity of retaining rhythmic synchronous firings, as shown in Fig. 6.8. The working mechanism is the network contains loops formed by low-degree neurons; a loop retains the activity seed between two synchronous firings; as the activity is propagated along a loop, few hub neurons are activated and synchronize the whole network. Thus, the length of a low-degree loop encodes the rhythm of synchronized firings of the network.

A scale-free network has large amounts of low-degree neuron and they form different lengths of low-degree loop (Fig. 6.9). So, the network has resources to maintain a broad range of rhythmic activities. When presented with a rhythmic external input, the neural system selects a low-degree loop from its reservoir with the loop size matching the input rhythm, and this matching operation can be achieved by a learning process.

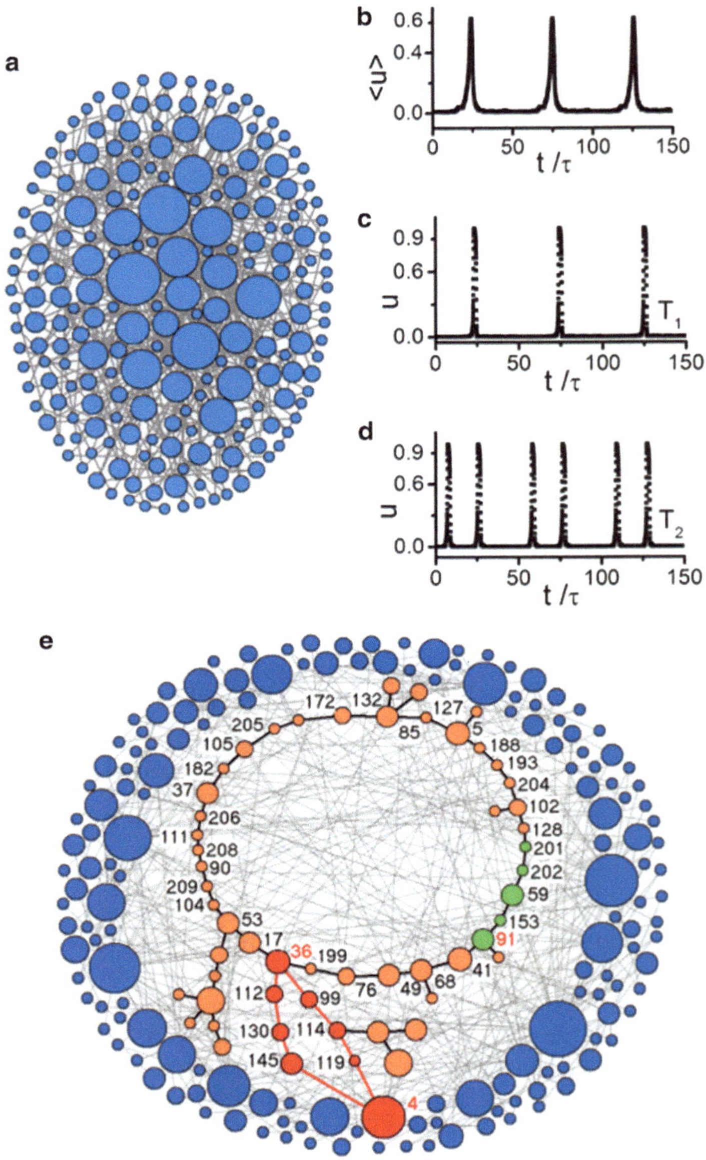

Fig. 6.9 (**a**) A network with scale-free topology $r = 3$, mean connectivity <k> =4, and $N = 210$. The diameter of a neuron is proportional to its connectivity. (**b**) The population activity of the network, which displays rhythmic synchronous firing. (**c** and **d**) Activity of the network, which displays rhythmic into two types: those that fire only once in a single period (T1 types, **c**) and neurons reorganized by placing all T2 neurons (*green and orange colored*) in the center and all T1 neurons (*blue colored*) outside. All T2 neurons have a low-degree connection and form a loop. Two red edges show the pathways along which activation of the loop is propagated to a hub neuron (Adapted with permission from Mi, Y et al. Proceedings of the National Academy of Sciences, Fig. 2)

6.4 Network Models for Decision-Making

Decision-making is a fundamental function of the brain. It refers to the cognitive process of choosing an opinion or an action among a set of two or more alternatives. Experimental studies have uncovered the neural signals related to decision-making at the single neuron level, yet the neural mechanism underlying decision-making remains far from clear. The present section reviews several computational models aiming to unveil the neural mechanism of decision-making.

6.4.1 Experimental Evidence

A classic experiment for studying decision-making is the task of random-dot motion (RDM) direction discrimination (Fig. 6.10). This task was first used to investigate the relationship between sensory encoding and perception. Later, a reaction-time (RT) version of the RDM task was developed to study decision-making.

In the task, the monkey made a choice between two possible (opposite) directions of motion displayed by random dots and was instructed to do so by saccade once a commitment to one of the alternatives is reached. The task difficulty was controlled by varying the percentage of coherently moving dots. The mean reaction time increased with the task difficulty. Neurons in LIP were recorded, and some of them displayed correlated activities with the decision behavior, in terms of that the neural activity is a graded function of the coherence level of moving dots (Fig. 6.11c).

6.4.2 Diffusion/Race Model

An abstract model, called race model, was developed to describe the decision-making process in the brain (Uchida et al. 2006). This model assumes that there exist decision variables which collect evidence to produce the corresponding choices, one for each alternative. It is like a race competition: accumulators representing different choice options build up their activities, and whichever is the first to reach a prescribed threshold produces the choice (Fig. 6.12).

The dynamics of the race model can be written as (two alternatives)

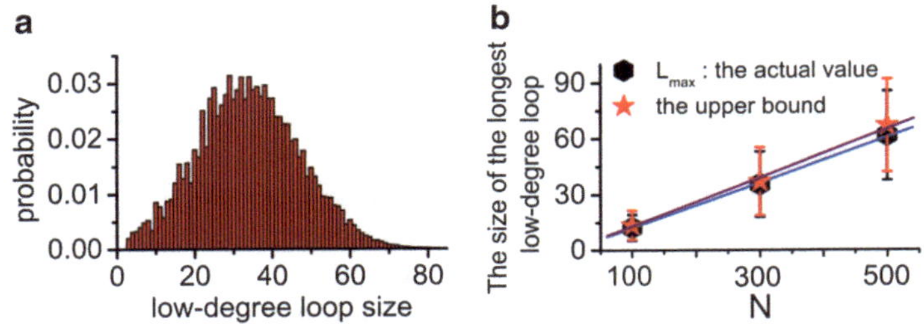

Fig. 6.10 (**a**) Distribution of the lengths of low-degree loops in a scale-free network of size $N = 300$. The result is obtained by averaging 100 randomly generated scale-free networks of the same size. (**b**) Length of the longest low-degree loop vs. the number of neurons in a network. Actual values are obtained through extensively searching all loops in each network, and the upper bounds are calculated theoretically. For each data point, the result is obtained by averaging over 100 randomly generated scale-free networks of the same size (Adapted with permission from Mi, Y et al. Proceedings of the National Academy of Sciences, Fig. 3)

$$\begin{aligned} e^{+/-}(t) &= \int_0^t s^{+/-}(t)dt, \\ s^+(t) &= \max(s(t), 0), \\ s^-(t) &= \max(-s(t), 0), \end{aligned} \tag{6.27}$$

where e(t) is the accumulator (decision variable) for one direction at time t, "+/−" represent different directions, and s(t) represents the stimulus on the corresponding direction.

A model similar to the race one was proposed to describe the decision-making process of LIP neurons. This model is called the diffusion model (Wang 2008), since the behavior of the accumulator is like a diffusion process (or random walk). Its dynamics for two alternative choices is written as

$$\frac{de}{dt} = s_L - s_R + noise \tag{6.28}$$

where *e(t)* is the evidence accumulator and s is the signal strength, with the subscripts "L" and "R" representing left and right directions, respectively.

Despite their simplicity, the race and diffusion models explain a large volume of experimental data, such as those in the RDM task (Fig. 6.11). To incorporate the concept of "forgetting" or integration time window, a linear leaky competing accumulator (LCA) model is developed (Wang 2008), which mimics the dynamics of real neurons by including a leakage term in evidence integration, whose dynamics is given by

$$\frac{de}{dt} = -\frac{e}{\tau} + s_L - s_R + noise, \tag{6.29}$$

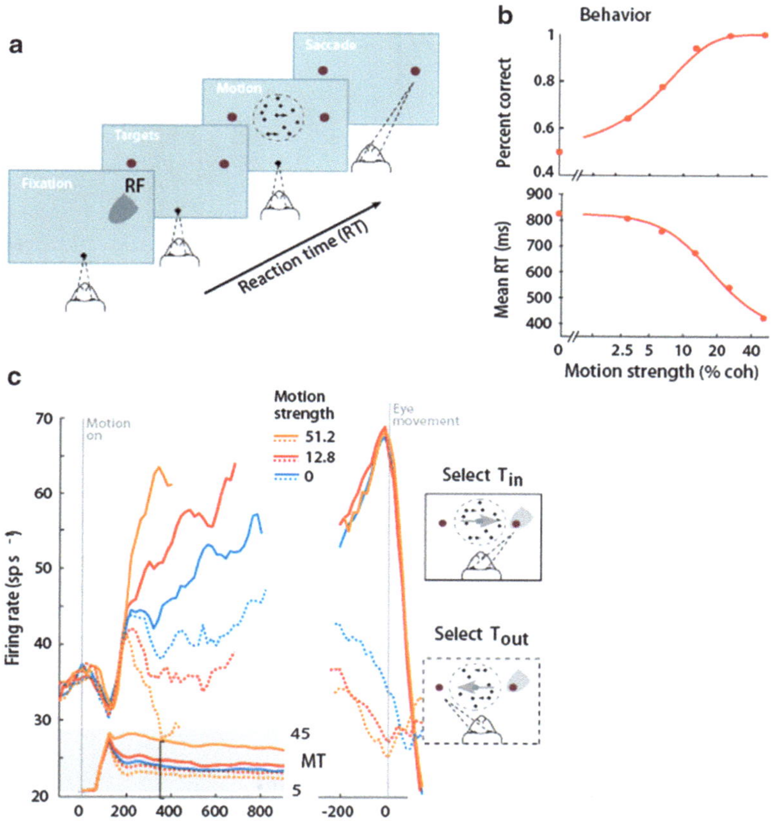

Fig. 6.11 (**a**) Schematic diagram of the random-dot motion direction discrimination task. (**b**) Effects of stimulus difficulty on accuracy and decision time. *Solid curves* are fits of the diffusion model. (**c**) Response of LIP neurons during decision formation. Average firing rate from 54 LIP neurons is shown for three levels of difficulty. Responses are grouped by motion strength and direction of choice, as indicated (Adapted with permission from Gold and Shadlen 2007, Fig. 4)

where τ is the time constant for leakage and determines the time window over which the stimulus evidence is accumulated. In reality, τ must be large (hundreds of milliseconds) in order for the model to reproduce the long accumulation process observed experimentally. But can this kind of long integration time constant be achieved in a realistic cortical network? The answer is "yes," which is the model to be introduced below.

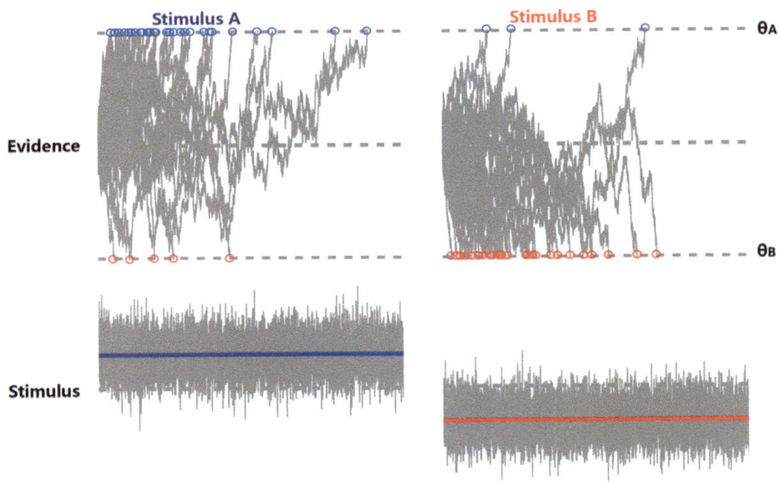

Fig. 6.12 An example of race model. The decision device receives stochastic input (*stimulus bottom*) representing evidence for two competing alternative choices (**a** and **b**). The integrator accumulates evidence until it reaches one of the two thresholds (θA, θB), at which point a decision is emitted. The graphs show the time course of accumulated evidence for several trials for two different stimuli with different mean signal strengths or drift rates (*slope of the arrows*). The input noise introduces stochasticity in the accumulation of evidence (*gray* traces represent different trials with different instantiations of noise). This creates trial-by-trial variability in response times and choices (*blue circles*, choice A; *red circles*, choice B) (Adapted with permission from Uchida et al. 2006, Fig. 1)

6.4.3 A Recurrent Neural Network Model

Wang et al. proposed a biologically more realistic network model to implement decision-making (Wang 2002). Inspired by the LCA model, this model realizes long-term evidence accumulation by using slow synapses mediated by NMDA receptors. Furthermore, the model exploits a combination of recurrent excitation and feedback inhibition to generate attractor dynamics, so that the subtle difference between conflicting inputs is amplified to get binary choices.

To simulate a two-choice decision task, the model consists of two neural groups, and each of them is selective to one of two motion directions (e.g., (a) left motion, (b) right motion; Fig. 6.13). The strong recurrent excitatory connections within a neural group are capable of generating self-sustained persistent activities. There also exists competition between two neural groups, mediated by the shared feedback from the jointly connected interneurons. All neurons receive background Poisson inputs and fire spontaneously when no stimulus input is presented.

During the discrimination task, both neural groups receive stochastic Poisson inputs at rates $s_A(t)$ and $s_B(t)$, respectively. The stimuli $s_A(t)$ and $s_B(t)$ vary in time; their distributions are Gaussian with a mean of μ_A and μ_B, respectively (Fig. 6.13b), and a standard deviation σ. The task difficulty is controlled by varying the percentage of coherently moving dots, which is measured by $c' = 100 \times \frac{\mu_A - \mu_B}{\mu_A + \mu_B}$.

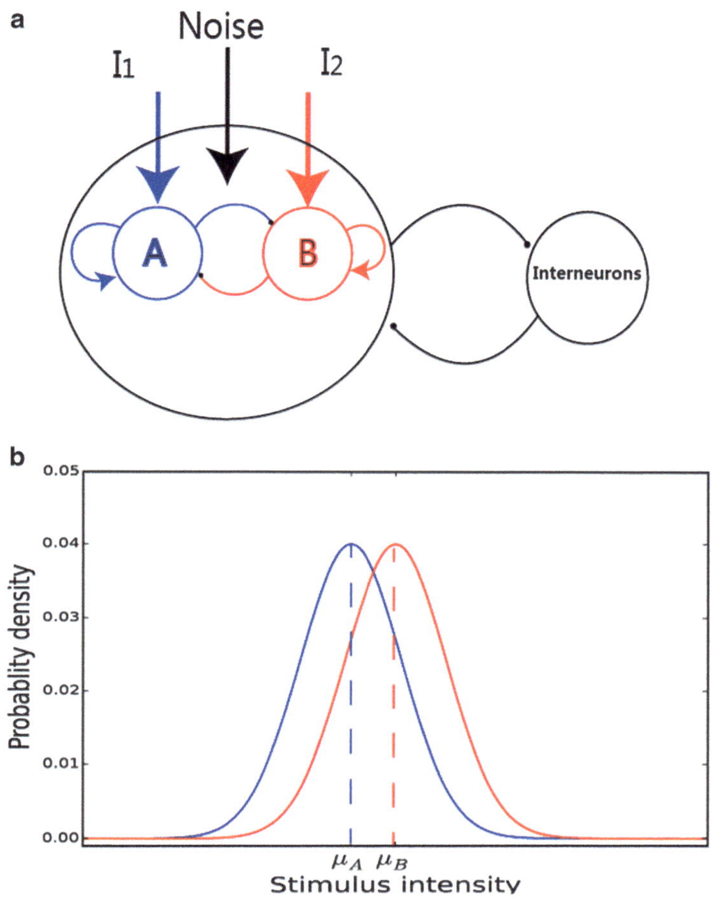

Fig. 6.13 (**a**) Schematic diagram of the model structure. (**a**) and (**b**) represent two groups of pyramidal neurons, each of which is selective to one of the two stimuli choice (*right* or *left*). Within each pyramidal neural group, neurons have strong recurrent excitatory connections that can sustain persistent activity triggered by a transient preferred stimulus. (**b**) The inputs are Poisson rates that vary in time and obey Gaussian distributions, with means μ_A and μ_B, respectively (Adapted with permission from Wang 2002, Fig. 1)

Thus, when the coherence is small, the two inputs to two neural groups are similar, and the discrimination task is difficult; otherwise, the task is easy.

The dynamics of a single neuron is given by

$$C_m \frac{dV(t)}{dt} = -g_L(V(t) - V_L) - I_{syn}(t), \tag{6.30}$$

where C_m is the membrane capacitance, $V(t)$ the membrane potential, V_L the resting potential, g_L the leak conductance, and I_{syn} the total synaptic current to the neuron.

The synaptic current to a neuron contains both excitatory and inhibitory elements, with the former mediated by AMPA (fast dynamics) and NMDA (slow dynamics) receptors and the latter mediated by GABA receptors. It is written as

$$I_{syn}(t) = I_{ext,AMPA}(t) + I_{rec,AMPA}(t) + I_{rec,NMDA}(t) + I_{rec,GABA}(t), \tag{6.31}$$

where

$$I_{ext,AMPA}(t) = g_{ext,AMPA}(V(t) - V_E)s^{ext,AMPA}(t), \tag{6.32}$$

$$I_{rec,AMPA}(t) = g_{rec,AMPA}(V(t) - V_E)\sum_j^{N_E} w_j s_j^{AMPA}(t), \tag{6.33}$$

$$I_{rec,NMDA}(t) = g_{rec,NMDA}(V(t) - V_E)\sum_j^{N_E} w_j s_j^{NMDA}(t), \tag{6.34}$$

$$I_{rec,GABA}(t) = g_{rec,GABA}(V(t) - V_E)\sum_j^{N_I} s_j^{GABA}(t), \tag{6.35}$$

where g is the conductance with subscript indicating the receptor type. Notice that g_{NMDA} is a function of membrane potential and $[Mg^{2+}]$ the other conductance are constant parameters. V_E and V_L are reverse potential of excitatory and inhibitory synapses, respectively. The parameter s is the gating variable, representing the fraction of open channels. For AMPA and GABA receptors, s is described by

$$\frac{ds_j^A}{dt} = -\frac{s_j^A}{\tau_j^A} + \sum_k \delta\left(t - t_j^k\right), \tag{6.36}$$

where the superscript is AMPA or GABA and τ is the time constant for AMPA or GABA, usually 2–5 ms. For NMDA receptors, s is described by

$$\begin{aligned} \frac{ds_j^{NMDA}(t)}{dt} &= -\frac{s_j^{NMDA}(t)}{\tau_{NMDA,decay}} + \alpha x_j(t)\left(1 - s_j^{NMDA}(t)\right), \\ \frac{dx_j(t)}{dt} &= -\frac{x_j(t)}{\tau_{NMDA,rise}} + \sum_k \delta\left(t - t_j^k\right), \end{aligned} \tag{6.37}$$

where $\tau_{NMDA,decay}$ is the time constant of NMDA, usually several hundreds of milliseconds, and $x_j(t)$ represents the fraction of open channels triggered by a spike.

The model reproduces the decision-making process. Figure 6.14 illustrates the process of evidence accumulation of the model for inputs at random coherence level. By assuming that a decision is made when the population activity of either of two neural groups reaches to a threshold, we can measure the reaction time. The simulation result is consistent with many aspects of the physiological data observed in the experiment (Fig. 6.15).

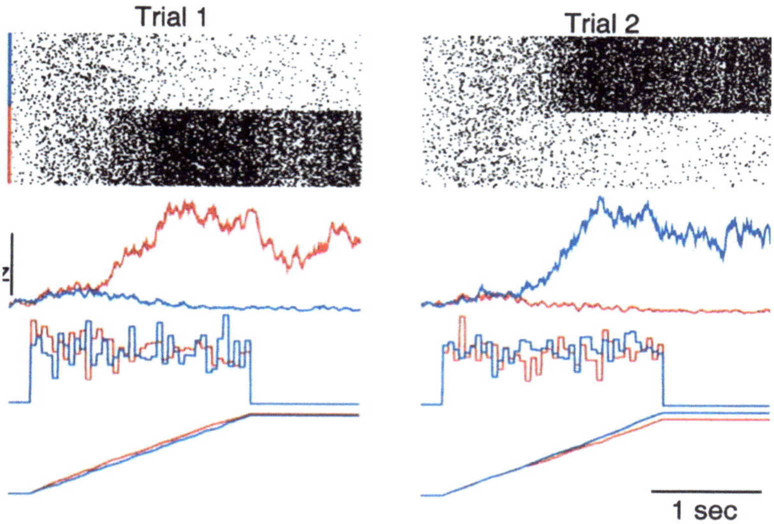

Fig. 6.14 Simulation results with input of zero coherence. Two trial simulations (*red*, neural group A; *blue*, neural group B). From *top* to *bottom*: raster, population firing rates, stochastic inputs, and time integrals of inputs (Adapted with permission from Wang 2002, Fig. 3)

6.5 Network Models with Dynamical Synapses

Neurons, synapses, and the networks they form are the fundamental units for the brain to implement various functions. In the conventional modeling studies, it is often assumed that the neuronal connection weight, which models the efficacy of firing of the presynaptic neuron on affecting the state of the postsynaptic neuron, is a constant. However, this is not true in a real biological system. In reality, neuronal synapses experience temporal changes depending on the firing history of the neurons, which is called short-term plasticity (STP) or dynamical synapses (Abbott and Regehr 2004). The dynamical synapses endow a neural system with new information processing capacities which are otherwise difficult to implement with static synapses (Tsodyks and Wu 2013). This section introduces the computational model of dynamical synapses and their effects on neural dynamics.

6.5.1 What Are Dynamical Synapses?

To start, we first briefly review the property of a neuronal synapse (Fig. 6.16). A synapse is a functional structure that passes an electrical or chemical signal from one neuron to another cell. Nevertheless, a synapse does this job in an ever-changing manner, which is dependent on the recent history at either or both sides

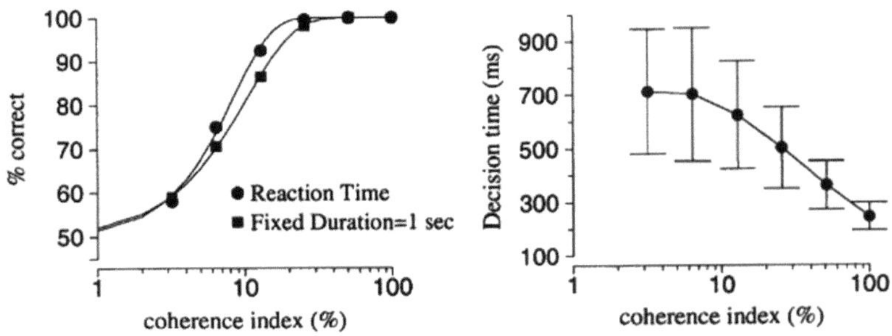

Fig. 6.15 Simulation results of the spiking neural network. *Left*: neurometric functions for the reaction time stimulation (*circle*) and with fixed stimulus duration of 1 s (*square*). *Right*: average decision time is linear in the logarithm of the coherence level. At very low coherence there is a saturation. Note the large standard deviation of decision time, especially at low coherence (Adapted with permission from Wang 2002, Fig. 5)

of the synapses. These activity-dependent changes can last from milliseconds to months and are collectively referred as synaptic plasticity.

Synaptic plasticity has been widely believed to be the substrate for learning and memory, and it also plays a significant role in the development of neural circuits. Synaptic plasticity can be divided into three broad categories: (1) long-term plasticity (LTP), which refers to changes, lasts for hours or longer and is crucial for learning and memory; (2) homeostatic plasticity is the capacity of neurons to regulate their own excitability relative to network activity, a compensatory adjustment that occurs over the time scale of days; and (3) short-term plasticity (STP), which is the main focus of this section, refers to a phenomenon that synaptic efficacy changes over time in a way reflecting the history of presynaptic activity. STP has shorter time scales compared to LTP, typically on the order of hundreds to thousands of milliseconds.

STP can be further divided into two opposite types, short-term depression (STD) and short-term facilitation (STF). STD is caused by depletion of neurotransmitters consumed during the synaptic signaling process at the axon terminal of a presynaptic neuron, whereas STF is caused by influx of calcium into the axon terminal after spike generation, which increases the release probability of neurotransmitters.

6.5.2 The Phenomenological Model of STP

The biological mechanisms giving rise to STP can be quite complex, so in the study of computational roles of STP, we rely mostly on a simplified phenomenological model. In the phenomenological model proposed by Tsodyks and Markram, STP is described as

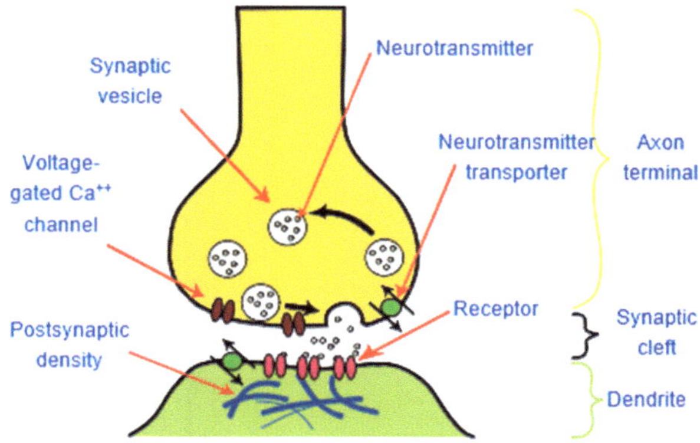

Fig. 6.16 Structure of a typical chemical synapse (From Wikipedia https://en.wikipedia.org/wiki/Synapse)

$$\frac{du}{dt} = -\frac{u}{\tau_f} + U(1 - u^-)\delta(t - t_{sp}), \tag{6.38}$$

$$\frac{dx}{dt} = -\frac{1 - x}{\tau_d} - u^+x^-\delta(t - t_{sp}), \tag{6.39}$$

$$\frac{dI}{dt} = -\frac{1}{\tau_s} + Au^+x^-\delta(t - t_{sp}) \tag{6.40}$$

where the STD effect is modeled by a normalized variable $x(0 \leq x \leq 1)$, denoting the fraction of resources that remain available after neurotransmitter depletion. The STF effect is modeled by a utilization parameter u, representing the fraction of available resources ready for use (release probability). τ_f and τ_d are the time constants of STF and STD, respectively. They determine how fast u and x is moving toward their equilibria. The t_{sp} denotes the spike time and U is the increment of u produced by a spike. We denote u^-, x^- as the corresponding variables just before the arrival of the spike, and x^+ refers to the moment just after the spike. I refers to the synaptic current. A denotes the response amplitude that would be produced by total release of all the neurotransmitter $(x = u = 1)$, called absolute synaptic efficacy of the connections. $\delta(t - t_{sp})$ is a Dirac delta function.

To get a more comprehensive understanding of this three-dimensional system, we first take a look at its stable equilibrium. Now considering when the system hasn't undergone any spikes for a long time, the stable equilibria of this system can be obtained by setting the left side of equations and the δ function to be zero. Thus, we have $u = 0,\ x = 1, I = 0$ as the equilibria. Next, we consider when a spike actually occurs; at time t_{sp}, from Eqs. (6.38), (6.39), and (6.40), we have $du = U, dx = -U, dI = AU$, which means u, the fraction of available resources ready for use, increases instantly by an amount of U, whereas x, the fraction of resources that remain available after neurotransmitter depletion, decreases instantly by an amount of U, and I, the synaptic current, increases by an amount of AU.

As the synaptic current is determined by the term u^+x^- (see Eq. 6.40), it's really the interplay between the dynamics of u and x that decides whether the current

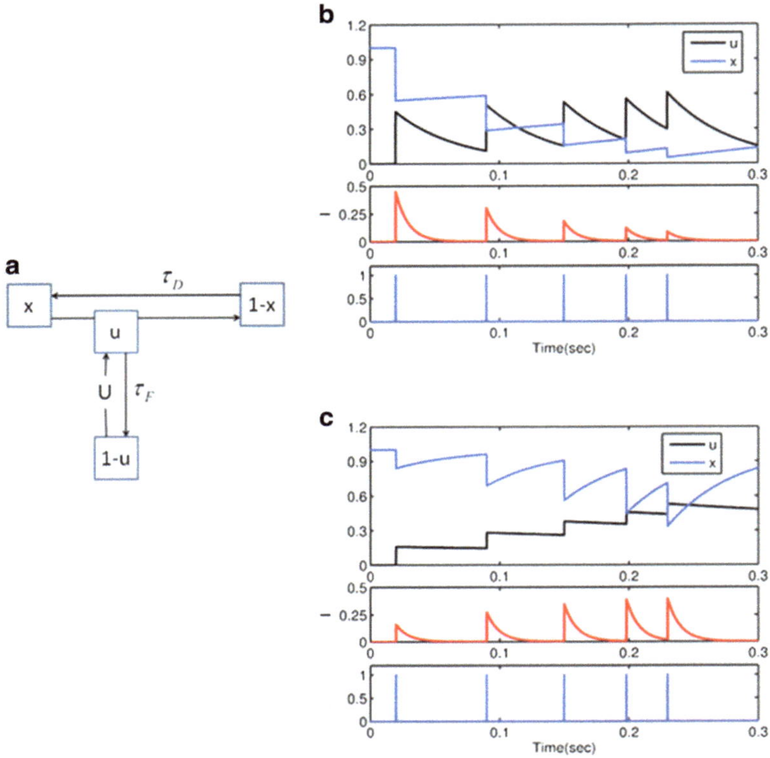

Fig. 6.17 (**a**) The phenomenological model for STP given by Eqs. (6.38), (6.39), and (6.40). (**b**) The postsynaptic current generated by an STD-dominated synapse. The neuronal firing rate $R = 15$ Hz. The parameters $A = 1$, $U = 0.45$, $\tau_s = 20$ms, $\tau_d = 750$ms, and $\tau_f = 50$ms. (**c**) The dynamics of a STF-dominating synapse. The parameters $U = 0.15$, $\tau_f = 750$ms, and $\tau_d = 50$ ms (From Scholarpedia http://www.scholarpedia.org/article/Short-term_synaptic_plasticity)

is dominated by depression or facilitation (Fig. 6.17). In the parameter regime of $\tau_d \gg \tau_f$ and large U, an initial spike incurs a large drop in x that takes a long time to recover; therefore, the synapse is STD dominated (Fig. 6.17b). In the regime of $\tau_f \gg \tau_d$ and small U, the synaptic efficacy is increased gradually by spikes, and consequently the synapse is STF dominated (Fig. 6.17c). This phenomenological model has successfully reproduced the kinetic dynamics of depressed and facilitated synapses observed in many cortical areas.

6.5.3 Effects on Network Dynamics

We first introduce the effects of STP on the dynamics of continuous attractor neural networks (CANNs) (Fung et al. 2012), and then we present its effects on other network models.

The dynamics of a one-dimensional CANN is written as

$$\tau\frac{\partial U(x,t)}{\partial t} = -U(x,t) + \rho\int_{-\infty}^{\infty} J\left(x,x'\right)f\left(x',t\right)p\left(x',t\right)r\left(x',t\right)dx' + I^{\text{ext}}(x,t), \tag{6.41}$$

$$J\left(x,x'\right) = \frac{J_0}{\sqrt{2\pi}a}\exp\left[-\frac{\left(x-x'\right)^2}{2a^2}\right], \tag{6.42}$$

$$r(x,t) = \frac{U(x,t)^2}{1 + k\rho\int_{-\infty}^{\infty} U\left(x',t\right)^2 dx'}, \tag{6.43}$$

where τ is the time constant and ρ the density of neurons in the network. $f(x, t)$ and $p(x,t)$ represent the effects of STF and STD, respectively. The parameter k controls the divisive normalization effect.

The dynamics of STP are given by

$$\tau_f\frac{\partial f(x,t)}{\partial t} = f_{\min} - f(x,t) + \alpha[1 - f(x,t)]r(x,t), \tag{6.44}$$

$$\tau_d\frac{\partial p(x,t)}{\partial t} = 1 - p(x,t) - \beta f(x,t)p(x,t)r(x,t), \tag{6.45}$$

where $f_{\min}$ is the initial value of $f(x, t)$. The parameters α and β determine how fast the STF and STD effects recover. τ_f and τ_d are the time scales of STF and STD, respectively.

6.5.4 Impact of STP on the Stationary States of a CANN

We first look at the impact of STP on the stationary states of a CANN when no external stimulus is applied. Figure 6.18a displays the individual impacts of STF and STD on the phase diagram of a CANN. Section 6.2 shows that when no STP is coupled with the CANN, a stationary bump can be held when and only when $k < k_c$, with $k_c \equiv \rho(f_{\min}J_0)^2/\left(8a\sqrt{2\pi}\right)$. We observe that STF broadens the region of CANNs to hold a static bump, whereas STD narrows down that region.

Note that in Fig. 6.18b, STD induces a new form of stationary state, called metastatic or moving state. In such a state, the network holds a spontaneously moving bump without relying on external drive. This dynamical property is intuitively understandable. Suppose that a bump is initiated at a position. Due to STD, those neurons which are most active undergo the strongest negative feedback, and

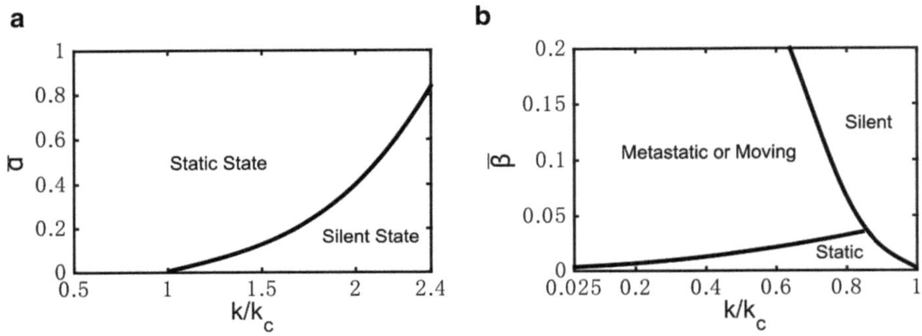

Fig. 6.18 Phase diagrams of a CANN with STF and STD. $\overline{\alpha}$ and $\overline{\beta}$ are scaled values of α and β. $\overline{\alpha}\equiv\alpha/\left(\rho^2J_0^2\right)$, $\overline{\beta}\equiv\beta/\left(\rho^2J_0^2\right)$. (**a**) The static state region enlarges as α increases. Parameters are $\tau_f = 50\,\text{ms}$, $\tau = 1\text{ms}$, $f_{\min} = 0.3$, and $a = 0.2$. 200 neurons uniformly distributed in $\left[-\pi,\pi\right)$. (**b**) The static state region shrinks as β increases. Parameters are $\tau_d = 50\text{ms}$, $\tau = 1\text{ms}$, $f_{\min} = 1$ (no STF) , and $a = 0.2$

their activity will be suppressed consequently. With the competition from the neighboring neurons which are less affected by STD, the bump tends to shift to the neighborhood; and at the new location, SFA tends to suppress neuronal responses again. Thus, the bump will keep propagating in the network like moving bump.

6.5.5 Prolonged Neural Response to a Transient Input

A CANN displays an interesting behavior in the marginally unstable region of the static bump. Since STD time scale is much longer than the synaptic time scale, a new feature STD can bring to the network dynamics is the prolongation of neural response to a transient input (Fig. 6.19). This stimulus-induced residual activity, called a slow decaying plateau, holds a memory trace of the input, lasting up to several hundred milliseconds (of the order of τ_d), and hence can serve as a buffer for information processing.

6.5.6 Tracking with Enhanced Mobility

Consider a CANN is receiving an external moving stimulus given by

$$I^{\text{ext}} = A\exp\left[-\frac{\left(x - z_0(t)\right)^2}{4a^2}\right],$$

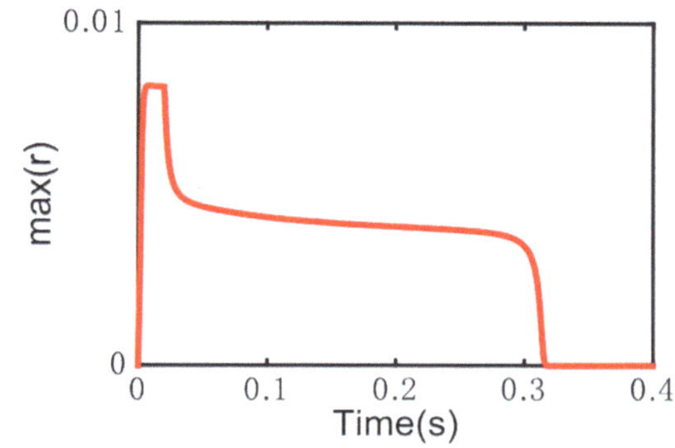

Fig. 6.19 A slow decaying plateau of CANN coupled with STD. External stimulus is removed at $t = 20\,\text{ms}$. Parameters are $\tau_d = 50$ ms, $\tau = 1$ ms, $a = 0.2$, $k = 7.43(0.94k_c)$, and $\beta = 10(\overline{\beta} = 0.0099)$

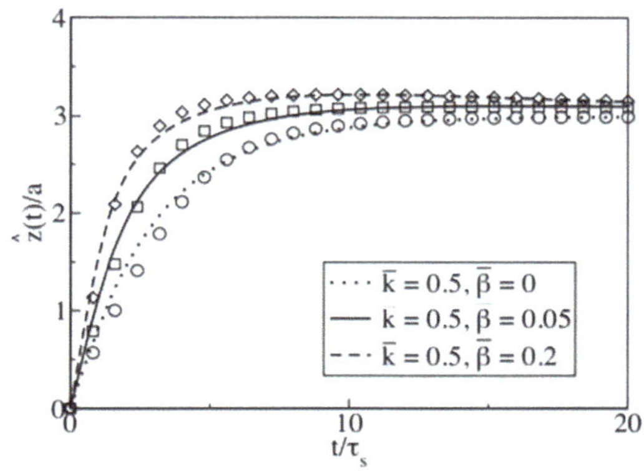

Fig. 6.20 The response of CANNs with STD to a stimulus that changed abruptly from $z_0/a = 0$ to $z_0/a = 3.0$ at $t = 0$ (Adapted with permission from Fung et al. 2012. Fig. 11)

where the stimulus position $z_0(t)$ is time dependent. Consider a tracking task in which the $z_0(t)$ abruptly changes from 0 at $t = 0$ to a new value at $t = 0$. Figure 6.20 shows the network response during the tracking process. Compared to the case without STD, we can see that the bump shifts to the new position faster. This is because STD tends to suppress localized neural activities and hence helps the network to catch up the moving input.

6.5.7 *Effects on Other Network Models*

STP also produces other interesting dynamical behaviors in network systems. For example, an STD-dominated network generates bursting synchronized neural activity, called population spike, in response to an external input. This is due to that after large-size synchronous firing in the network, the synapses are weakened by STD, reducing neuronal interactions rapidly, and consequently the network activity returns to the baseline immediately.

STP can also be used to carry out valuable computations. Considering a network that holds multiple attractor states competing with each other, STD destabilizing one of them can incur the network to switch to another attractor state. This property has been linked to spontaneous transition between up and down states of cortical neurons, the binocular rivalry phenomenon, and enhanced discrimination capacity for superimposed ambiguous inputs.

6.6 Summary and Advanced Readings

Neural circuits formed by a large number of neurons and synapses are the computational substrates of brain functions. The computation of a neuron is rather simple, which basically integrates synaptic inputs and generates action potentials. It is the computational power of neural networks that endow the brain with the capacity of implementing various cognitive functions.

This chapter is devoted to introduce some fundamental network models and their associated capabilities of information processing.

Section 6.1 introduces the classical Hopfield model for associative memory. The Hopfield considers that the neuronal state is a binary variable and that neuronal connections satisfy the Hebbian learning rule and ignores all other biological details. In the Hopfield model, a memory is stored as a stationary state (attractor) of the network. Given only partial knowledge of a memory, the network will evolve to the memory state emergently, achieving associative memory.

Section 6.2 is focused on continuous attractor neural networks (CANNs), a network model that has been successfully applied to describe the encoding of continuous stimuli in neural systems. A key structure of a CANN is the translational invariance of neuronal recurrent interactions in the stimulus feature space. This property enables a CANN to hold a continuous family of stationary states, and they form a sub-manifold in the network state space on which the network is neutrally stable. This neutral stability endows a CANN with the capacity of tracking moving inputs smoothly. Moreover, when slow negative feedback modulation is included in the neural dynamics, the network response can track a moving input anticipatively.

Section 6.3 presents a type of network model called reservoir networks. This type of network captures a few fundamental features of real neural systems, such as that the number of neurons is enormous and single neuron properties (including synapses) exhibit large diversity. The key idea of a reservoir network is that there is no dedicated encoding strategy, and in response to an external input, the stimulus information is straightforwardly embedded in the extremely high-dimensional space of the network. The information extraction occurs only at the reading-out stage and can be implemented easily by a single layer network through learning. A reservoir network with a large-size scale topology can have extra computational capacities, for instance, a scale-free network can extract the rhythm information of periodic inputs.

Section 6.4 introduces several computational models aiming to unveil the underlying mechanism of decision-making in neural systems, a fundamental function of the brain. These models include the phenomenological diffusion/race models and the biology-based recurrent networks.

Section 6.5 focuses on introducing some new computational properties brought by dynamical synapses to network models. In reality, neuronal synapses are not fixed, but rather experience constant changing depending on the firing history of neurons, which is called short-term plasticity (STP). STP endows a neural system with new information processing capacities which are otherwise difficult to implement with static synapses.

6.7 Advanced Readings

Dayan P, Abbott LF (2001) Theoretical neuroscience – computational and mathematical modeling of neural systems, The MIT Press

Hertz J, Krogh A, Palmer R-G (1995) Introduction to the theory of neural computation, Addison Wesley

References

Abbott LF, Regehr WG. Synaptic computation. Nature. 2004;431(7010):796–803.

Amari S. Dynamics of pattern formation in lateral–Inhibition type neural fields. Biol Cybern. 1977;27:77–87.

Amit DJ, Gutfreund H, Sompolinsky H. Storing infinite numbers of patterns in a spin-glass model of neural networks. Phys Rev Lett. 1985;55(14):1530.

Amit DJ, Gutfreund H, Sompolinsky H. Information storage in neural networks with low levels of activity. Phys Rev A. 1987;35(5):2293.

Benda J, Herz AVM. A universal model for spike-frequency adaptation. Neural Comput. 2003;15:2523–64.

Ben-Yishai R, Bar-Or RL, Sompolinsky H. Theory of orientation tuning in visual cortex. Proc Natl Acad Sci. 1995;92:3844–8.

Berry MJ, Brivanlou IH, Jordan TA, Meister M. Anticipation of moving stimuli by the retina. Nature. 1999;398:334–8.

Blair HT, Sharp PE. Anticipatory head direction signals in anterior thalamus: evidence for a thalamocortical circuit that integrates angular head motion to compute head direction. J Neurosci. 1995;15:6260–70.

Buonomano DV, Maass W. State-dependent computations: spatiotemporal processing in cortical networks. Nat Rev Neurosci. 2009;10(2):113–25.

Fung CC, Wong KYM, Wu S. A moving bump in a continuous manifold: a comprehensive study of the tracking dynamics of continuous attractor neural networks. Neural Comput. 2010;22:752–92.

Fung CA, Wong KM, Wang H, Wu S. Dynamical synapses enhance neural information processing: gracefulness, accuracy, and mobility. Neural Comput. 2012;24(5):1147–85.

Gold JI, Shadlen MN. The neural basis of decision making. Annu Rev Neurosci. 2007;30:535–74.

Goodridge JP, Touretzky DS. Modelling attractor deformation in the rodent head-direction system. J Neurophysiol. 2000;83:3402–10.
Gutkin B, Zeldenrust F. Spike frequency adaptation. Scholarpedia. 2014;9(2):30643.
Heeger DJ. Normalization of cell responses in cat striate cortex. Vis Neurosci. 1992;9:181–97.
Hertz JA, Krogh A, Palmer RG. Introduction to the theory of neural computation, volume 1 of Santa Fe institute studies in the sciences of complexity: lecture notes. 1991.
Hopfield JJ. Neural networks and physical systems with emergent collective computational abilities. Proc Natl Acad Sci. 1982;79(8):2554–8.
Kanter I, Sompolinsky H. Associative recall of memory without errors. Phys Rev A. 1987;35 (1):380.
Maass W. Liquid state machines: motivation, theory, and applications. In: Computability in context: computation and logic in the real world, 2010;275–96.
Maass W, Natschläger T, Markram H. Real-time computing without stable states: a new framework for neural computation based on perturbations. Neural Comput. 2002;14(11):2531–60.
Mi Y, Liao X, Huang X, Zhang L, Gu W, Hu G, Wu S. Long-period rhythmic synchronous firing in a scale-free network. Proc Natl Acad Sci. 2013;110(50):E4931–6.
Samsonovich A, McNaughton BL BL. Path integration and cognitive mapping in a continuous attractor neural network model. J Neurosci. 1997;17:5900–20.
Sato TK, Nauhaus I, Carandini M. Travelling waves in visual cortex. Neuron. 2012;75:218–29.
Sussillo D, Abbott LF. Generating coherent patterns of activity from chaotic neural networks. Neuron. 2009;63(4):544–57.
Tsodyks M, Wu S. Short-term synaptic plasticity. Scholarpedia. 2013;8(10):3153.
Uchida N, Kepecs A, Mainen ZF. Seeing at a glance, smelling in a whiff: rapid forms of perceptual decision making. Nat Rev Neurosci. 2006;7(6):485–91.
Wang XJ. Probabilistic decision making by slow reverberation in cortical circuits. Neuron. 2002;36(5):955–68.
Wang XJ. Decision making in recurrent neuronal circuits. Neuron. 2008;60(2):215–34.
Yuanyuan Mi, Alan Fung CC, Michael Wong KY, Si Wu. Spike frequency adaptation implements anticipative tracking in neural systems. Adv Neural Info Process Syst. 2013.
Zhang K. Representation of spatial orientation by the intrinsic dynamics of the head-direction cell ensemble: a theory. J Neurosci. 1996;16:2112–26.

MIX
Papier aus verantwortungsvollen Quellen
Paper from responsible sources
FSC® C105338

If you have any concerns about our products,
you can contact us on
ProductSafety@springernature.com

In case Publisher is established outside the EU,
the EU authorized representative is:
Springer Nature Customer Service Center GmbH
Europaplatz 3, 69115 Heidelberg, Germany

Printed by Libri Plureos GmbH
in Hamburg, Germany